International Series in Operations Research & Management Science

Volume 308

The book series *International Series in Operations Research and Management Science* encompasses the various areas of operations research and management science. Both theoretical and applied books are included. It describes current advances anywhere in the world that are at the cutting edge of the field. The series is aimed especially at researchers, doctoral students, and sophisticated practitioners. The series features three types of books:

- *Advanced expository* books that extend and unify our understanding of particular areas.
- *Research monographs* that make substantial contributions to knowledge.
- *Handbooks* that define the new state of the art in particular areas. They will be entitled

Recent Advances in (name of the area). Each handbook will be edited by a leading authority in the area who will organize a team of experts on various aspects of the topic to write individual chapters. A handbook may emphasize expository surveys or completely new advances (either research or applications) or a combination of both.

The series emphasizes the following four areas: *Mathematical Programming*: Including linear programming, integer programming, nonlinear programming, interior point methods, game theory, network optimization models, combinatorics, equilibrium programming, complementarity theory, multiobjective optimization, dynamic programming, stochastic programming, complexity theory, etc. *Applied Probability*: Including queuing theory, simulation, renewal theory, Brownian motion and diffusion processes, decision analysis, Markov decision processes, reliability theory, forecasting, other stochastic processes motivated by applications, etc. *Production and Operations Management*: Including inventory theory, production scheduling, capacity planning, facility location, supply chain management, distribution systems, materials requirements planning, just-in-time systems, flexible manufacturing systems, design of production lines, logistical planning, strategic issues, etc. *Applications of Operations Research and Management Science*: Including telecommunications, health care, capital budgeting and finance, marketing, public policy, military operations research, service operations, transportation systems, etc.

More information about this series at http://www.springer.com/series/6161

Chiwoo Park • Yu Ding

Data Science for Nano Image Analysis

Springer

Chiwoo Park
Florida State University
Tallahassee, FL, USA

Yu Ding
Industrial & Systems Engineering
Texas A&M University
College Station, TX, USA

ISSN 0884-8289 ISSN 2214-7934 (electronic)
International Series in Operations Research & Management Science
ISBN 978-3-030-72821-2 ISBN 978-3-030-72822-9 (eBook)
https://doi.org/10.1007/978-3-030-72822-9

This Springer imprint is published by the registered company Springer Nature Switzerland AG
The registered company address is: Gewerbestrasse 11, 6330 Cham, Switzerland

To our parents:

Chanbum Park and Yongok Lee
Jianzhang Ding and Yujie Liu

and to our families:

Sun Joo Kim, Layna Park and Leon Park
Ying Li and Alexandra Ding

Foreword

The ability of (scanning) transmission electron microscopy (S/TEM) to obtain images on the nanometre to atomic scale has revolutionised our scientific understanding of structure-property relationships in structural, electronic and biomedical materials. More recently, there has been a significant expansion in the use of in-situ/operando gas and liquid stages that moves beyond static samples and/or images and now permits (S)TEM images to be acquired at defined intervals during precisely controlled kinetic reactions. When these in-situ/operando experiments use most advanced CMOS detectors, there is the potential to image dynamic functionality on the scale of individual molecular interactions and for unprecedented fundamental insights into a wide range of next-generation materials to be developed. While it is always tempting for any experimentalist to believe that the imaging challenge is complete when the hardware is designed, installed and commissioned, this book on data science for nano image analysis by Chiwoo Park and Yu Ding shows clearly and explicitly that the hardware is actually just the start of the scientific process. There are tremendous advances in our understanding of dynamics that image analytics can bring to any problem that is being studied in a (S)TEM.

The comprehensive description of the use of data analytical methods for image analysis in this book contains the authors' combined insights from 15 years of pioneering the use of analytics for dynamic imaging in (S)TEM. The contents of the book are structured to take the reader from the basic mathematical definitions of an image and its information content through to the analysis of the main changes taking place in any dynamic sequence: morphology, spacing, temporal evolution and motion/interactions. For students/experimentalists just getting started on the use of analytics, the book follows through the mathematics for a series of examples that are important to imaging both inorganic and organic systems, such as nucleation and growth, coalescence, corrosion and decay. With the current rapid advancement in the use of artificial intelligence (AI) to power scientific discovery, the methods in this book are ideally suited to educate researchers in the benefits of AI for their research and help them with their initial implementation.

This book successfully integrates data science with imaging in a way that allows a new practitioner of the methodology to work through concepts established in each field (data science and imaging) without getting bogged down by differing terminologies and complex mathematical concepts. As such, it is an ideal reference for graduate courses in this area and for scientists who maybe many years from graduate school but now wish to enhance their understanding of new methods for image analysis. This book will be a key reference for me, my research group and our collaborators going forward, and I am grateful for the authors' major accomplishment in completing this text.

Liverpool, UK
January 2021

Nigel Browning

Preface

Scientific imaging is one of the crucial steps in biological and materials sciences to see and understand small-scale phenomena that cannot be seen with human eyes. In the earlier years when the resolution of scientific imaging instruments was not very high and the process of imaging was not automated, the volume of the resulting images was manageable. Analyzing the scientific images was by and large manual in nature, yet the manual process was not considered a big burden then. When we first worked on material images nearly 15 years ago, the high-end electron microscope that we could access produced 500 × 500-pixel images with the resolution of several nanometers per pixel. The time spent from sample preparation to the actual microscope imaging session was over several hours. In a single imaging session, roughly tens of microscopic images were produced. Despite all the complexities in the resulting images and the labor-intensive process of obtaining them, the manual analysis of material images was the norm and default in those days.

Driven by scientific curiosity to see smaller scale objects and fast-changing phenomena and aided with technological prowess and innovations, the improvement in spatial and temporal resolutions of scientific images has seen a dramatic acceleration in the past decade, and the pace of the improvement follows the Moore's Law in the semiconductor industry. Consequently, the data volume and generation rate of scientific images increase at an unprecedented speed. Nowadays the spatial resolution of electron microscopes goes easily below sub-nanometers, resolving every single atom. Millions of such high-resolution images could be produced every second while capturing interesting physical, chemical, and biological phenomena. Manual analysis of such huge volumes of scientific images, easily reaching terabytes per one imaging session, is no longer feasible. If indeed attempted, the manual analysis could take months of dedicated work by a team of experts to obtain a partial analysis. There is a pressing need for automated image analysis of scientific images in order to keep up with the fast pace of data generation.

In spite of its importance, automated material image analysis has progressed relatively slowly, if one compares it to, say, the progress of bio-image analysis. There is a void in terms of research monographs and textbooks dedicated to the topic of automated material image analysis. This void motivated us to write this book covering a broad aspect of nano image analysis problems as well as recent research developments. While the contents of the book are presented in the specific context of nanomaterials, many methods covered herein can in fact be applicable to general material image analysis.

This book describes comprehensively the necessary steps in material image analysis, including mathematical representation of material images, material object detection, separation, and recognition, size and shape analysis, spatial pattern recognition at both local and global scales, time-series modeling of temporally resolved material images, visual tracking of a population of or individual materials for understanding or quantifying their changes, and image super-resolution that enhances the quality of raw images and benefits the downstream analyses.

We intend for this book to be a good reference both for material scientists who are interested in automated image analysis for accelerating their research and for data scientists who are interested in developing machine learning/data science solutions for materials research.

Tallahassee, FL, USA Chiwoo Park
May 2021

College Station, TX, USA Yu Ding
June 2021

Acknowledgments

We would like to acknowledge the contribution of our co-authors and collaborators to the research work that forms the backbone of this book: Patricia Abellán, Lawrence Drummy, Ali Esmaieeli, Rolland Faller, Jianhua Huang, Jim Ji, Xiaodong Li, Xin Li, Bani Mallick, Layla Mehdi, Trevor Moser, Lucas Parent, Joseph Patterson, Yanjun Qian, Brian Smith, Anuj Srivastava, Mollie Touve, Garret Vo, David Welch, and Jiaxi Xu.

A special thanks goes to the fellow microscopists: Nigel Browning, James Evans, Nathan Gianneschi, Hong Liang, and Taylor Woehl. The collaborations with them motivated most of the research reported in this book. They provided material images that are hard to get elsewhere and shared their knowledge and experiences in materials image analysis. Our research and contribution would not have been possible without their collaborations.

We would also like to acknowledge the generous support from our sponsors, particularly, the Dynamic Data and Information Processing (DDIP) program (formerly the Dynamic Data-Driven Applications Systems, or the DDDAS program) of the Air Force Office of Scientific Research (AFOSR) and the Cyber-Physical System (CPS) program of the National Science Foundation. We extend a special thanks to Dr. Frederica Darema, former Director of AFOSR, for her appreciation in our work and heartfelt encouragement when we faced difficulties. The countless stimulating discussions with Dr. Darema drove our research to new levels.

The planning for this book started when Camille Price, Series Editor, Springer International Series in Operations Research and Management Science, reached out to us in 2018. We appreciate Camille for her patience and guidance at the book-proposal stage and for meeting with us a couple of times. In the ensuing years during the book-writing process, the Springer team did a fantastic job in managing this project and assisting us in numerous occasions.

Chiwoo is grateful to Florida State University for supporting a year-long sabbatical leave. That time period gave him more freedom and allowed him to concentrate on writing this book. The book's writing would have taken much longer to complete without this support. Thanks to his former students, Xin Li, Garret Vo, and Ali Esmaieeli, for some of the excellent works in this book.

Yu is grateful to his former students and co-authors, Yanjun Qian and Jiaxi Xu, for arranging the image data and computer code for Chaps. 5, 7, 9, and 11. Yu expresses his gratitude to Mike and Sugar Barnes for their generous endowment that supports his work in this important field. Yu is also indebted to his Ph.D. advisor, Dr. Jianjun Shi, at the Georgia Institute of Technology for bringing him to the data science world and for teaching him to be an independent researcher.

Last but not least, Chiwoo would like to express sincere thanks to his wife Sun Joo Kim who has been very supportive of this year-long book-writing project during the difficult time of the global outbreak of COVID-19. Without Sun Joo's sacrifice, Chiwoo could not have been able to concentrate on writing the book. Yu would like to thank his wife, Ying, and their daughter, Alexandra, for their love and support.

Contents

Acronyms

ADMM	Alternating direction multiplier method
AIC	Akaike information criterion
AMISE	Asymptotic mean integrated squared error
ANOVA	Analysis of variance
ARL	Average run length
ARMA	Autoregressive moving average
BIC	Bayesian information criterion
BSP	Binary segmentation process
CSR	Complete spatial randomness
DP	Dirichlet process
EDSR	Enhanced deep-residual networks super-resolution
EM	Expectation maximization
fps	Frames per second
GPU	Graphics processing unit
HPRC	High performance research computing
ID	Index of dispersion
IEEE	Institute of Electrical and Electronics Engineers
i.i.d	Independently, identically distributed
IG	Inverse gamma
LASSO	Least absolute shrinkage and selection operator
LB	Library-based
LP	Linear programming
NLM	Non-local mean
MAP	Maximum a posterior
MCMC	Markov chain Monte Carlo
MCMC-DA	Markov chain Monte Carlo data association
MLE	Maximum likelihood estimation or maximum likelihood estimator
MOTA	Multi-object tracking analysis
MSE	Mean squared error
MWDA	Multi-way minimum cost data association
NN	Nearest neighborhood

NPSD	Normalized particle size distribution
OA	Orientated attachment
OR	Ostwald ripening
PC	Principal component(s)
PCA	Principal component analysis
pdf	Probability density function
PELT	Pruned exact linear time
PSD	Particle size distribution
PSNR	Peak signal-to-noise ratio
RCAN	Residual channel attention networks
ReLU	Rectified linear unit
ScSR	Sparse-coding based super-resolution
SE	Shannon entropy
SEM	Scanning electron microscope
SISR	Single image super-resolution
SK	Skewness
SNR	Signal-to-noise ratio
SSIM	Structural similarity index measure
SPC	Statistical process control
SPM	Scanning probe microscope
SR	Super resolution
SRCNN	Super-resolution convolutional neural network
SRSW	Super-resolution by sparse weight
SQC	Statistical quality control
SSE	Sum of squared errors
STEM	Scanning transmission electron microscope
STM	Scanning tunneling microscope
SVM	Singular value decomposition
TEM	Transmission electron microscope
VDSR	Very deep convolutional neural network for super resolution
UE	Ultimate erosion
UECS	Ultimate erosion for convex set
WBS	Wild binary segmentation

Chapter 1
Introduction

Materials, according to Merriam-Webster online, are "*the elements, constituents, or substances of which something is composed or can be made.*" Invention or discovery of new materials has significantly influenced the course of civilization. For example, the discovery of metals prompted the transition from the Stone Age to the Bronze Age, changing drastically the ways to grow crops and enabling people to settle in cities and countries. Various metals and metal processing techniques brought out a series of transitions from the Bronze Age through the Iron Age to the industrial revolution, fundamentally altering human societies. The discovery of electrons and silicon technology led to the development of vacuum tubes, transistors, and semiconductors. More recent materials discoveries include nanomaterials and biomaterials, both of which are regarded as promising materials to shape the future of our lives.

Nanoscience and engineering deal with a specialty material—nanomaterials, defined as materials with nanoscale structures, i.e., either some external dimensions or internal structures are in the nanoscale. By *nanoscale*, we refer to a length scale between one and one hundred nanometers. Figure 1.1 presents a few such examples, of which a nanoparticle is a nanomaterial with all of its external dimensions in the nanoscale, a nano-fiber (nanotudes as hollow nanofibers and nanorods as solid nanofibers) has two of its dimensions in the nanoscale, and a nanoplate has only one of its dimensions in the nanoscale.

The basis of modern materials science and engineering is a trilogy of processing, structure, and properties of materials. The trilogy starts off with studying the structure of a material and its relation to the properties and performance of the material. Understanding of and insights into the structure-property correlation lead to designing the right structure as well as developing new processing or synthesis for making the structure. As such, the desired properties for intended uses can be materialized. Studying the processing-structure-property interrelation appears to be the major paradigm of the modern materials science and engineering (Reid 2018). This major paradigm certainly influences the study of nanoscience and engineering.

C. Park, Y. Ding, *Data Science for Nano Image Analysis*, International Series in Operations Research & Management Science 308,
https://doi.org/10.1007/978-3-030-72822-9_1

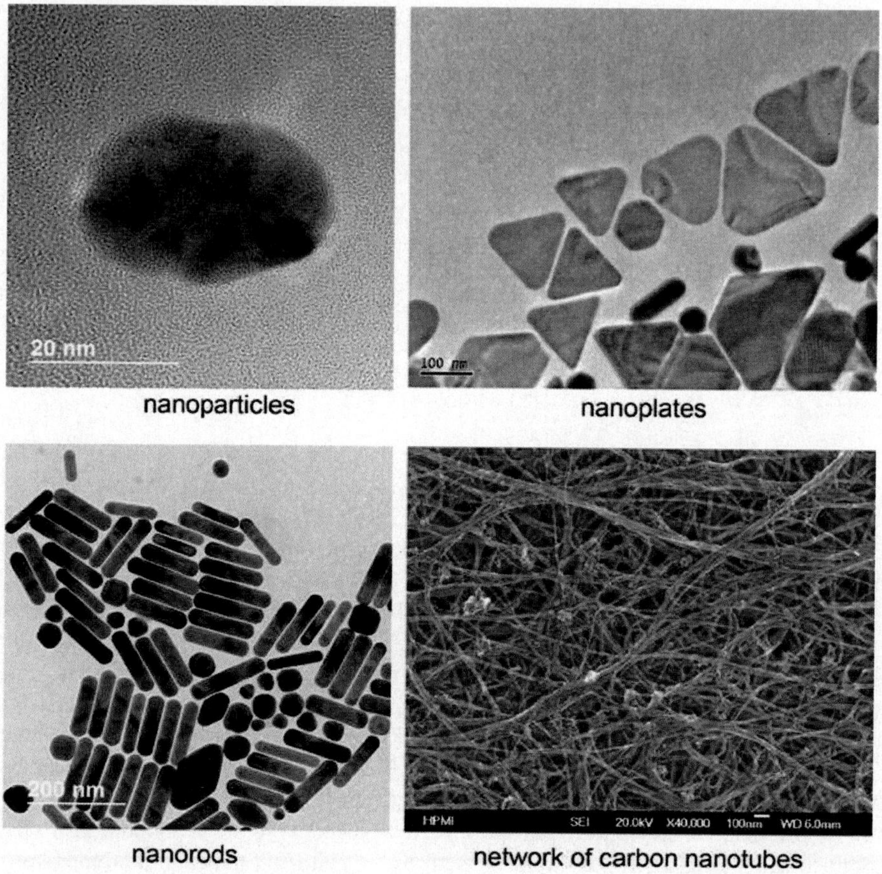

Fig. 1.1 Examples of nanomaterials

In the nanoscale, the structure of a material is more influential to the material's properties or performance, than in traditional scales, making the nanoscale structure an even more crucial factor to be researched and characterized.

While the processing-structure-properties correlation has been studied with a combination of theory, computation, and experiments, metrology and processing techniques play a pivotal role in advancing the capability of materials characterization (Fig. 1.2). There are two major techniques for nanoscale structural characterization—the microscopy imaging techniques and spectroscopy techniques. Both techniques produce imagery data. The microscopy techniques produce real-space images, i.e., images in a spatial domain, while the spectroscopy techniques produce reciprocal-space images, i.e., images in the frequency domain. Theoretically, a reciprocal-space image can be converted to a real-space image by taking the inverse Fourier transform or something equivalent. For many instruments in use,

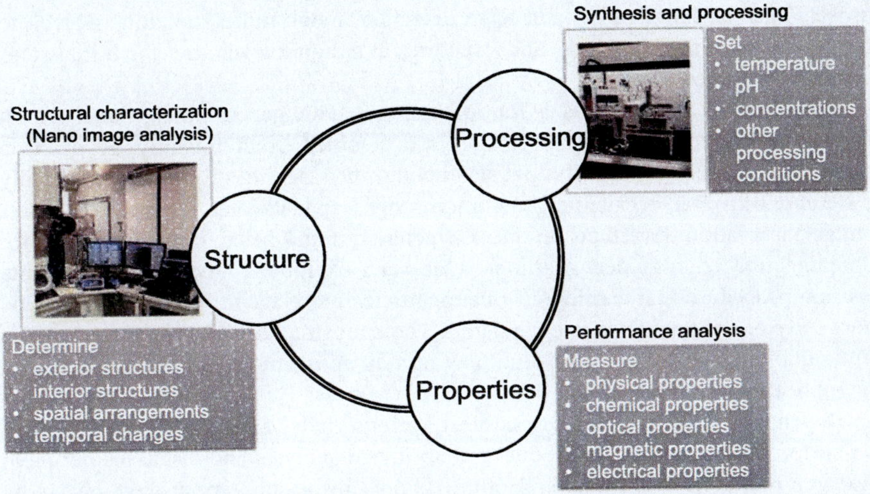

Fig. 1.2 The role of nano image analysis in nanomaterials research, which fulfills principally the characterization of nanoscale structures of materials synthesized under applied processing conditions

however, the reciprocal-space images are taken only in a single dimension, which hampers such inverse conversion. In other words, the reciprocal-space image are mostly in a functional form, rather than in imagery forms.

In this book, the real-space images taken from nanomaterials by microscopes are referred to as nano images. Most of the nano images are understandably two-dimensional, but with recent advances in microscopy, three-dimensional or four-dimensional images are also made available. The three dimensions can be either just three spatial dimensions or two spatial dimensions plus one temporal dimension. Likewise, the four dimensions can be three spatial dimensions plus one temporal dimension, or two spatial dimensions in the real space plus the other two dimensions in the reciprocal space. Chapter 2 will provide a detailed account of the existing microscopes and the corresponding image formats.

Broadly speaking, the purpose of nano image analysis, to be discussed in this book, is to analyze nano images and extract the structural information of nanomaterials from the images. With the processing-structure-property paradigm in mind, it is not difficult to see the significant role that nano image analysis plays, which is a key part in structural characterization and the important enabler for many subsequent actions in materials discovery, design, and synthesis. We want to stress that nano image analysis, like the general image analysis, is basically a data science, considering the rich amount of data, embedding all sorts of complexities, that the nano images bring about. There are certainly specialized tools used for image analysis and processing, but the readers will discover that many modeling tools used in analyzing image data nowadays are also commonly used for other data science problems. In this book, while touching upon basic image processing aspects such as binarization and edge detection, much more attention is given to how to model material objects and structures at a higher level after the initial steps of image preprocessing.

Nano image analysis has a lot in common with generic image analysis. It also has its uniqueness. Nano images are obtained predominantly by electron microscopes, meaning that the physics behind the imaging process is profoundly different from that of visible light microscopes and cameras. The nano objects under observation have their unique characteristics and behaviors, which calls for domain knowledge-guided solutions. One such example is the tailored ultimate erosion procedure that can identify nanoparticles more accurately when the shape convexity of nanoparticles is considered. The nanostructures also presents a range of unique research problems, which may or may not be prevalent in other domains or applications.

Depending on the structures of interest, various kinds of nano image analyses can be performed. Section 1.1 introduces nano image analysis and discusses pertinent research problems by examples. Section 1.2 presents a summary of the topics to be covered in this book and how they are organized. Section 1.3 describes how this book can be used for various groups of intended readers. The last section of this chapter lists the information for online supplementary materials, including datasets and example codes.

1.1 Examples of Nano Image Analysis

The structural features of interest in nano image analysis include individualized features such as morphology (size and shape), facets and interior organization (interior atomic arrangements), as well as ensemble features such as the distribution of the individualized features of a group of nano objects and the local and global spatial arrangements of individual objects (dispersion and network). When a time-resolved material imaging capability is available, the temporal evolution of these structural features for a history of the applied processing conditions is of great interest as well.

In the sequel, we describe a number of nano image analyses for the purpose of delivering concrete examples and guiding the reading of this book. Four examples are presented, as shown in Fig. 1.3, exemplifying the four main categories of nano image analysis. The structural information of interest is highlighted in Fig. 1.3 for each category of the analysis, which is, from top-left to bottom-right, the morphology (size and shape), spatial arrangements (location, dispersion and lattice), distribution changes, and motion and interaction.

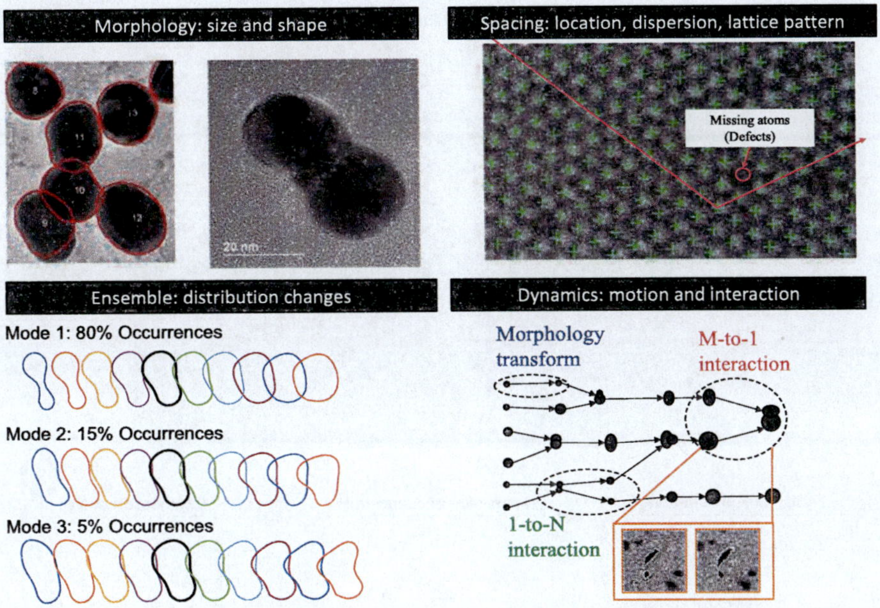

Fig. 1.3 Four main categories of nano image analysis to be discussed in this book

1.1.1 Example 1: Morphology

Our first example is the morphology analysis of nanoparticles. Morphology analysis is to analyze the sizes, shapes and other structural parameters related to the exterior outlines of materials. The morphology analysis of nanoparticles is an essential step in the scientific studies of the morphology-property correlation of the nanoparticle-embedding materials. It is also an integral component when people want to check whether the sizes and shapes of synthesized nanoparticles are within the desirable ranges or design intents, an important question to be addressed for quality monitoring of nanoparticle fabrications.

The input data for morphology analysis is a microscope image of nanoparticles. As illustrated in Figure 1.4, the exterior outlines of nanoparticles shown in the input image are extracted (the top plot). The size and shape information is quantified using the extracted outlines. Then, the size and shape information can be statistically analyzed to provide important statistics such as the probability distributions of sizes, shapes, aspect ratios, and so on. Below summarizes the data analysis sequence:

$$\text{image} \Rightarrow \text{outlines} \Rightarrow \text{sizes, shapes, other structure parameters} \Rightarrow \text{statistics}$$

The step of the analysis encounters technical challenges, including how to identify material objects and their outlines buried under noises, how to recover outline occlusions due to overlaps among materials, how to define and model the

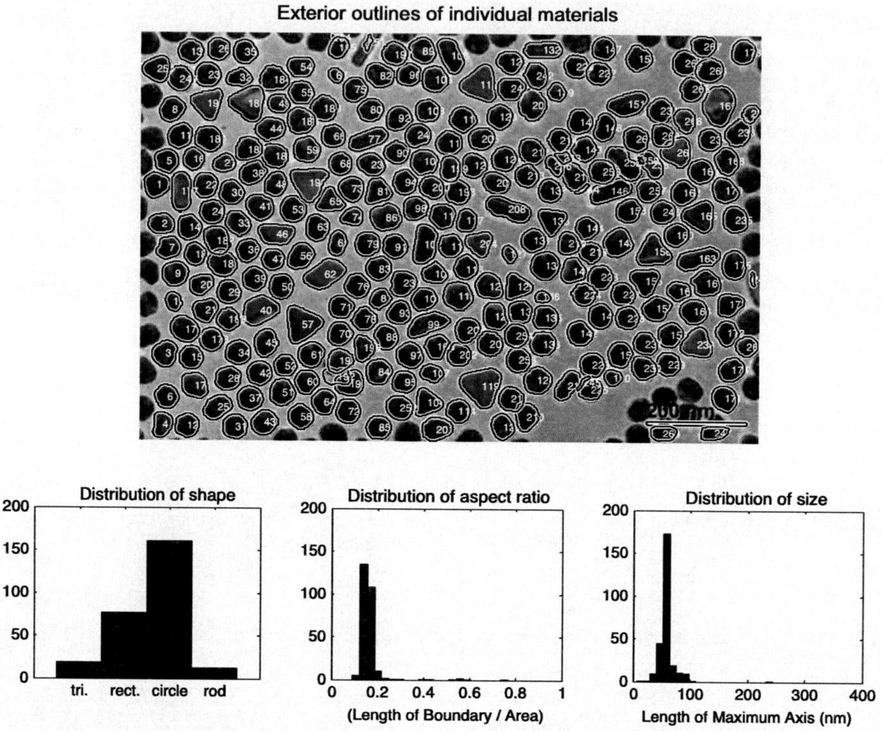

Fig. 1.4 Morphology analysis

sizes and shapes of the outlines mathematically, how to define the probability spaces of the sizes and shapes, how to cluster or group by sizes and shapes and calculate the mean shape and its covariance for a group of materials, and how to perform statistical inferences on a shape space such as model parameter estimation and hypothesis testing. Methodologies addressing these challenges come from the studies of image segmentation in computer vision, statistical shape analysis in statistics, and computational geometry in applied mathematics.

1.1.2 Example 2: Spacing

Our second example is about the interior micro-patterns of materials—specifically, the spatial positioning and arrangements of smaller scale elements within materials. Figure 1.5 presents an example, which is the crystallographic structure of Mo-V-M-O catalysts captured by a high-angle annular dark field scanning transmission electron microscope (HAADF-STEM). The top figure in Fig. 1.5 shows the STEM image taken at an atomic resolution, where each atom is imaged as a white spot over the dark background. Given the input image, many research questions naturally follow, for example, how atoms are spaced, whether the spacing is symmetric (the

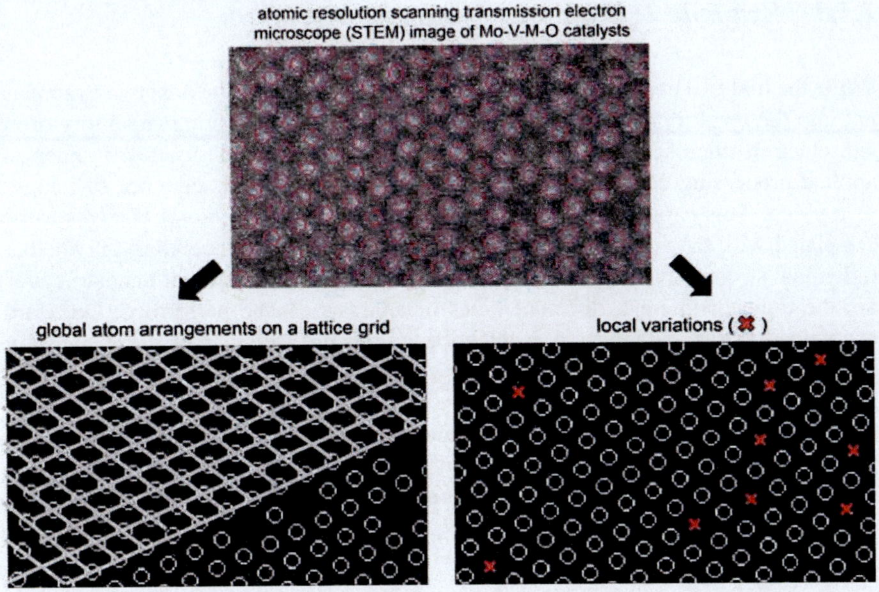

Fig. 1.5 Spatial positioning and arrangements of smaller scale elements within materials

bottom-left figure), and whether there are local variations that break the symmetry (the bottom-right figure). The analysis of such arrangements could yield insights concerning functionalities of materials.

The major steps for this analysis is to first identify all white spots in the image and then analyze the centroid locations of the white spots for identifying possible symmetries or symmetry violations. The first step is well known as the spot detection problem in image processing (Hughes et al. 2010). The spatial symmetry in the second step is mathematically represented as a lattice in geometry, the symmetry group of discrete translational symmetric in $\mathbb{R}^d$, i.e., the d-dimensional space. A point on a lattice in $\mathbb{R}^d$ is represented by a weighted sum of d linear independent vectors in $\mathbb{R}^d$ with integer weights. The basis vectors need to be estimated using the information associated with the detected spots. Not every white spot conforms to the lattice, causing local non-conformity to the global lattice. While many earlier approaches attempt to solve the first and second steps sequentially, more recent attentions are given to integrated approaches, in which spot detection and spatial symmetry analysis are solved together for more robust solutions.

1.1.3 Example 3: Temporal Evolution

While the first two examples use still (snapshot) nano images, the next two examples get into the territory of dynamic imaging. With the time-resolving capability of in situ electron microscopes, the temporal evolution of material structures under an applied processing condition can be imaged in the form of a sequence of images or a video. This capability provides the potential for the structural changes to be correlated with the applied processing condition over time. Depending on whether individual structural changes or the ensemble changes for a group of nanostructures are the topic of interest, different types of analysis can be performed. Our third example, illustrated via Fig. 1.6, is intended to show the need to track the changes for a group of nanostructures in the form of a probability distribution of the structural characteristics.

In the example of Fig. 1.6, a silver nanoparticle growth process is studied. The growth of silver nanoparticles is initiated by an electron beam shot on the growth solution. The growth process is video-imaged by an in situ scanning transmission electron microscope. The top row in Fig. 1.6 presents a few image frames taken

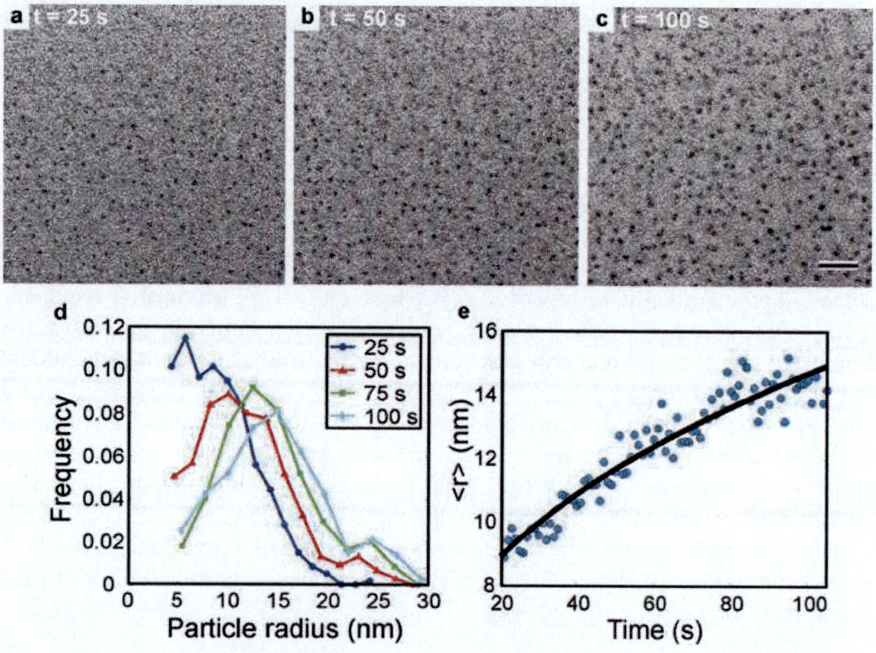

Fig. 1.6 Temporal evolution in particle size distribution. Panels (**a**)–(**c**) show three image snapshots taken from the video of a silver nanoparticle growth process. Panel (**d**) shows the particle size distributions at different times. Part (**e**) shows the trend of the average particle sizes over time. (Reprinted with permission from Woehl et al. 2013)

from the video. The goal of the nano image analysis here is to track the probability distribution of particle sizes over time and understand the particle growth behavior such as the mean growth rate. Figure 1.6d shows the particle size distributions at a few selected time points, whereas Fig. 1.6e shows a polynomial curve fitted to the mean particle size versus time. The exponent of the polynomial quantifies the nanoparticle growth rate. Apparently, a prerequisite for this dynamic image analysis is the morphology analysis of material objects in each image frame.

1.1.4 Example 4: Motions and Interactions

In addition to distribution changes, which are associated with a group of nano objects or structures, individual motions and interactions are also of great interest in materials research. Figure 1.7 shows our fourth example, in which the topic of interest is to understand the orientations of two nanoparticles when they aggregate or merge. Such insights are important to studying how larger nanoparticles are formed and how nanoparticles grow. The in situ scanning transmission electron microscopy, essential to the study in Sect. 1.1.3, is also used for this study. A video sequence of particle aggregation events is captured and then analyzed.

A number of nanoparticle aggregation events are identified and extracted from the video by multi-object tracking algorithms. Figure 1.7a shows two such examples, both of which are aggregations involving two elongated nanoparticles. Figure 1.7b illustrates how the aggregation events are modeled. The orientations of the two elongated objects in the merge are defined as the center coordinates of the merge zone, which is $c_{X,Y}$ in the figure, with respect to the respective coordinate frames of the two objects. The orientations of the two objects are denoted by θ_X and θ_Y, respectively. Figure 1.7c presents the probability distributions describing the orientations. The top-left panel of Fig. 1.7c is the joint distribution of θ_X and θ_Y, fit to the observed data of θ_X and θ_Y, The bottom-left and top-right panels of Fig. 1.7c present the marginal distributions of θ_X and θ_Y, respectively. The bottom-right panel of Fig. 1.7c depicts the mean orientation of θ_X and θ_Y in the fitted distributions. In this example, the bivariate von Mises distribution is used to model the joint distribution. Once the joint distribution is fit, many statistical analyses can be performed; for example, a hypothesis test can be conducted to check whether the two samples of nanoparticle aggregations exhibit the same mean orientations or not.

1.2 How This Book Is Organized

We organize this book into two large parts of nano image analysis: the pre-processing and the main analysis (Fig. 1.8).

The pre-processing analysis is a set of image processing methods that can be performed on any images before the main analysis. This includes image

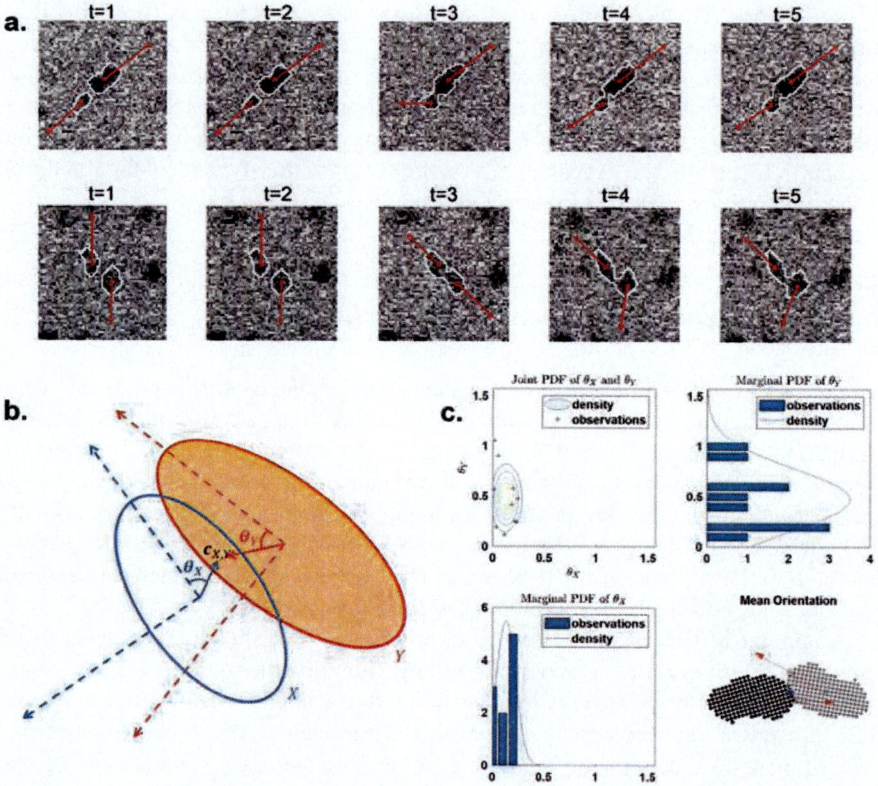

Fig. 1.7 Analysis of particle motions and inter-particle interactions. Panel (**a**) shows two sequences of smaller nanoparticles merged into a larger one. Panel (**b**) presents the diagram illustrating how the orientations of the primary particles involved in the merge are mathematically modeled. Panel (**c**) shows the empirical distributions of the orientations of the two particles observed in multiple merging events, as well as the bivariate angular distribution fitted to the image data

representation, image super-resolution enhancement, and image segmentation. As shown in the top panel of Fig. 1.8, the topics of pre-processing are covered in the following three chapters:

- **Chapter** 2 introduces various kinds of nanoscale imaging instruments, i.e., different types of microscopes, and explains the mathematical representations of the resulting material images. Image representations lay the foundation for mathematical formulations and technical solutions in the subsequent nano image analyses.
- **Chapter** 3 describes image segmentation, essential for separating the material objects and structures of interest from the noisy background. Image segmentation is the prerequisite to almost all subsequent higher level image analyses.

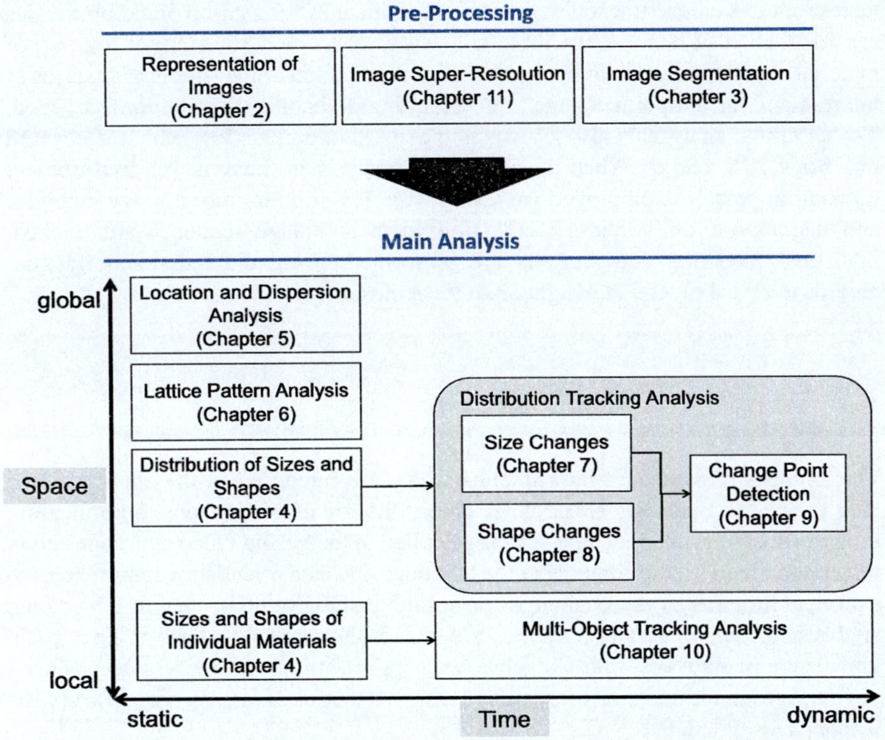

Fig. 1.8 The organization of the topics in the book

- **Chapter** 11 discusses an image enhancement approach, known as super-resolution. Image super-resolution is to improve the resolution of raw microscope images by exploiting the relation between a pair of low- and high-resolution images. It is an optional step before image segmentation and the main analysis. But if multi-resolution image data are indeed available, super-resolution actions can definitely help deliver better quality in later analyses.

The organization of main analysis is based on the structural and temporal aspects of nano image analysis, as shown in the bottom panel of Fig. 1.8.

When the topics of static analysis are covered in Chaps. 4 through 6, the focus of the analysis is the static structural features of nanomaterials at the last stage of processing or a certain chosen time point. The structural features of interest include morphology of the exterior outlines of nanomaterials (in **Chap.** 4), spatial arrangements of nanoscale objects in a bulk material (in **Chap.** 5), and spatial arrangements of atoms in the interior of a nanomaterial (in **Chap.** 6).

When the topics of dynamic analysis are covered in Chaps. 7 through 10, the temporal changes of nanostructures are the focus. The modeling and analysis concerning temporal changes are colloquially referred to as *tracking*—individual nanostructures can be tracked or the ensemble changes for a group of nanomaterials can be tracked. When the ensemble analysis is concerned, the state of a group of material objects is described by a probability distribution of the structural features of interest, and the temporal change in the resulting probability distribution is tracked. The ensemble analysis is also known as the distribution tracking, which is covered in **Chaps.** 7, 8, and 9. When the individual analysis is concerned, a multi-object tracking approach is employed (in **Chap.** 10). The motion, morphology changes, and interactions of individual nano objects in an image sequence are tracked. Statistical modeling and analysis are performed, using the individual tracking records to reveal profound insights into the nanoscale world.

1.3 Who Should Read This Book

This book is designed for data scientists who are interested in the application of data science to materials research. It covers the topics of statistics, optimization, image processing, and machine learning applied to image and video data captured in materials research. The contents of the book provide data scientists a comprehensive overview in terms of the challenges and state-of-the-art approaches in image data analysis for nanomaterials research. We expect the readers to have at least basic knowledge in statistics, linear algebra, and optimization. The readers do not need any background in materials research, because the book is largely written from the perspectives of a data scientist.

Materials scientists and practitioners who are interested in artificial intelligence and machine learning (AI/ML) could also benefit from this book. In the past few years, we witness a fast increasing presence of AI/ML sessions and workshops in major materials research conferences, such as the Materials Research Society (MRS)'s Spring and Fall Meetings and the Microscopy & Microanalysis (M&M) conference. We also noticed that many students from the materials science departments are taking, or have taken, machine learning courses. With the basic knowledge of statistics and machine learning, materials scientists can branch out to more advanced topics by using this book, which builds upon examples and case studies taken from materials research. Practitioners can run the algorithms and methods through the code and examples accompanying the book and apply the solutions for their own benefit.

1.4 Online Book Materials

Either one of the following book websites has the example codes and datasets used in the book:

https://www.chiwoopark.net/book/dsnia.
https://aml.engr.tamu.edu/book-dsnia.

The contents on both websites are the same. Readers please feel free to access either of them.

References

Hughes J, Fricks J, Hancock W (2010) Likelihood inference for particle location in fluorescence microscopy. The Annals of Applied Statistics 4(2):830–848

Reid R (2018) Inorganic Chemistry. ED Tech - Press, Waltham Abbey Essex, UK

Woehl TJ, Park C, Evans JE, Arslan I, Ristenpart WD, Browning ND (2013) Direct observation of aggregative nanoparticle growth: Kinetic modeling of the size distribution and growth rate. Nano Letters 14(1):373–378

Chapter 2
Image Representation

2.1 Types of Material Images

2D Material Images Most material imaging techniques produce two-dimensional material images over a lateral plane of a material specimen as shown in Fig. 2.1. These material imaging techniques are largely grouped into optics-based microscopes and probe-based microscopes. The optics-based microscopes emit a beam of lights (photons) or charged particles (electrons and ions) to a material specimen of interest and exploit the interaction of the light or charged particle beam with the specimen to produce images. Depending on the beam sources, we have various types of microscopes, such as the optical microscopes (visible light beam), X-ray microscopes (X-ray beams), electron microscopes (electron beams), and ion-beam microscopes (ion beams). The wavelengths of the beam sources influence the resolutions of these microscopes. In the optics-based microscopes, various modes of beam-material interactions such as diffraction, reflection and refraction have been exploited for imaging. For example, in a transmission electron microscope (TEM), an electron beam of uniform current density irradiates one side of a thin material specimen, and the electron current density on the other side of the specimen is imaged through multiple stages of magnification lens (Reimer 2013). A scanning electron microscope (SEM) has a focused beam of electrons probe a raster over a material specimen, and the secondary electrons emitted for each raster spot are then detected (Reimer 2013). The second electron yield is related to the topography of the specimen. The electron probe size is 0.5–2 nm diameter if a field emission gun is used, which determines the spatial resolution of imaging. A scanning transmission electron microscope (STEM) uses a focused electron beam like SEM to probe over a raster of a specimen but measures the transmitted beam intensity like TEM (Pennycook and Nellist 2011).

C. Park, Y. Ding, *Data Science for Nano Image Analysis*, International Series in Operations Research & Management Science 308,
https://doi.org/10.1007/978-3-030-72822-9_2

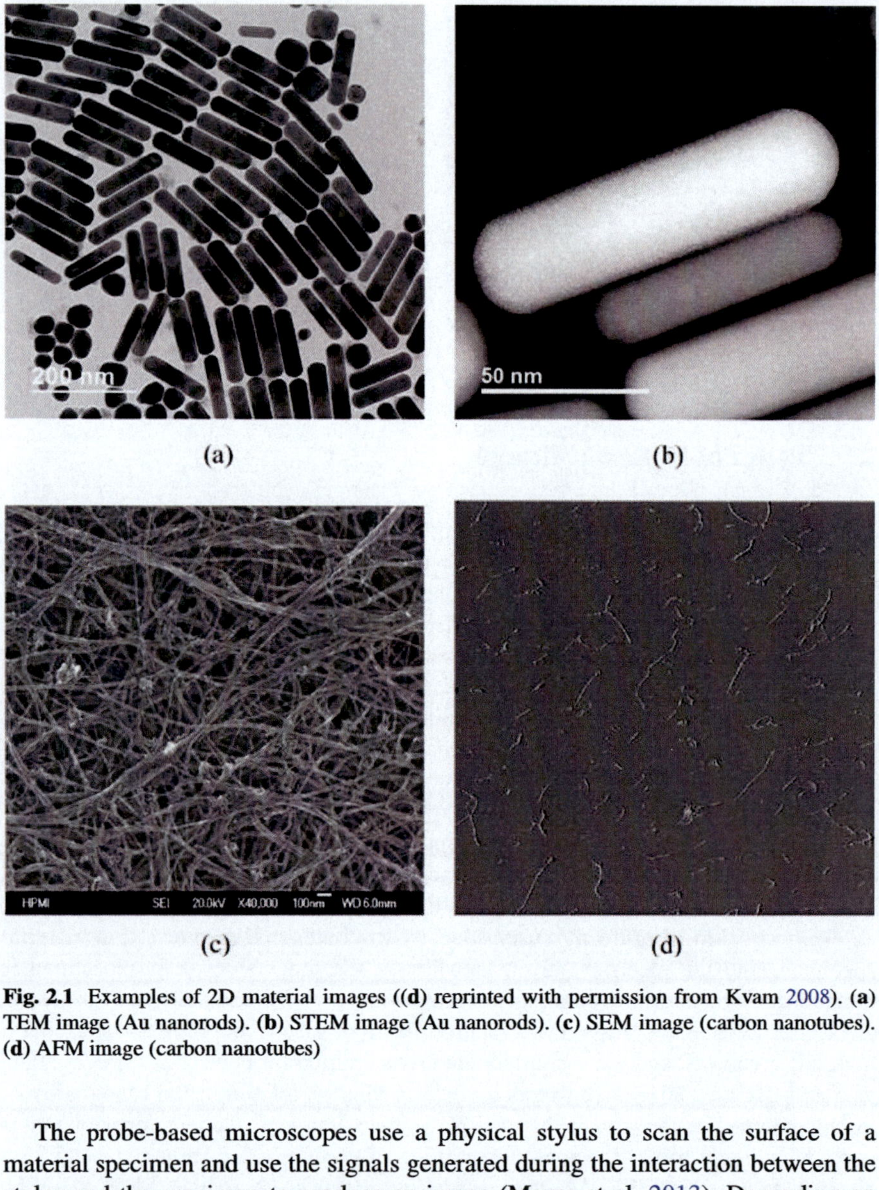

Fig. 2.1 Examples of 2D material images ((**d**) reprinted with permission from Kvam 2008). (**a**) TEM image (Au nanorods). (**b**) STEM image (Au nanorods). (**c**) SEM image (carbon nanotubes). (**d**) AFM image (carbon nanotubes)

The probe-based microscopes use a physical stylus to scan the surface of a material specimen and use the signals generated during the interaction between the stylus and the specimen to produce an image (Meyer et al. 2013). Depending on the types of the interactions exploited, the probe-based microscopes can be subcategorized. In a scanning tunneling microscope (STM), a conductive tip is brought very close to the surface of a specimen, and the tip position is moved over the surface. The tunneling current in the tip and the specimen at each probing location is measured and used to produce images. The tunneling current is a function of

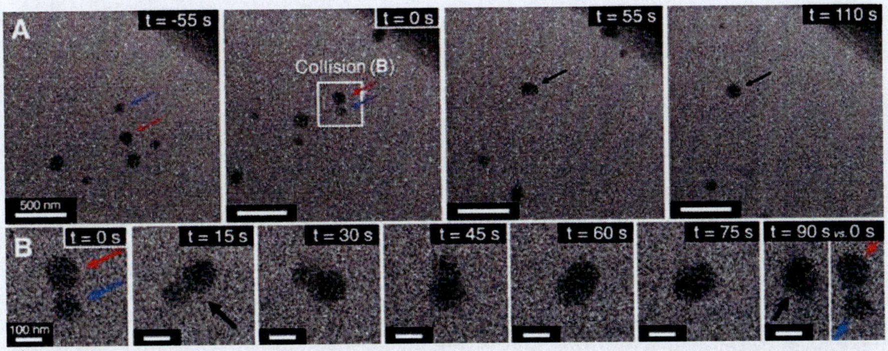

Fig. 2.2 Examples of 2D material video: Micelle fusion process (reprinted with permission from Parent et al. 2017)

the surface height and the local material density of the material specimen (Binnig and Rohrer 2000). Atomic force microscope (AFM, Fig. 2.1d) can work in three modes: contact mode, non-contact mode, and tapping mode (Binnig et al. 1986). In the contact mode, a stylus tip scans the specimen in close contact with the surface, and the deflection of the tip is measured. The tapping mode is similar to the contact mode except that the tip moves up and down, like taping on the surface, to prevent the tip from being stuck to the surface. In the non-contact mode, the tip scans the specimen surface with a constant distance above the surface, and the Van der Waals force between the tip and the surface is detected to construct the topographic image of the surface.

2D Material Videos In situ microscopy (Ross 2007) is the imaging technique to capture the change of a material specimen in response to chemical or physical stimuli in real time. The imaging technique yields a sequence of 2D images recorded over the period of measurements as illustrated in Fig. 2.2. Most of the conventional electron microscopes and some scanning probe microscopes had not been equipped with this real-time imaging capability, mainly because of the nature of their sample preparation process. In order for the conventional electron microscopes to image properly, the sample specimen should be dried first and then placed in a high-vacuum environment. But once dried, the material samples can no longer change or evolve, as most of materials can only change in liquid or gas phases. In situ microscopy uses a special sample holder attached onto the conventional microscopes, which allows the natural phases of a sample to be placed for imaging. For examples, liquid-cell transmission electron microscopes use a liquid cell holding a very thin layer of liquid samples that are sandwiched by two silicon or graphene windows (Ross 2016). Such in situ capabilities were first applied to TEMs. The reason is that SEM rasters over a specimen and acquires its image pixel by pixel. In the earlier days, the raster speed is not fast enough to make the in situ capability meaningful. Unlike the scanning electron microscopes, TEM acquires all its image pixel simultaneously from a single electron beam shot and its imaging process is thus much faster. The scanning speed has since improved greatly, which makes the in situ capability also feasible for the scanning electron microscopes; see the example of in situ STEM (Woehl et al. 2013).

3D Material Images Material samples are inherently 3D. The 2D material images produced by the conventional microscopes are the results of projecting the 3D structures onto a 2D plane perpendicular to the beam irradiation direction or the results of simply scanning the surface of the 3D structures. While these 2D images provide important structural information, they have very little ability in revealing the 3D structures of material specimens. Three-dimensional imaging capabilities have been developed with three major tomography techniques. The first one is the electron tomography, in which a sample is rotated around a single axis, and a series of 2D TEM images are sequentially taken at different rotation angles. The series are analytically combined into a 3D image by a tomography reconstruction method (Mu and Park 2020). The second technique is the X-ray tomography, similar to the electron tomography, except for its use of X-ray photon as the beam source (Salvo et al. 2003). The third technique is the focused ion beam (FIB) tomography, which exploits SEM and a focused ion-beam microscope for achieving the 3D imaging capability. The tomography technique uses an ion beam to peel off the surface of a material specimen little by little via a sputtering process, while the SEM uses an electron beam to scan the new surface once the old surface is peeled off (Holzer and Cantoni 2012). This process creates a series of 2D surface images at different layers of the material specimen, and the 3D structure of the material can be reconstructed by stacking these image series.

4D Material Images Electron tomography produces a static 3D image of material structures: either the state of a solidified material or a snapshot of an evolving material. Four-dimensional electron tomography refers to the imaging technique that creates a series of 3D tomograms representing the dynamic changes of the 3D material structures (Kwon and Zewail 2010). 4D electron tomography first obtains a series of 2D images at a number of projection angles and times using ultrafast electron pulses and then reconstructs the series into 4D tomograms. Another 4D material image data comes from 4D STEM. Unlike the conventional STEM that yields one scattering intensity at each probe location over the raster lines, a 4D STEM yields 2D scattering patterns at each probe location; see an example in Fig. 2.3. The 2D scattering pattern is known as the convergent beam electron diffraction, which contains important local structure information.

The material images so far discussed are mostly available in the form of digital images in two to four dimensions. The digital image data can be modeled through various mathematical abstractions, including a function, tensor, graph and set. The choice of mathematical abstraction of the image data impacts considerably the subsequent image analysis. In the subsequent sections, we describe a few commonly used mathematical abstractions of digital images and the related analysis approaches.

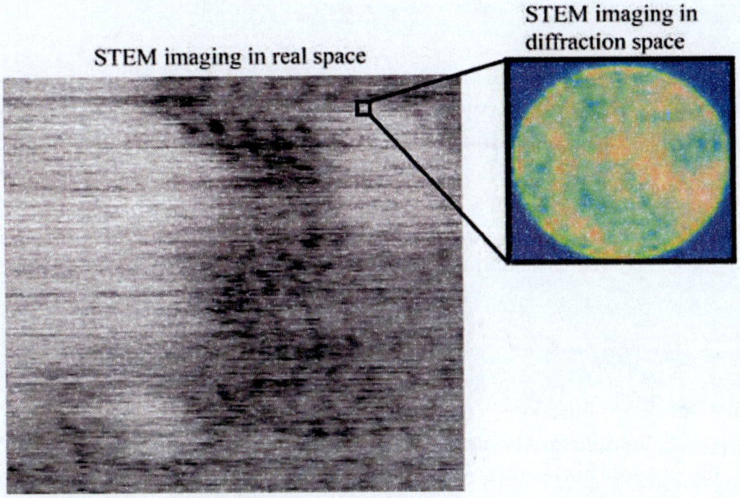

Fig. 2.3 Example 4D material images

2.2 Functional Representation

One can model an image as a function, $f : \mathcal{X} \mapsto \mathcal{Y}$, where $\boldsymbol{x} \in \mathcal{X}$ represents a location of an imaging space $\mathcal{X}$, and $f(\boldsymbol{x}) \in \mathcal{Y}$ represents the image intensity at the location (Sonka et al. 2014). The domain $\mathcal{X}$ represents the imaging space of two to four dimensions, typically rectangular and modeled as the product of respective real intervals, i.e., $\prod_{j=1}^{d}[0, W_j]$, where d is the dimension of the imaging space and $W_j > 0$ specifies the size of the domain for the j-th dimension. The co-domain $\mathcal{Y}$ is often a single-dimensional space of non-negative scalars for a grayscale image or a multi-dimensional space of vector quantities for color or multi-channel images.

For a digital image, both of $\boldsymbol{x}$ and $f(\boldsymbol{x})$ are digitized, implying that both the domain and co-domain are discretized to finite sets. The domain $\mathcal{X}$ is uniformly partitioned into grids of equal sizes, and the set of the grids defines the discrete domain for f. Each cubicle (in 3D) or quadrat (in 2D) of the smallest element in the X-domain partition is called a pixel. The physical dimension of a pixel is naturally known as the pixel size, e.g., a pixel of the size of one nm^2. The pixel size determines the spatial resolution of the image. Similarly, the value of the co-domain $\mathcal{Y}$ is also discretized into a discrete set. The carnality of the co-domain set is called the pixel depth, which determines the resolution of image intensities. For instance, a grayscale image with an 8-bit depth uses 2^8 integer values in the range of [0, 255] to represent the pixel intensity from the lightest to the darkest.

Using the functional representation, some basic image processing can be expressed mathematically as applying either a point-wise transformation or a convolution to f. The point-wise transformation is to apply a function g on the image output f point-wise, denoted as

$$g \circ f(x).$$

For example, the transformation function g can be a thresholding function, i.e., $g(y) = \mathbb{1}_{y \geq \tau}$, where $\mathbb{1}$ is the indicator function and τ is the threshold for binarization. such point-wise transformation converts a grayscale image to a binary image.

A convolution to f is a linear combination of f weighted by a filter function g. Many popular image filtering operations are represented by the convolution of an image $f: \mathcal{X} \mapsto \mathcal{Y}$, which is defined as

$$g * f(x) = \int f(x')g(x' - x)dx'. \tag{2.1}$$

Naturally, the filter function g defines the weights used in the convolution of f. The support of the filter function is typically symmetric around a zero-origin, so the result of the convolution is often a linear combination of the intensities of the pixels neighboring the evaluation location x. This is referred to as a spatial filter, since it is based on image pixels surrounding the evaluation location.

A commonly used spatial filter is the Gaussian filter, which takes the convolution of f with a Gaussian density function,

$$g(x) = (\pi h^2)^{d/2} \exp\{-x^T x / h^2\},$$

where $h > 0$ is the standard deviation of the Gaussian density. With this filter, the value of $g(x' - x)$ monotonically decreases as the Euclidean distance from x' to x increases. Therefore, the convolution (2.1) takes the weighted average of the intensities of the pixels neighboring x, with more weights given to the pixels closer to the evaluation location x, and produces a local estimate of $f(x)$. The Gaussian filter is popularly used for image denoising.

Another popular spatial filter is the mean filter

$$g(x) = \frac{1}{\int_{\mathcal{X}} dx}.$$

The filter function is a constant function, with which the convolution (2.1) is just the simple arithmetic average of the neighboring pixel intensities.

There are spatial filters that cannot be represented in the convolution form (2.1), like the rank statistics-based filters such as the median filter and the spatially weighted median filter (Mitra and Sicuranza 2000). These rank-based filters are more robust to outliers, working better on denoising Poisson noises.

There are non-spatial image filters such as the non-local mean filter (Buades et al. 2004), the bilateral filter (Elad 2002), and other variants (Lou et al. 2009), where the filter function does not only depend on the spatial distance $x' - x$ but also depends on $f(x') - f(x)$. Therefore, a non-spatial filtering cannot be represented in the convolution form but can be represented in a more general form of

$$g \otimes f(\boldsymbol{x}) = \frac{1}{C_g(\boldsymbol{x})} \int f(\boldsymbol{x}')g(\boldsymbol{x}, \boldsymbol{x}')d\boldsymbol{x}', \tag{2.2}$$

where $g(\boldsymbol{x}, \boldsymbol{x}')$ is a non-spatial filter function, and $C_g(\boldsymbol{x}) = \int g(\boldsymbol{x}, \boldsymbol{x}')d\boldsymbol{x}'$. For example, the non-local mean filter uses the following filtering function,

$$g(\boldsymbol{x}, \boldsymbol{x}') = \exp\{-b(\mu_f(\boldsymbol{x}) - \mu_f(\boldsymbol{x}')^2)\},$$

where $b > 0$ is a filter parameter, and μ_f is the local mean of $f(\boldsymbol{x})$ that can be obtained by applying the mean filter on f. These non-spatial filters are more effective to preserve local intensity jumps than spatial filters that tend to blur out sharp features.

Image reconstruction and estimation are more advanced image processing than the basic operations like image binarization or filtering explained above. The problem of image estimation is to estimate or infer a pixel's intensity based on either noisy pixel measurements or pixel measurements elsewhere. Using the functional representation of image data, the problem can be formulated as a regression analysis. The underlying noise-free image f is linked with its noisy measurements at N grid locations, $\{(\boldsymbol{x}_i, y_i) : i = 1, \ldots, N\}$, through the following model,

$$y_i = f(\boldsymbol{x}_i) + \epsilon_i,$$

where ϵ_i is an independent random noise with zero mean and variance σ_ϵ^2. The image de-noising problem can be formulated as a regularized learning problem, which is to find f that minimizes the sum of a loss function L and a regularization term, i.e.,

$$\underset{f}{\text{minimize}} \sum_{i=1}^{N} L(f(\boldsymbol{x}_i), y_i) + \lambda R(f), \tag{2.3}$$

where $R(f)$ is the regularization term that penalizes the complexity of function f, and the regularization term is added for avoiding the overfitting of f. A popular choice for the loss function is the L_2 norm defined by the squared difference $||f(\boldsymbol{x}_i) - y_i||_2^2$, as used in a least squares regression, and another choice is the L_1 norm defined by the absolute difference $|y_i - f(\boldsymbol{x}_i)|$, as used in robust regression.

The function f is typically represented in either a semi-parametric or a non-parametric form to accommodate complex patterns of images. In the literature, f is commonly represented by a basis expansion with a pre-determined basis function set, i.e.,

$$f(\boldsymbol{x}) = \sum_{k=1}^{K} \beta_k \phi_k(\boldsymbol{x}),$$

where $\phi_k(\boldsymbol{x}) : \mathcal{X} \mapsto \mathcal{Y}$ is the kth basis function, β_k is the respective coefficient that are to be estimated, and K is the number of the basis functions. Popular basis sets for a two-dimensional domain are the Fourier basis or wavelet basis in signal processing (Figueiredo and Nowak 2003) and the spline basis in statistics (Unser 1999). For more than two dimensions, a product basis is popularly used (Verhaeghe et al. 2007). The product basis implies that the basis function for a higher dimensional domain is constructed by taking the product of basis functions defined on low dimensional subspaces. For example, when $\mathcal{X} = \mathcal{X}_1 \times \mathcal{X}_2$ and $\phi_k^{(l)} : \mathcal{X}_l \mapsto \mathcal{Y}$ is the basis function on $\mathcal{X}_l$, the basis function on $\mathcal{X}$ is defined by the product form,

$$\phi_{k,k'}(\boldsymbol{x}) = \phi_k^{(1)}(\boldsymbol{x}_1)\phi_{k'}^{(2)}(\boldsymbol{x}_2)$$

for $\boldsymbol{x} = (\boldsymbol{x}_1, \boldsymbol{x}_2)$.

With the basis functions defined and chosen, the space of f is now spanned by the basis functions. As such, the function is uniquely decided once the coefficients of $\boldsymbol{\beta} = (\beta_1, \ldots, \beta_K)^T$ are estimated. The complexity function, $R(f)$, is defined as a function of $\boldsymbol{\beta}$. The popular choices are the L_p norms of $\boldsymbol{\beta}$ for using $p = 1$ or $p = 0$ (although $p = 0$ induces only a pseudo norm). With the orthogonal basis $\{\phi_k\}$ and the L_1 norm as the complexity function, solving the problem (2.3) produces soft-thresholding on an image, while the L_0 norm as the complexity function, the solution produces hard-thresholding with the threshold of λ.

A popular non-parametric representation of f is based on a local linear kernel smoother (Takeda et al. 2007), where the value of f is locally approximated around a neighborhood of $\boldsymbol{x}$ by a simple polynomial function,

$$f(\boldsymbol{x}) \approx \alpha_{\boldsymbol{x}} + \boldsymbol{\gamma}_{\boldsymbol{x}}^T \boldsymbol{x},$$

and with its coefficients $\alpha_{\boldsymbol{x}}$ and $\boldsymbol{\gamma}_{\boldsymbol{x}}$ estimated for each $\boldsymbol{x}$ by

$$(\hat{\alpha}_{\boldsymbol{x}}, \hat{\boldsymbol{\gamma}}_{\boldsymbol{x}}) = \arg\min \sum_{i=1}^{n} K_h(\boldsymbol{x} - \boldsymbol{x}_i) || y_i - \alpha_{\boldsymbol{x}} - \boldsymbol{\gamma}_{\boldsymbol{x}}^T \boldsymbol{x}_i ||_2^2,$$

where $K_h(\boldsymbol{x} - \boldsymbol{x}_i)$ is a kernel function with the bandwidth h, and n is the number of pixels in the prescribed neighborhood of $\boldsymbol{x}$. The estimated $\hat{\alpha}_{\boldsymbol{x}}$ becomes the estimate of f at $\boldsymbol{x}$. The kernel function K_h determines the weighting function of the ith data location $\boldsymbol{x}_i$ neighboring $\boldsymbol{x}$, which is typically defined as an isotropic function depending on the distance of $||\boldsymbol{x} - \boldsymbol{x}_i||_2$. The isotropic smoothing gives a consistent estimator of f when f is continuous and smooth around $\boldsymbol{x}$. When f is discontinuous or has directional changes around $\boldsymbol{x}$, locally adaptive kernel or one-sided kernel functions are used to accommodate the anisotropy (Takeda et al. 2008) or the discontinuity (Qiu 2009).

2.3 Matrix Representation

The observations of a function f at grid locations can be represented as a vector, a matrix, or more generally, a tensor. Such representation can be viewed as a discrete version of the functional representation. For example, a 2D grayscale image observed at $m \times n$ grid locations can be represented as an $m \times n$ matrix,

$$\boldsymbol{F} = \begin{bmatrix} f(1,1) & f(1,2) & \cdots & f(1,n) \\ f(2,1) & f(2,2) & \cdots & f(2,n) \\ \vdots & \vdots & \ddots & \vdots \\ f(m,1) & f(m,2) & \cdots & f(m,n) \end{bmatrix},$$

where $f(i, j)$ is the image intensity at the (i, j)th grid location. A c-channel 2D image is then represented by an $m \times n \times c$ tensor, and more generally, a d-dimensional image by a d-dimensional tensor.

With the matrix representation, image processing can be expressed as matrix operations. Unsurprisingly many types of digital image processing are defined as a matrix convolution, which is a discrete version of the convolution of two functions f and w on a two-dimensional domain (x, y). Recall that the convolution of f and w are expressed as

$$w * f(x, y) = \int_{-a}^{a} \int_{-b}^{b} w(s, t) f(x - s, y - t) ds dt,$$

where $[-a, a] \times [-b, b]$ is the support of w. Its discrete version is then

$$f_w(i, j) := w * f(i, j) = \sum_{s=-a}^{a} \sum_{t=-b}^{b} w(s, t) f(i - s, j - t),$$

where w is called a filter function of the convolution, and $f_w(i, j)$ is the image intensity of the convoluted image at the (i, j)th grid location. Using a matrix representation, the discrete convolution can be written as $\boldsymbol{W} * \boldsymbol{F}$, where $\boldsymbol{W}$ is the matrix representation of the filter function,

$$\boldsymbol{W} = \begin{bmatrix} w(-a,-b) & w(-a,-b+1) & \cdots & w(-a,b) \\ w(-a+1,-b) & w(-a+1,-b+1) & \cdots & w(-a+1,b) \\ \vdots & \vdots & \ddots & \vdots \\ w(a,-b) & w(a,-b+1) & \cdots & w(a,b) \end{bmatrix}.$$

Popular filter matrices and the result of their applications to a material image example are shown in Fig. 2.4.

Singular value decomposition (SVD) is another population matrix operation used in image analysis, which is to decompose an $m \times n$ matrix of rank p as the product of three matrices,

$$\boldsymbol{F} = \boldsymbol{U}\boldsymbol{D}\boldsymbol{V}^T,$$

where $\boldsymbol{U}$ and $\boldsymbol{V}$ are $m \times p$ and $n \times p$ unitary matrices respectively, and $\boldsymbol{D}$ is a $p \times p$ diagonal matrix with non-negative real numbers on the diagonal. When SVD is applied to a matrix $\boldsymbol{F}$ representing an image, many diagonal elements of $\boldsymbol{D}$ are close to zero, for which one can approximate the matrix by its low-rank version,

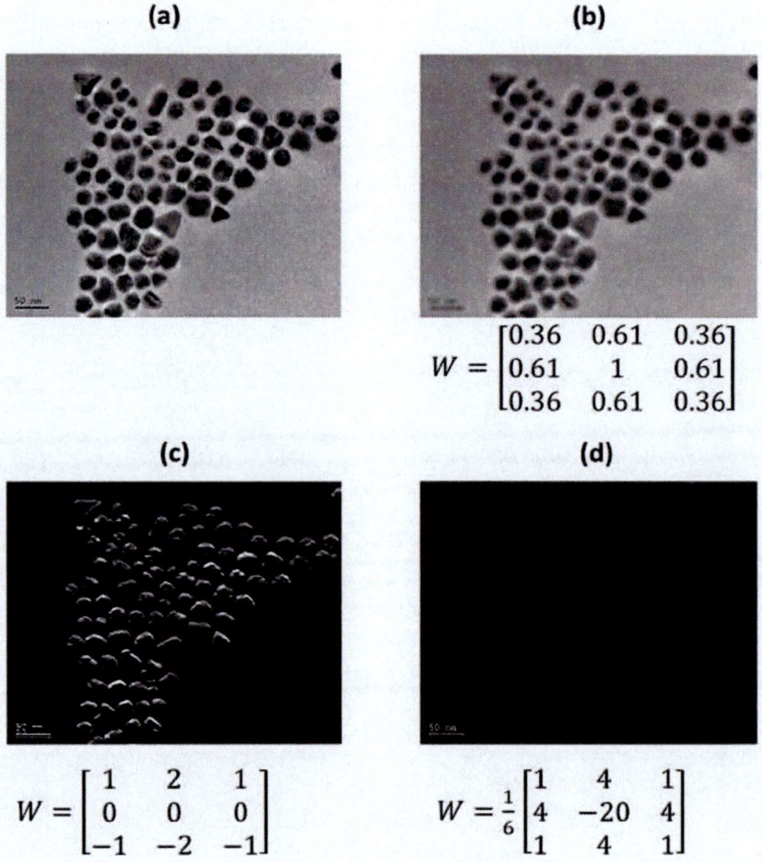

Fig. 2.4 Popular image filters and their exemplary applications. (**a**) Original image. (**b**) Gaussian filter. (**c**) Sobel filter. (**d**) Laplacian filter

$$U_q D_q V_q^T,$$

where U_q and V_q are the $m \times q$ and $n \times q$ matrices with the q columns of U and V that corresponds to the largest q diagonal elements of D, and D_q is a $q \times q$ diagonal matrix with the q largest diagonal elements of D. It can be shown

$$||F - U_q D_q V_q^T||_F^2 = \text{trace}(D) - \text{trace}(D_q),$$

where $|| \cdot ||_F$ is the Frobenius norm. The above expression says that the approximation error is the sum of the $q - p$ minor diagonal elements. This result is very useful, as it allows one to quantify information loss when compressing images or removing noises from images. Figure 2.5 illustrates different rank approximations to a 1000×1000 image matrix of rank 1000. Its rank-100 approximation is very close to the original image but it contains less image noise than the original image. The low-rank approximation can be used for more advanced image analysis such as subtracting the background and separating the foreground (Vo and Park 2018).

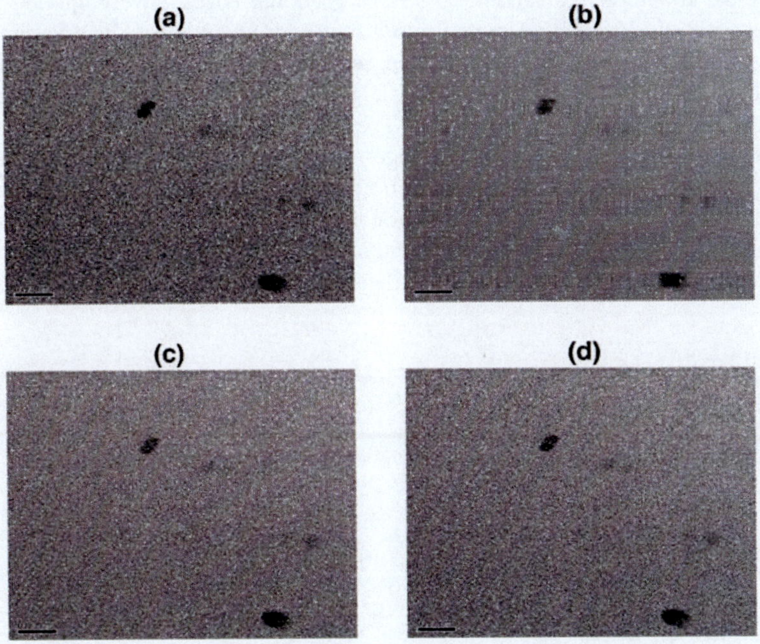

Fig. 2.5 Low-rank approximations of an image. (**a**) Original image (rank 1000). (**b**) Rank-10 approximation. (**c**) Rank-100 approximation. (**d**) Rank-200 approximation

2.4 Graph Representation

Digital images have been also represented by graph networks of different topologies (Sanfeliu et al. 2002). The simplest form is a graph of individual image pixels connected by a grid network as shown in Fig. 2.6a, where each pixel serves a node in the graph, and the edge links the neighboring pixels on a grid of pixel locations (Boykov and Funka-Lea 2006). A cost can be assigned to each of the edges. The choice of the edge cost is dependent on the types of analysis to be performed on images. For example, the edge cost can be assigned to the intensity difference between the source pixel and the target pixel for the purpose of edge detection and image segmentation (Shi and Malik 2000).

Another graph representation of images is a tree-based representation, including a tree-pyramid and a quad-tree (Sonka et al. 2014). As shown in Fig. 2.6b, a tree-pyramid represents an image as a balanced tree, where a leaf node on the bottom layer corresponds to an image pixel. The leaf nodes are then grouped into blocks of 2×2 neighboring pixels, and each of the blocks becomes a parent node of the four leaf nodes within the block; doing this to all leaf nodes create the layer of parents immediate above the leaf nodes. The same grouping is recursively applied to the blocks until all blocks are grouped into a root node. A quad-tree is a variant of a tree-pyramid, but it is an unbalanced tree, while a tree-pyramid is a balanced tree.

When an image is in the form of a graph, many graph algorithms can be employed to facilitate image analysis. The most noteworthy application of graph algorithms in image processing is image segmentation. The idea is to formulate image segmentation as a graph partitioning problem which finds minimal cuts to partition a graph into connected sub-graphs (Shi and Malik 2000; Felzenszwalb and Huttenlocher 2004). To describe the idea more specifically, let $G = (V, E)$ represent a graph representation of a digital image f, where an vertex $v \in V$

(a)

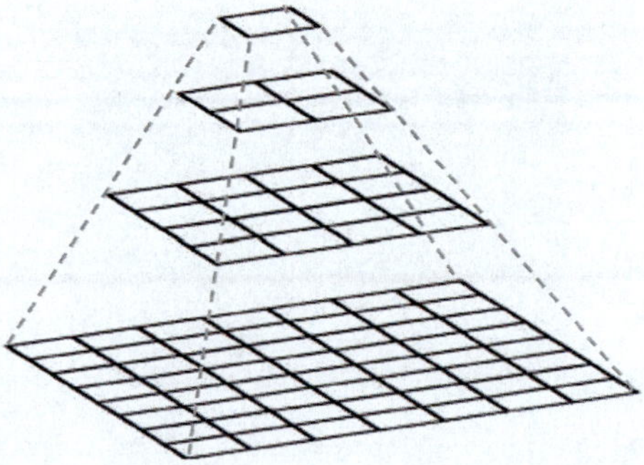

(b)

Fig. 2.6 Graph representations of images. (**a**) Rank-200 approximation. (**b**) Tree-pyramid representation corresponds to an image pixel in f, and an edge $e \in E$ defines the link between two neighboring pixels. A graph cut to G is defined as a set of the edges in E linking two disjoint sets A and B of the vertex set V,

$$cut(A, B) = \{e \in E : s(e) \in A \text{ and } t(e) \in B \text{ or } s(e) \in B \text{ and } t(e) \in A\},$$

where $s(e)$ denotes the source node of edge $e \in E$ and $t(e)$ is the destination node. The cut cost is defined by

$$w(A, B) = \sum_{e \in cut(A,B)} c(e),$$

where $c(e)$ is the edge cost for $e \in E$. The edge cost is defined by the intensity similarity between the image pixels associated with the edge. The intensity-based image segmentation seeks for a partition of an image into regions so that the intensity level would be homogeneous within each region. The problem can be formulated as an optimization problem of finding a graph cut that minimizes the associated cut cost. Shi and Malik (2000) find that directly minimizing the cut cost promotes many small subgraphs being segmented out, creating the over-segmentation of an image, because the cut costs for smaller subgraphs are less than those for larger subgraphs. Shi and Malik (2000) propose a normalized cut idea, which is to normalize the cut cost by the sizes of the subgraphs. The major shortcoming of this graph cut is its high computational expense. Felzenszwalb and Huttenlocher (2004) propose a faster graph partitioning method that recursively merges individual nodes into subgraphs by adding high cost edges incrementally.

2.5 Set Representation

The third image representation introduced in this chapter is the set representation, used especially in morphological image analysis (Soille 2013). In the morphological image analysis, an image is viewed as a set of image pixels. The extension of the set theory is used to define operations on an image.

A digital binary image has pixels on an integer grid with each pixel having only two possible intensity values, 0 for white and 1 for black. Using the set representation, the digital binary image is represented by the set of the pixel coordinates having intensity values equal to one; see the examples in Fig. 2.7. Elementary morphological image operations are defined using some basis set operations. Let $\mathbb{E} = \mathbb{Z}^2$ denote a two-dimensional integer grid, and I denotes the set representation of an image on $\mathbb{E}$, i.e., $I \subset \mathbb{E}$. The translation of the image by a coordinate $z \in \mathbb{E}$ is denoted by

$$I_z = \{x + z : x \in I\}.$$

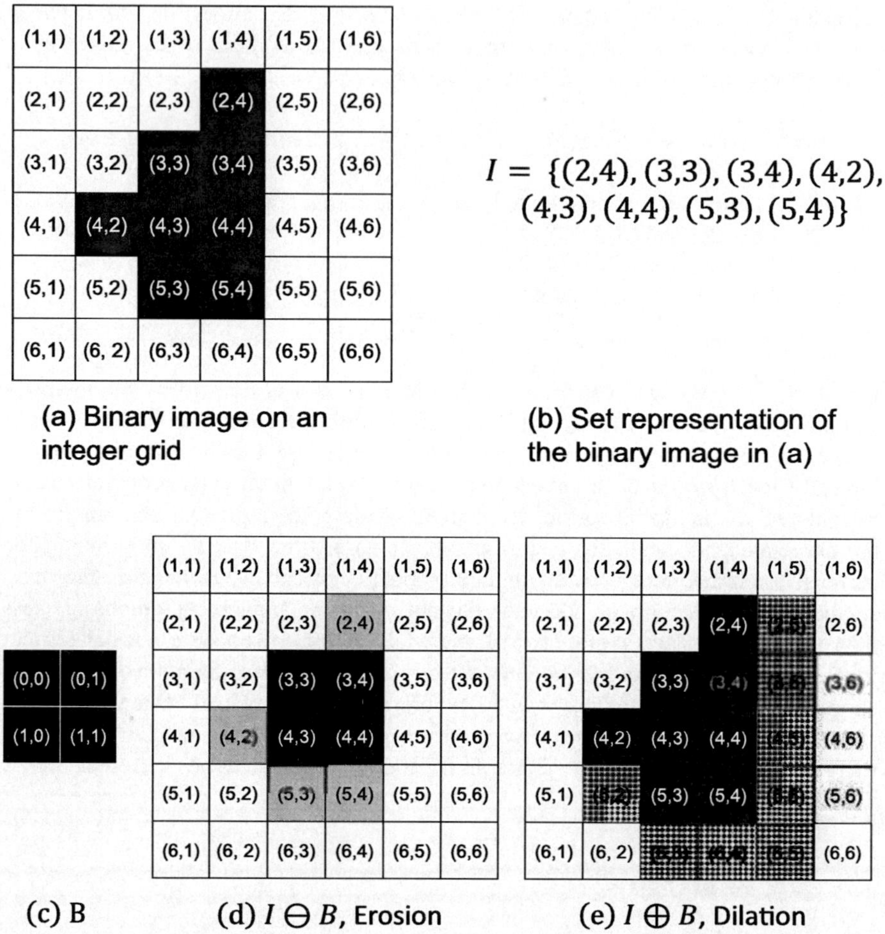

Fig. 2.7 Set representation of an binary image. In (**a**), the black boxes represent image pixels having intensity one. In (**d**), the pattern-filled pixels are the ones subtracted from I. In (**e**), the patterned fills are the ones added to I

The morphological erosion of I with respect to a structural element B is

$$I \ominus B = \bigcap_{z \in B} I_{-z},$$

where the structural element is a binary image of a simple shape, also a subset of the integer grid $\mathbb{E}$. For example, a simple box structural element is $B = \{(0,0), (0,1), (1,0), (1,1)\}$. The morphological erosion with a structural element functions is like a subtraction of B from the boundary of I. The operator opposite to the morphological erosion is the morphological dilation, which adds B to I from the boundary of I,

$$I \oplus B = \bigcup_{z \in B} I_z.$$

Figure 2.7c, d illustrates the application of the two operators on the image in Fig. 2.7a.

Based on the two basis morphological operations, many advanced operations were defined. The morphological opening is defined by

$$I \star B = (I \ominus B) \oplus B,$$

which is useful to smooth out some sharp features in I such as the outline roughness due to noises and hairy features. Another useful application of the morphological opening operator is to separate touching objects when the touching objects are bridged by image features whose thickness is less than the size of B. The morphological closing operator is opposite to the opening operator, and defined by

$$I \diamond B = (I \oplus B) \ominus B.$$

This operator is useful for bridging narrow breaks and eliminating small interior holes. Edges of I can be extracted by an operator of

$$I - (I \ominus B),$$

which is referred to as the morphological skeleton operator. This operator generates a skeleton of a binary image I, which is defined by

$$B_n(I) = (I \ominus nB) - (I \ominus nB) \cdot B,$$

where n is a positive integer, and $nB = \underbrace{B \oplus \cdots \oplus B}_{n \text{ times}}$.

In the morphological image analysis, a grayscale image is represented by a pixel intensity function f, defined on an integer grid $\mathbb{E}$, and a grayscale structural element is represented by a function g defined on a subset of the integer grid $\mathbb{E}$. The morphological image operations for grayscale images are defined on the functions. The grayscale morphological erosion of an image f with respect to the structural element g is

$$f \ominus g(x) = \inf\{f(y) - g(y - x) : y \in \mathbb{E}\} \quad \forall x \in \mathbb{E},$$

where $\inf\{A\}$ is the infimum of a set A. The grayscale morphological dilation is defined by

$$f \oplus g(x) = \inf\{f(y) + g(x - y) : y \in \mathbb{E}\} \quad \forall x \in \mathbb{E}.$$

With the definition of the two basic morphological operations on the grayscale images, advanced operations are defined similarly to those on binary images. For example, the grayscale morphological closing and opening are defined as $(f \oplus g) \ominus g$ and $(f \ominus g) \oplus g$, respectively. For a grayscale image, a morphological gradient operation can be defined in the form of

$$G(f) = (f \oplus g) - (f \ominus g),$$

where g is a flat structural element such that $g(x) = 0$ when $|x| \leq 1$ and $-\infty$ otherwise.

2.6 Example: Watershed Segmentation

Image segmentation is a very useful image processing technique that partitions an image into homogeneous regions, within which certain image features of interest such as image intensities are similar or even identical. In this section, we describe a popular watershed image segmentation, which serves also as a demonstration of the applications of the aforementioned concepts of image representations and operations.

Here we consider a grayscale image, which is defined as an intensity function that maps a point x on a discrete grid $\mathbb{E}$ to the intensity value of $f(x)$. A watershed transform of f converts the input image f to the *watershed lines* that divide the image into sub regions, each of which is a connected subset of $\mathbb{E}$ that contains only one local minimum of f, also known as a marker. A local minimum or a marker represents the core of a dark foreground object, and the watershed lines are the lines separating the cores. Here we would like to formally define the *watershed lines*. For doing so, we first define some basic set and image operations.

The level set of f is defined as

$$L_f(i) = \{x \in \mathbb{E} : f(x) \leq i\}.$$

For an arbitrary set $A \subset \mathbb{E}$, one can construct a neighborhood graph G_A, where each element of A serves a node, and two nodes are linked with an edge of weight one if and only if the two nodes are neighboring to each other. The geodesic distance of two elements in A, x and y, is defined by the shortest graph distance between them over G_A; we denote the geodesic distance by $dist(x, y)$. A connected subset of A can be defined by a subset of A that has a finite geodesic distance for every pair of the elements in the subset. For a connected subset $B \subset A$, one can define a reachable subset of A from B,

$$R_A(B) = \{a \in A : \exists b \in B \text{ such that } dist(a, b) < \infty\}.$$

Suppose that $B_1, \ldots, B_K$ represent K connected subsets of A that partition B, i.e., $\bigcup_{k=1}^{K} B_k = B$ and $B_k \cap B_l = \emptyset$ for $k \neq l$. The influence zone of B_k within A is defined by a subset of A whose elements are reachable from B_k with the shorter geodesic distance than the other connected subsets, which can be formally written as

$$IZ_A(B_k) = \{a \in R_A(B_k) : \min_{b \in B_k} d(a, b) < \min_{c \in \bigcup_{j \neq k} B_j} dist(a, c)\},$$

and for convenience, we denote

$$IZ_A(B) = \bigcup_{k=1}^{K} IZ_A(B_k).$$

The watershed lines of f are the lines that divide $\mathbb{E}$ into the influence zones of the local minima of f. We denote the local minima at the intensity level i by $m_f(i) = \{x \in \mathbb{E} : f(x) = i, x \text{ is a local minimum of } f\}$. As illustrated in Fig. 2.8b–d, the influence zones of the local minima are iteratively updated, starting with their influence zones in the smallest level set $L_f(0) = \emptyset$ and expanding them to those in the upper level sets through the iteration,

$$W_f(i) = \left[IZ_{L_f(i)}(L_f(i-1))\right] \cup m_f(i).$$

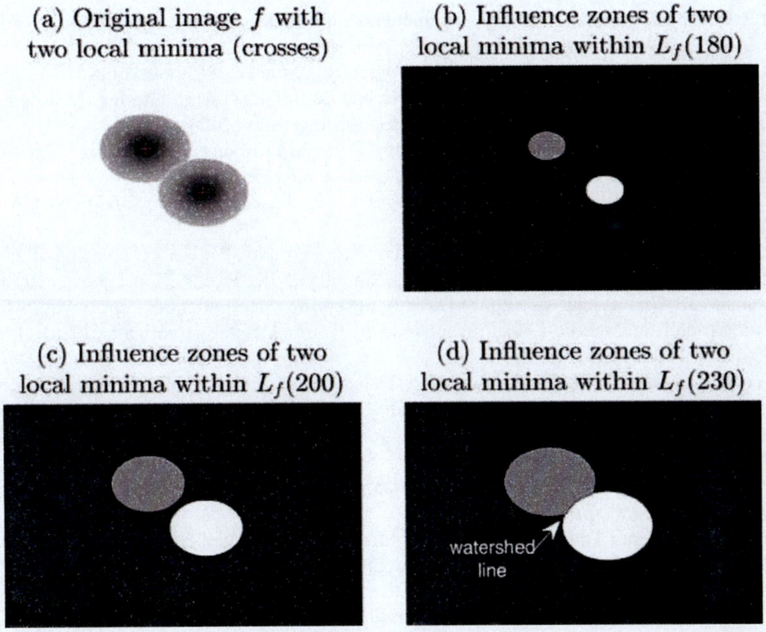

Fig. 2.8 Watershed transformation explained by a simple example. In **(a)**, the original image f is shown, where two red crosses represent two local minima of f. In **(b)** to **(d)**, the influence zones of the two local minima are presented for different level sets of f. The last three plots show how the influence zones expand as the level of the water flowing from the two local minima changes

The watershed lines are the ones dividing the influence zones at the final iteration,

$$WL(f) = \mathbb{E} - W_f(N),$$

where $N = \max_{x \in \mathbb{E}} f(x)$ is the maximum intensity of f.

References

Binnig G, Rohrer H (2000) Scanning tunneling microscopy. IBM Journal of Research and Development 44(1/2):279

Binnig G, Quate CF, Gerber C (1986) Atomic force microscope. Physical Review Letters 56(9):930

Boykov Y, Funka-Lea G (2006) Graph cuts and efficient N-D image segmentation. International Journal of Computer Vision 70(2):109–131

Buades A, Coll B, Morel JM (2004) On image denoising methods. CMLA Preprint 5

Elad M (2002) On the origin of the bilateral filter and ways to improve it. IEEE Transactions on Image Processing 11(10):1141–1151

Felzenszwalb P, Huttenlocher D (2004) Efficient graph-based image segmentation. International Journal of Computer Vision 59(2):167–181

Figueiredo MA, Nowak RD (2003) An EM algorithm for wavelet-based image restoration. IEEE Transactions on Image Processing 12(8):906–916

Holzer L, Cantoni M (2012) Review of FIB-tomography. In: Nanofabrication using Focused Ion and Electron Beams: Principles and Application, Oxford University Press, New York, NY, USA, chap 11, pp 410–435

Kvam P (2008) Length bias in the measurements of carbon nanotubes. Technometrics 50(4):462–467

Kwon OH, Zewail AH (2010) 4D electron tomography. Science 328(5986):1668–1673

Lou Y, Favaro P, Soatto S, Bertozzi A (2009) Nonlocal similarity image filtering. In: International Conference on Image Analysis and Processing, Springer, pp 62–71

Meyer E, Hug HJ, Bennewitz R (2013) Scanning Probe Microscopy: The Lab on a Tip. Springer Science & Business Media, New York, NY, USA

Mitra S, Sicuranza G (2000) Nonlinear Image Processing. Academic Press, San Diego, CA 92101, USA

Mu C, Park C (2020) Sparse filtered sirt for electron tomography. Pattern Recognition 102:107253

Parent LR, Bakalis E, Ramírez-Hernández A, Kammeyer JK, Park C, De Pablo J, Zerbetto F, Patterson JP, Gianneschi NC (2017) Directly observing micelle fusion and growth in solution by liquid-cell transmission electron microscopy. Journal of the American Chemical Society 139(47):17140–17151

Pennycook SJ, Nellist PD (2011) Scanning Transmission Electron Microscopy: Imaging and Analysis. Springer Science & Business Media, New York, NY, USA

Qiu P (2009) Jump-preserving surface reconstruction from noisy data. Annals of the Institute of Statistical Mathematics 61(3):715–751

Reimer L (2013) Transmission Electron Microscopy: Physics of Image Formation and Microanalysis. Springer, New York, NY, USA

Ross FM (2007) *In situ* Transmission Electron Microscopy. Springer, New York, NY, USA

Ross FM (2016) Liquid Cell Electron Microscopy. Cambridge University Press, New York, NY, USA

Salvo L, Cloetens P, Maire E, Zabler S, Blandin JJ, Buffière JY, Ludwig W, Boller E, Bellet D, Josserond C (2003) X-ray micro-tomography an attractive characterisation technique in materials science. Nuclear Instruments and Methods in Physics Research (Section B: Beam Interactions with Materials and Atoms) 200:273–286

Sanfeliu A, Alquézar R, Andrade J, Climent J, Serratosa F, Vergés J (2002) Graph-based representations and techniques for image processing and image analysis. Pattern Recognition 35(3):639–650

Shi J, Malik J (2000) Normalized cuts and image segmentation. IEEE Transactions on Pattern Analysis and Machine Intelligence 22(8):888–905

Soille P (2013) Morphological Image Analysis: Principles and Applications. Springer Science & Business Media, New York, NY 10013, USA

Sonka M, Hlavac V, Boyle R (2014) Image Processing, Analysis, and Machine vision. Cengage Learning, Independence, KY, USA

Takeda H, Farsiu S, Milanfar P (2007) Kernel regression for image processing and reconstruction. IEEE Transactions on Image Processing 16(2):349–366

Takeda H, Farsiu S, Milanfar P (2008) Deblurring using regularized locally adaptive kernel regression. IEEE Transactions on Image Processing 17(4):550–563

Unser M (1999) Splines: A perfect fit for signal and image processing. IEEE Signal Processing Magazine 16(6):22–38

Verhaeghe J, D'Asseler Y, Staelens S, Vandenberghe S, Lemahieu I (2007) Reconstruction for gated dynamic cardiac PET imaging using a tensor product spline basis. IEEE Transactions on Nuclear Science 54(1):80–91

Vo GD, Park C (2018) Robust regression for image binarization under heavy noise and nonuniform background. Pattern Recognition 81:224–239

Woehl TJ, Park C, Evans JE, Arslan I, Ristenpart WD, Browning ND (2013) Direct observation of aggregative nanoparticle growth: Kinetic modeling of the size distribution and growth rate. Nano Letters 14(1):373–378

Chapter 3
Segmentation

3.1 Challenges of Segmenting Material Images

Material images has some unique characteristics that need to be considered while segmenting images. First, material images are mostly grayscale except for some fluorescence microscopic imaging of biological materials with color dyes. At the first glance, image segmentation of grayscale images may appear simpler than that of color images. However, achieving image segmentation of good accuracy for grayscale images is in fact more challenging than the same task for natural-color images. This is because image intensities are the only information that can be exploited for segmenting grayscale images, whereas multi-channel color information (red, green and blue channels) is available for segmenting color images.

Another unique factor existing in material images is the intensity and pattern of background image noises. The noises for microscopes are mainly due to random intensity variations in the radiation beams of light or charged particles, and this type of noises follow a Poisson distribution, instead of a Gaussian distribution (Marturi et al. 2014). This is particularly true for low dose microscopes where the amount of the radiation beam is low. When imaging with a higher dose, the noise distribution, namely a Poisson distribution, approaches a Gaussian distribution, but when the dose is low, the Gaussian approximation of Poisson distribution is poor. Low dose microscopy is rather common in material experiments, particularly for biological material samples (Fujiyoshi 2013), as a strong beam could damage material specimens too much.

In addition, the signal-to-noise ratios (SNRs) of material images are much lower than most of the daily life images. SNR is the intensity distinction relative to the background noise and typically quantified by the squared foreground-background intensity difference over the variance of background noises. Under the constraint that the total radiation of the light or particle sources is limited to certain amount, the SNR of the resulting image decreases as the spatial and temporal resolutions of

C. Park, Y. Ding, *Data Science for Nano Image Analysis*, International Series in Operations Research & Management Science 308,
https://doi.org/10.1007/978-3-030-72822-9_3

imaging increase, because the total beam amount needs to be split over more spatial and temporal points of measurement. In other words, a higher resolution means a lower dosage of radiation per point. For the low-dose microscopic images, the SNRs could be as low as 0.1 (Baxter et al. 2009).

In addition to these unique challenges, material images share some common complicating factors with image segmentation tasks developed from other application domains, e.g., uneven illumination and uneven background pattern, and overlaps among foreground objects. A "non-flat" image background could create a big headache in separating the image background from the foreground of material interest. When materials are overly crowded in a sample and their images partially overlap, a simple segmentation of an image into the background and foreground regions gives only the image regions of the overlapped materials instead of the image regions of individual materials. Splitting the overlaps into individual material regions is pressingly needed as a key image processing step.

3.2 Steps for Material Image Segmentation

Park et al. (2013) discuss the necessary steps to segment material images considering the challenges described in the previous section. Figure 3.1 summarizes the three main steps for material image segmentation. In the first step, as illustrated in Fig. 3.1b, a raw image is partitioned into the background region and multiple regions of foreground materials. The partitioning is treated as a binary classification problem that classifies the image pixels into, respectively, the background group and the foreground group. For this reason, this step is also referred to as image binarization.

When materials are not overlapping with each other, image segmentation is completed with the first step conducted, and each of the foreground regions identified corresponds to the region of individual material components. When

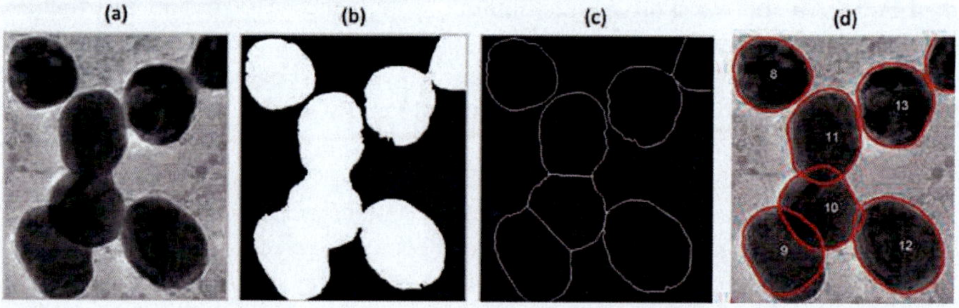

Fig. 3.1 Three main steps for material image segmentation. **(a)** Original image. **(b)** Step 1. Image binarization. **(c)** Step 2. Foreground segmentation. **(d)** Step 3. Shape inference (Reprinted with permission from Park et al. 2013)

materials are overlapping with each other, certain foreground regions need to be further segmented into sub-regions as shown in Fig. 3.1c. This step is referred to as foreground segmentation, and it is an essential step to further extract the regions of individual material components out of the aggregating agglomerate.

Depending on the degrees of the overlaps, subsequent adjustment to the segmented foreground may be necessary because certain important structural information could be hidden by the overlaps. The hidden parts of the separated regions are missing in the process of foreground segmentation, resulting incomplete shapes or contours. They thus need to be estimated in order to recover the whole shape and achieve enhanced and practical segmentation outcome. Please compare the shape differences between Fig. 3.1c and d. This step is referred to as the shape inference step.

The three steps can be performed sequentially, or some of the steps may be combined and solved simultaneously. Performing all three steps in a single integrative framework is still very challenging, due to either computational or analytical complexity. Combing some of the steps may not be very synergistic, either, owing to different nature of their own processing step. For example, the image binarization step exploits the distinction in terms of image intensity between the background and foreground regions, but when it comes to the foreground segmentation step, the intensity distinction among foreground regions is very minor. Instead, the foreground segmentation step rely mostly on the prior knowledge of material structures such as circularity and convexity. In order to solve the problems of image binarization and foreground segmentation in a unified framework, we need to include both of the intensity and shape profiles of the segmented regions in a problem formulation. While the method of active contour with shape priors (Chen et al. 2001) is capable of handling this type of problem, actually solving it is too computationally expensive. Hence, a combined approach is more appropriate for locating a small number of complex foregrounds rather than identifying numerous overlapping foregrounds. Combining the second and third steps, on the other hand, is much more feasible. Several approaches have shown successful combinations of the two steps based on different shape priors; for example, convex shape prior (Park et al. 2013) or elliptical shape prior (Zafari et al. 2015; Qian et al. 2016).

3.3 Image Binarization

Image binarization exploits the image intensity distinction between the background and foreground regions in a noisy image and creates the binary silhouette of foreground objects. A good binarization solution lays a solid foundation for downstream image analysis. In particular, many foreground segmentation algorithms (Park et al. 2013; Schmitt and Hasse 2009; Zafari et al. 2015; Qian et al. 2016) require the binary silhouette as inputs to find markers of individual foreground objects. Naturally, the quality of the foreground segmentation is significantly affected by the accuracy of the binary silhouette input. In this section, we review some image binarization approaches and their applications to material images.

3.3.1 Global Image Thresholding

The simplest approach of image binarization is the global thresholding (Otsu 1979), which is to apply a single intensity threshold to an image and divides its pixels into two groups: the foreground and the background. The choice of the threshold is mostly based on the histogram of pixel intensities; comprehensive reviews can be found in Sezgin and Sankur (2004) and Stathis et al. (2008). When the SNR of an image is high, the histogram has two distinct modes as illustrated in Fig. 3.2b. One of the modes is the mode of the intensity distribution of foreground pixels, whereas the other is the mode of background pixels. Then the valley in between is chosen as the threshold. As the SNR decreases, however, the distinction between the two distributions becomes less and less pronounced, as shown in Fig. 3.2d, f, which makes the choice of a threshold hard. Please note that material images do have low SNRs, especially for those low-dose microscopic images and the in situ images. Consequently, the global thresholding approach is less likely effective.

Another challenge for gloabl thresholding may be caused by sparsity of foregrounds. Some material images contain very few foreground materials. If foreground objects are very sparse in the image, the histogram of image intensities can hardly show the presence of two modes, distinguishing the foreground distribution from the background distribution, even for high SNR cases; see the example in Fig. 3.3. This is because one of the two modes in the histogram would be very minor, thereby being buried in the tail part of the other mode.

3.3.2 Local Image Thresholding

Many material images also have uneven backgrounds, which means that the image intensities of the background region have a long-range spatial correlation on top of the pure random variation. The unevenness is resulted owing to many factors, including lack of uniformity of the host materials, and thickness variation in material specimens. For the images with nonuniform backgrounds, using a single, global threshold to divide image pixels would not work, because the intensity distribution changes from location to location. Instead, a localized analysis is more suitable, pointing to a local image thresholding. A local thresholding approach selects a threshold for each pixel, based on the local statistics within a neighborhood centered at the pixel. Depending on the types of the statistics used, there are several approaches to carry out local thresholding.

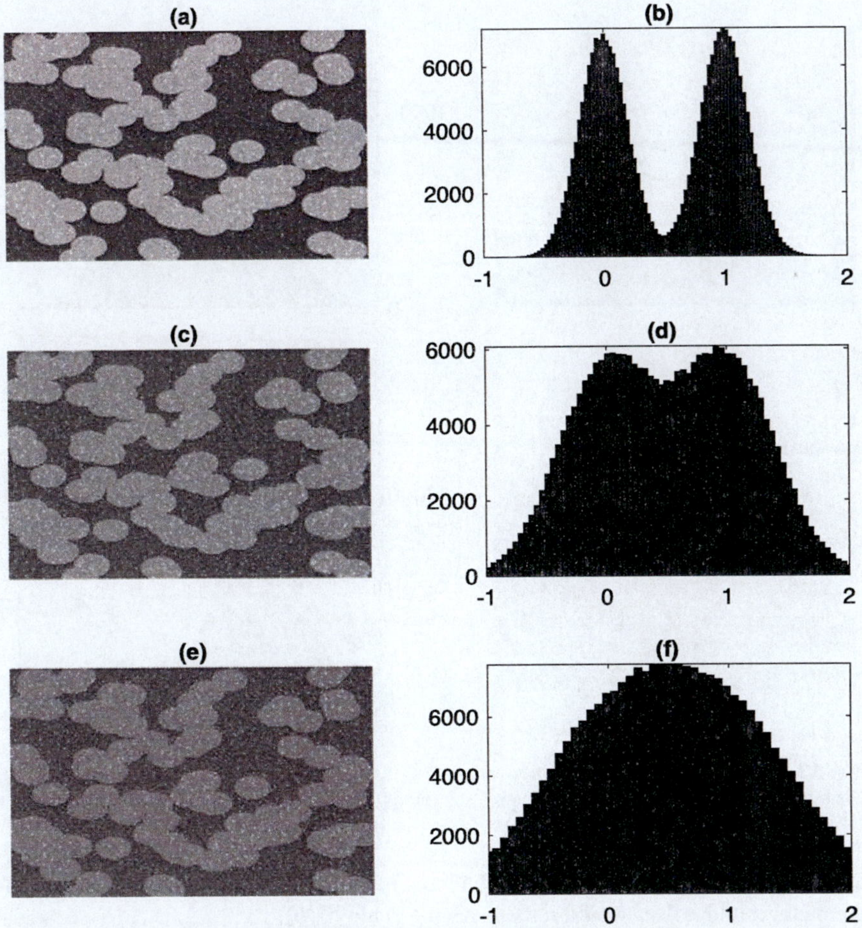

Fig. 3.2 Histograms of image foreground and background under different SNRs. (**a**) SNR = 25. (**b**) Histogram of image intensities. (**c**) SNR = 6.2. (**d**) Histogram of image intensities. (**e**) SNR = 2.8. (**f**) Histogram of image intensities

Earlier versions of local thresholding use simple local statistics. Niblack (1985) uses the local mean M and local standard deviation S to choose the local threshold as

$$M + k \times S,$$

where k is a user-defined constant, often fixed to -0.2. A weakness of this local approach is that the background noise still remains in the resulting binary image. If

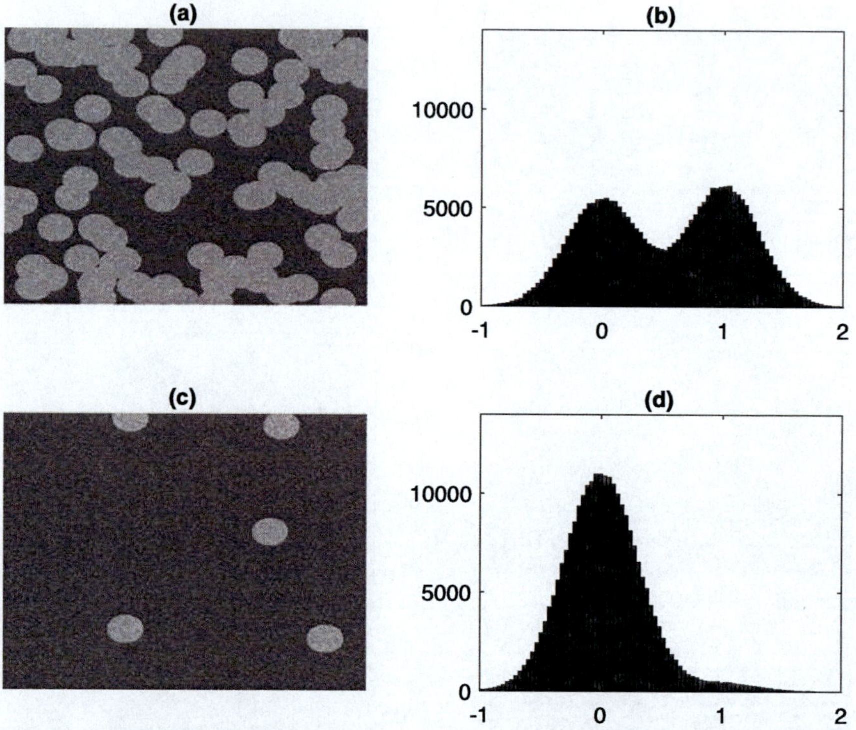

Fig. 3.3 Histogram of image intensities versus sparsity of foreground. (**a**) Dense foregrounds. (**b**) Histogram of image intensities. (**c**) Sparse foregrounds. (**d**) Histogram of image intensities

the foreground objects are sparse in the local neighborhood, a significant amount of the background noises would remain (Gatos et al. 2006).

Another local thresholding approach is the contrast threshold method, which uses local image contrasts to decide the belonging of a pixel. The local image contrast at a pixel location is defined as the range of the intensities of the pixel's neighborhood. If the contrast is high, the pixel likely belongs to a transition area between foregrounds and backgrounds; otherwise the pixel would belong to either the foreground or the background.

Following this idea, researchers would first categorize the local image contrasts into high and low contrasts, and then classify the low contrast regions as either foreground or background. Existing local contrast methods differ largely in terms of how they define the local contrast. For instance, Bernse (1986) defined the local contrast at the (i, j)th pixel location as

$$C(i, j) = I_{max}(i, j) - I_{min}(i, j),$$

where $I_{max}(i, j)$ and $I_{min}(i, j)$ represent the maximum and minimum intensities within a local neighborhood window centered around the (i, j)th pixel location. Because the afore-defined local contrast is influenced by the local background intensity level in the neighborhood, Su et al. (2010) further normalizes the local contrast with the local intensity level, such as

$$\tilde{C}(i, j) = \frac{I_{max}(i, j) - I_{min}(i, j)}{I_{max}(i, j) + I_{min}(i, j) + \epsilon},$$

where ϵ is a small positive constant and its inclusion makes the evaluation feasible even for the case of $I_{max}(i, j) + I_{min}(i, j) = 0$.

Edge-level thresholding approach is similar to the contrast-based thresholding. The basic principle of this approach is to use an existing edge detector to locate the edge pixels that delineate the image foreground regions, and to estimate a local threshold at a pixel location using the image intensities of the nearby edge pixels. The existing methods differ in terms of how edges are detected and how the local threshold is achieved.

For instance, Parker (1991) and Parker et al. (1993) use the Shen-Castan (ISEF) edge detection algorithm to locate edge pixels. To determine a local threshold at the (i, j) pixel location, the authors solve the local linear kernel regression to interpolate the intensity values of the nearby edge pixels. Let $N_r(i, j)$ denote the set of all pixels within their Euclidean distance to the pixel location (i, j) less than r, and let $E_r(i, j)$ represents a subset of $N_r(i, j)$, which contains only the edge pixels. To estimate the local threshold, a local linear kernel regression approach uses the intensity levels of the edge pixels in the neighborhood, such as

$$(a^*, b^*, c^*) = \text{argmin} \sum_{(i', j') \in E_r(i, j)} \omega(|i' - i|, |j' - j|)\{I(i', j') - T(i', j')\}^2.$$

where $\omega(x, y)$ is the kernel function whose value diminishes as $|i' - i|$ or $|j' - j|$ increases, $I(i, j)$ represents the image intensity at the pixel location (i, j), and $T(i, j) = ai + bj + c$ represents the linear function locally fitted. Let $\hat{T}(i, j) = a^*i + b^*j + c^*$ represent the fitted solution, which then serves as the local threshold at the pixel location.

Ramírez-Ortegón et al. (2010) generalize what constitutes edge pixels using the concept of transition pixels. The authors defined a pixel located at (i, j) as a r-transition pixel if its neighborhood $N_r(i, j)$ contains foreground and background pixels. When $r = 1$, $R_1(i, j)$ becomes an edge pixel. Ramírez-Ortegón et al. (2010) introduced a transition function $F(i, j)$ to determine whether the pixel (i, j) is a r-transition pixel. A suggested transition function is a min-max function, defined by

$$F(i, j) = I_{max}(i, j) + I_{min}(i, j) - 2I(i, j),$$

where $I_{max}(i, j)$ and $I_{min}(i, j)$ represent the maximum and minimum intensities within a local neighborhood window centered around the (i, j)th pixel location.

A set of all transition pixels can be identified as $\mathcal{T}(i, j) = \{(i', j') \in N_r(i, j) : t^- \leq F(i', j') \leq t^+\}$ with proper choices of $t^+ \geq t^- \geq 0$, which is referred to as a transition set. Ramírez-Ortegón et al. (2010) showed that if $(i, j) \in \mathcal{T}$, the corresponding pixel is almost surely a transition pixel. After the transition set is identified, the foreground and background pixels are separately identified as $\mathcal{T}^- = \{(i', j') \in N_r(i, j) : F(i', j') < t^-\}$ and $\mathcal{T}^+ = \{(i', j') \in N_r(i, j) : F(i', j') > t^+\}$, respectively. The histograms of foreground intensities and background intensities can then be attained, which can be in turn used to define the local threshold.

3.3.3 Active Contour

The active contour method (Kass et al. 1988) is a popular approach for image binarization. The method identifies the contours of foregrounds that divide the foreground and background regions. The interiors of the contours are regarded as foregrounds, and the exteriors are regarded as the background. In the framework of active contour, an energy function is defined for each contour. Then the contour is evolving to minimizing the energy functional until the local optima is found; doing so is supposed to produce the optimal separation.

There are many variants of the active contour methods with different representations of a contour and different choices of the energy functional. In the earlier version of the active contour (Kass et al. 1988), a contour is defined as a closed curve, represented by a parametric curve $\boldsymbol{v}(s) = (x(s), y(s))$, where $x(s)$ and $y(s)$ map a parameter $s \in [0, 1]$ to the x and y coordinates of a point on the closed curve, respectively. The energy functional comes in two parts: internal energy functional $\mathcal{E}_{internal}$ and image energy functional $\mathcal{E}_{image}$,

$$\mathcal{E}(\boldsymbol{v}(s)) = \int_0^1 \mathcal{E}_{internal}(\boldsymbol{v}(s)) + \mathcal{E}_{image}(\boldsymbol{v}(s))ds.$$

The internal energy functional is related to the inherent properties of the contour $\boldsymbol{v}(s)$, such as continuity and smoothness. For example, it can be defined as

$$\mathcal{E}_{internal}(\boldsymbol{v}(s)) = \alpha(s)||\nabla\boldsymbol{v}(s)||^2 + \beta(s)||\nabla^2\boldsymbol{v}(s)||^2,$$

where $\nabla\boldsymbol{v}(s) = \left(\frac{\partial x(s)}{\partial s}, \frac{\partial y(s)}{\partial s}\right)$, $\nabla^2\boldsymbol{v}(s) = \left(\frac{\partial^2 x(s)}{\partial s^2}, \frac{\partial^2 y(s)}{\partial s^2}\right)$, and $||\cdot||$ is the Euclidean norm. In the internal energy, the first term is related to the continuity of the curve and the second term is related to the smoothness of the curve. The $\alpha(s)$ and $\beta(s)$ are user-defined weights associated with the two terms. The image energy functional quantifies the fitness of the contour $\boldsymbol{v}(s)$ to the evidences from images. The image energy functional is based on edge information, intensity information, or both. For example, a popular edge-based image energy functional is

$$\mathcal{E}_{image}(\boldsymbol{v}(s)) = -||\nabla I(\boldsymbol{v}(s))||^2,$$

which decreases around edge pixels.

Another way to represent a contour is to use the level-set of a continuous function (Chan and Vese 2001). The r-level set of a continuous function f on a domain $\mathcal{X}$ is defined as

$$L_r(f) = \{\boldsymbol{x} \in \mathcal{X} : f(\boldsymbol{x}) = r\}.$$

If r is in the range of the co-domain of f, the level-set is composed of one or more closed curves. As f changes, the closed curves represented by the level set change. The level-set-based active contour is optimized likewise using an energy functional. Similar to other active contour approaches, the energy functional for f comes in two parts: the internal energy function $\mathcal{E}_{internal}$ and the image energy functional $\mathcal{E}_{image}$:

$$\mathcal{E}_{energy}(f) = \mathcal{E}_{internal}(f) + \mathcal{E}_{image}(f). \tag{3.1}$$

Chan and Vese (2001) use an intensity-based image energy functional, which assesses the uniformity of the image intensities within the interior and exterior regions of the active contours. Let $IN_r(f) = \{\boldsymbol{x} \in \mathcal{X} : f(\boldsymbol{x}) > r\}$ and $OUT_r(f) = \{\boldsymbol{x} \in \mathcal{X} : f(\boldsymbol{x}) < r\}$ represents the interior and exterior regions, respectively. The image energy function is defined as

$$\mathcal{E}_{image}(f) = -\int_{c_1,c_2 \geq 0} \int_{IN_r(f)} |c_1 - I(\boldsymbol{x})|^2 d\boldsymbol{x} - \int_{OUT_r(f)} |c_2 - I(\boldsymbol{x})|^2 d\boldsymbol{x}.$$

The internal energy functional for f is defined as

$$\mathcal{E}_{internal}(f) = \beta \int_{\mathcal{X}} |\nabla H(f(\boldsymbol{x}))| d\boldsymbol{x} + \alpha \int_{\mathcal{X}} H(f(\boldsymbol{x})) d\boldsymbol{x},$$

where $H(\cdot)$ is the heavyside function. The first term of the internal energy functional quantifies the perimeter of the active contours, the second term quantifies the interior area of the active contours, and α and β are user-defined constants to weigh the two terms.

Many later works (Allili and Ziou 2007; Zhang et al. 2010; Tian et al. 2013) consolidate these assumptions and design their own version of active contours, which, to certain extent, makes uses of both intensity-based and gradient-based information. There are a few difficulties confronting the active contour method when it is applied to low-contrast and noisy microscopic images, in which the intensity distinction between foregrounds and the background is not so pronounced. A lot of sharp image noises existing in the material microscopic images create many high-gradient pixels randomly spread over the image, so that the gradients at the foreground boundaries versus those at other interior regions are not so different, leading to a large number of local optima in the active contour's optimization procedure. Consequently, the effectiveness of the active contour methods highly depends on the choice of the initial contour (also known as a mask). The active contour method proposed by Tian et al. (2013) uses both intensity and gradient information and is hence more robust, as this method has a better convergence property than its counterparts that use only the intensity information or only the gradient information (Chan and Vese 2001; Caselles et al. 1997).

Even with a capable method like in Tian et al. (2013), the choice of the initial contour still has a profound impact on the outcomes of contour detection in low contract electron microscopic images. Qian et al. (2016) reported such sensitivity; please see Fig. 3.4 as an example. They start the active contour method from either a large mask or a small mask of the foreground. Then the active contour algorithm can shrink the large mask or expand the small one to get the estimated contour. Both of the small and large masks are created using the Otsu's method (Otsu 1979), which estimates a global threshold τ. The pixels of intensities above τ are classified as foreground, whereas those below τ are regarded as the background. The large mask of the foreground is created by increasing the Ostu's threshold by a constant, while the small mask is created by decreasing the threshold by a constant. Qian et al. (2016) observed that using different initial contours did lead to different final outcomes and stated that using the small mask as the initial contour in the chosen active contour method is more effective for low contrast microscopic images.

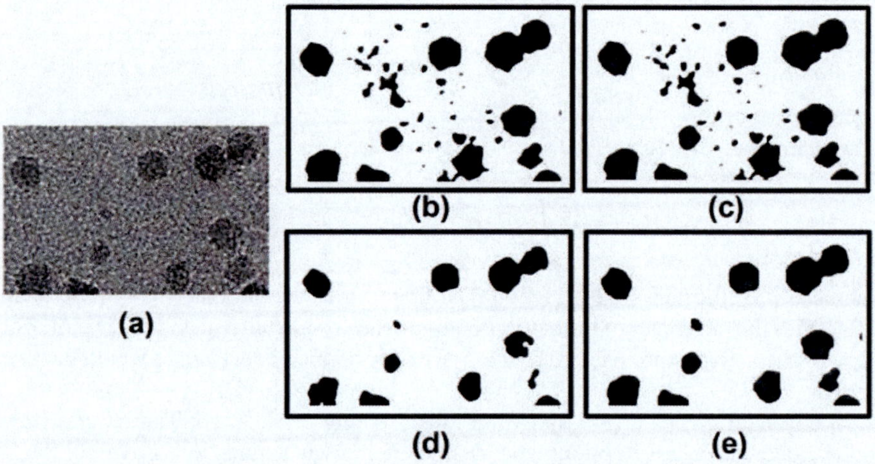

Fig. 3.4 The outcomes of active contour with different initial contours: (**a**) The original image; (**b**) Large mask as initial contours; (**c**) Result of active contour with the large mask in (**b**); (**d**) Small mask as initial contours; (**e**) Result of active contour with the small mask in (**d**). (Reprinted with permission from Qian et al. 2016)

3.3.4 Graph Cut

Graph-cut methods are based on a graph representation of an image, which was explained in much more details in Sect. 2.4. Once a graph representation of the image is constructed, a graph $G = (V, E)$ represents a digital image f, where an vertex $v \in V$ corresponds to an image pixel in f, and an edge $e \in E$ defines the link between two neighboring pixels. The image binarization problem can be formulated as finding cuts of the graph that would partition the image into foreground and background regions. This section reviews two representative works of the graph cut approach.

In the graph theory, a graph cut to G is defined as a set of the edges in E linking two disjoint subsets A and B of the vertex set V. This cut set is denoted by $cut(A, B)$. The cut cost is typically defined as the total cost of the edges that belong to the cut set, i.e.,

$$w(A, B) = \sum_{e \in cut(A,B)} c(e),$$

where $c(e)$ is the cost associated with $e \in E$. For the image binarization problem, the edge cost is defined by the intensity similarity between the image pixels connected by the edge. Shi and Malik (2000) pointed out that finding a cut that minimizes the cut cost would prompt choosing a cut set containing a small number of edges that segment an image into smaller pieces. Shi and Malik (2000) proposed to normalize the cut cost by computing the cut cost as a fraction of all edges from the disjoint sets,

$$\tilde{w}(A, B) = \frac{w(A, B)}{a(A, V)} + \frac{w(A, B)}{a(B, V)},$$

where $a(A, V) = \sum_{e \in E: s(e)=A, t(e)=V} c(e)$ with $s(e)$ and $t(e)$ as the start and terminal nodes of edge e quantifies the total cost of all the edges that are connected to a vertex in A. The authors proposed to minimize the normalized cut cost to partition an image into two regions. This dyadic partitioning can be applied recursively to produce multiple disjoint sets. Solving the minimization problem for a normalized cut is an NP-complete problem. Shi and Malik (2000) proposed a continuous relaxation of the problem, so that the approximate solution can be achieved analytically by solving a generalized eigenvalue problem. The generalized eigenvalue problem, however, is still computationally expensive for large images because the computational cost increases quadratically in the number of image pixels. Felzenszwalb and Huttenlocher (2004) proposed another cut cost,

$$w_{min}(A, B) = \min_{e \in cut(A,B)} c(e),$$

and devised a heuristic algorithm to solve a graph cut problem much more efficiently.

3.3.5 Background Subtraction

The idea of background subtraction, as it is applied to image binarization, is to estimate the intensity profile of an image's background and employ some threshold to subtract the background from the input image. There are different strategies in carrying out the idea. Gatos et al. (2006) split an input image into background and foreground pixels and interpolate the background pixel intensities for defining the background intensity profile, which is further used for creating the corresponding local threshold policy. The major weakness of this approach is that achieving a good initial split to the foreground and background regions is difficult when the input image has uneven background and heavy noise.

Vo and Park (2018) proposed a background subtraction approach that does not require any rough estimate of the foreground and background regions. In their work, Vo and Park (2018) use the functional representation of images, which we described in Sect. 2.2. Let $Y(i, j)$ denote the image intensity at an image pixel location (i, j), where $i = 1, ..., m$ and $j = 1, ..., n$. The input image $Y(i, j)$ is decomposed into the background $L(i, j)$, the foreground $F(i, j)$, and the noise $E(i, j)$, such that

$$Y(i, j) = L(i, j) + F(i, j) + E(i, j), \tag{3.2}$$

Fig. 3.5 presents an example illustrating this decomposition. Suppose that background $L(i, j)$ has already been estimated, then a global threshold τ can be applied to the background subtracted image for image binarization:

$$B(i, j) = \begin{cases} 1 & \text{for } Y(i, j) - L(i, j) \leq \tau \\ 0 & \text{for } Y(i, j) - L(i, j) > \tau, \end{cases} \tag{3.3}$$

where $B(i, j) = 1$ indicates that the pixel at (i, j) part of the foreground and $B(i, j) = 0$ says it is in the background.

A more challenging part is of course how to estimate $L(i, j)$. For that, Vo and Park (2018) develop a robust regression approach, which proceeds as follows. Vo and Park (2018) took a note of the difference between the background pixels and the foreground ones, namely, $Y(i, j) = L(i, j) + E(i, j)$ for background versus $Y(i, j) = L(i, j) + F(i, j) + E(i, j)$ for foreground. It is not strange to view the foreground pixels as outliers far deviating from the unknown background function $L(i, j)$. Prompted by this observation, the authors formulate the estimation problem of $L(i, j)$ as a robust regression problem, i.e., they intend to estimate $L(i, j)$ from $Y(i, j)$ by shielding off the effect of outliers as much as possible.

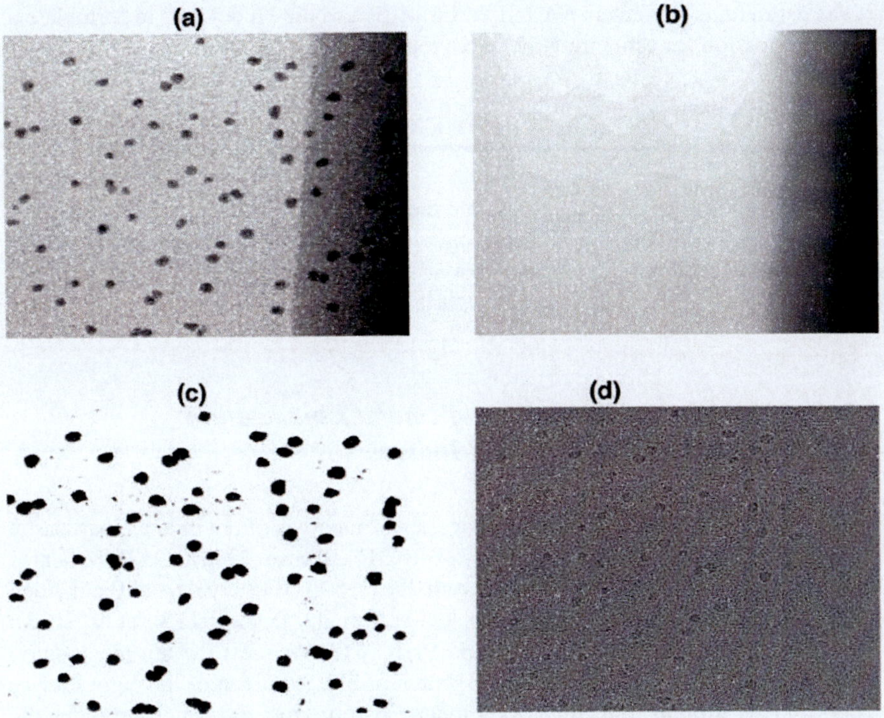

Fig. 3.5 Modeling an image as addition of foreground, background, and noise functions. (**a**) image: Y(i, j). (**b**) background: L(i, j). (**c**) foreground: F(i, j). (**d**) noise: E(i, j)

Research in robust statistics (Rousseeuw and Leroy 2005) reveals that a more robust estimation (or function fitting) is to use a weighted squared error loss (Rousseeuw and Leroy 2005), i.e.,

$$\rho(Y(i,j), L(i,j)) = W(i,j)(Y(i,j) - L(i,j))^2,$$

where the weighting coefficient, $W(i, j)$, ought to be designed to lower the weights where an outlier locates. As such, smaller weights on outlier regions make the outcome of the regression less affected by outlying pixels. A popular choice for the weight factor is the Huber loss weight (Rousseeuw and Leroy 2005),

$$W(i,j) = \begin{cases} 1 & \text{for } |Y(i,j) - L(i,j)| \leq \delta \\ \frac{\delta}{|Y(i,j) - L(i,j)|} & \text{otherwise,} \end{cases}$$

where $\delta = 1.346$ is the default value. The Huber loss places lower weights for higher absolute difference $|Y(i, j) - L(i, j)|$, so that the effect of extreme outliers on the estimate is mitigated. Vo and Park (2018) use the Huber loss to formulate a robust regression for estimating the background function of an image as:

$$\min \sum_{i=1}^{m} \sum_{j=1}^{n} \rho(Y(i, j), L(i, j)) + \lambda \sum_{i=2}^{m-1} \sum_{j=2}^{n-1} |\nabla^2 L(i, j)|^2, \tag{3.4}$$

where the first term is the Huber loss, and the second term is the regularization term that penalizes the large value of the second derivative $\nabla^2 L(i, j)$. The regularization term ensures smoothness in the background intensity function $L(i, j)$. Vo and Park (2018) proposed an efficient iterative algorithm to solve the minimization problem.

3.3.6 *Numerical Comparison of Image Binarization Approaches for Material Images*

In this section, we present the image processing outcomes of ten image binarization methods: NIBLACK (Niblack 1985), BERNSE (Bernse 1986), GATOS (Gatos et al. 2006), BRADLEY (Bradley and Roth 2007), SAUV (Sauvola and Pietikäinen 2000), PHAN (Phansalkar et al. 2011), LU (Lu et al. 2010), SU (Su et al. 2013), HOWE (Howe 2013), and BGS (Vo and Park 2018). We test the ten methods on the eight TEM images of nanoparticles shown in Fig. 3.6. Each of the eight images comes with a ground truth binary sil/houette of foregrounds, which provides the basis for comparison with the outcomes of the binarization methods.

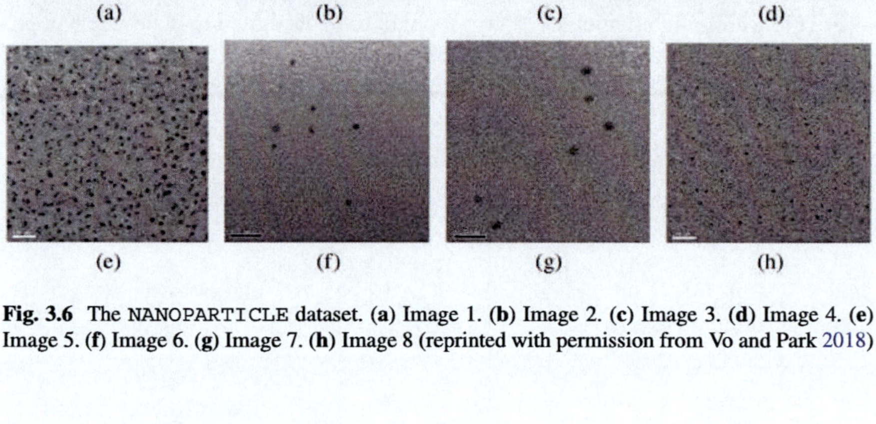

Fig. 3.6 The NANOPARTICLE dataset. (**a**) Image 1. (**b**) Image 2. (**c**) Image 3. (**d**) Image 4. (**e**) Image 5. (**f**) Image 6. (**g**) Image 7. (**h**) Image 8 (reprinted with permission from Vo and Park 2018)

To evaluate image processing outcomes quantitatively, four performance metrics are commonly used, including the F-measure (FM), peak signal-to-noise ratio (PSNR), distance reciprocal distortion metric (DRD), and misclassification penalty metric (MPM).

To define FM, let us first define the pixel-wise bianrization recall (RC) and precision (PR). Consider a binary image B of size $M \times N$ resulting from an image binarization approach, and define by G the ground truth binary image of the same size. The recall is the count of the pixels, whose intensities equal to one in both B and G, as a fraction of the count of the pixels whose intensities in G equal to one,

$$\mathrm{RC} = \frac{\sum_{i=1}^{M}\sum_{j=1}^{N} B(i,j)G(i,j)}{\sum_{i=1}^{M}\sum_{j=1}^{N} G(i,j)}.$$

The precision is the count of the pixels, whose intensities equal to one in both B and G, as a fraction of the count of the pixels whose intensities in B equal to one,

$$\mathrm{PR} = \frac{\sum_{i=1}^{M}\sum_{j=1}^{N} B(i,j)G(i,j)}{\sum_{i=1}^{M}\sum_{j=1}^{N} B(i,j)}.$$

The FM is a combined measure based on both pixel-wise binarization recall and precision, i.e.,

$$FM = \frac{2 \times RC \times PR}{RC + PR}.$$

The PSNR is defined as $10\log(1/\mathrm{MSE})$, where MSE refers the mean square difference between B and G, i.e.,

$$\mathrm{MSE} = \frac{1}{MN}\sum_{i=1}^{M}\sum_{j=1}^{N}(B(i,j) - G(i,j))^2.$$

The DRD has been used to measure the visual distortion of an estimated binary image from its ground truth counterpart. To define DRD, let $Q = \{(i,j) : B(i,j) \neq G(i,j)\}$ represent the set of all the pixels where the pixel intensities of B and G do not match. The DRD is first calculated for each pixel in the set of Q: for $(i,j) \in Q$, let $N_m(i,j) = \{(i',j') : |i' - i| \leq m, |j' - j| \leq m\}$ represents the local neighborhood of size $(2m+1) \times (2m+1)$ for a positive integer m, and the DRD at the pixel of (i,j) is defined as

$$DRD(i,j) = \frac{\sum_{(i',j') \in N_m(i,j)} \omega_{(i,j)}(i',j')|B(i',j') - G(i',j')|}{\sum_{(i',j') \in N_m(i,j)} \omega_{(i,j)}(i',j')},$$

where $\omega_{(i,j)}(i', j') = 1/\{(i - i')^2 + (j - j')^2\}^{1/2}$ if $i \neq i'$ or $j \neq j'$ and $\omega_{(i,j)}(i', j') = 0$ otherwise. We set $m = 3$ for our analysis. The overall DRD measure is defined as

$$DRD = \frac{\sum_{(i,j)\in Q} DRD(i, j)}{NURB},$$

where the $NURB$ is the cardinality of the set, $\{(i, j) : \sum_{(i',j')\in N_m(i,j)} G(i', j') = 0$ or $1\}$.

To define MPM, let $FN = \{(i, j) : B(i, j) = 0, G(i, j) = 1\}$, $FP = \{(i, j) : B(i, j) = 1, G(i, j) = 0\}$ and $P = \{(i, j) : G(i, j) = 1\}$.

$$MPM = \frac{\sum_{(i,j)\in FN} d_h((i, j), P) + \sum_{(i,j)\in FP} d_h((i, j), P)}{D},$$

where $d_h((i, j), P) = \min_{(i',j')\in P}\{(i - i')^2 + (j - j')^2\}^{1/2}$, and D is a scaling constant; we set $D = \sum_{(i,j)\in P} d_h((i, j), P)$.

When using these metrics, a greater FM or PSNR is better, whereas a lower DRD or MPM is desirable. Table 3.1 presents the outcomes of the performance metrics when the ten methods are applied to the NANOPARTICLE dataset. For images having low SNR values, local image contrasts are significantly affected by image noise, causing the local contrast-based methods such as Su et al. (2013) less competitive. When foreground sizes are large and noise are severe, it is challenging to estimate the background accurately, causing some of the background subtraction methods such as Lu et al. (2010) and Gatos et al. (2006) not to work as effectively. Under those difficult circumstances, Vo and Park (2018) is still able to capture the background rather robustly. As an illustration, a few image backgrounds estimated by Vo and Park (2018) are presented in Fig. 3.7.

Table 3.1 Four performance metrics of ten image binarization methods applied to the NANOPARTICLE dataset

Dataset: NANOPARTICLE	FM	PSNR	DRD	MPM
NIBLCK (Niblack 1985)	30.57	4.70	965.97	0.1912
BERNSE (Bernse 1986)	29.86	4.09	1118.00	0.2325
GATOS (Gatos et al. 2006)	39.09	10.28	205.80	0.0728
BRADLEY (Bradley and Roth 2007)	35.92	6.26	760.91	0.1274
SAUV(Sauvola and Pietikäinen 2000)	40.80	8.14	380.86	0.0772
PHAN (Phansalkar et al. 2011)	39.61	7.82	472.26	0.0802
LU (Lu et al. 2010)	24.60	3.08	920.42	0.3463
SU (Su et al. 2013)	25.08	5.50	719.45	0.1389
HOWE (Howe 2013)	37.54	11.20	229.17	0.0354
BGS (Vo and Park 2018)	80.77	17.68	10.90	0.0036

Source: Vo and Park (2018) with permission

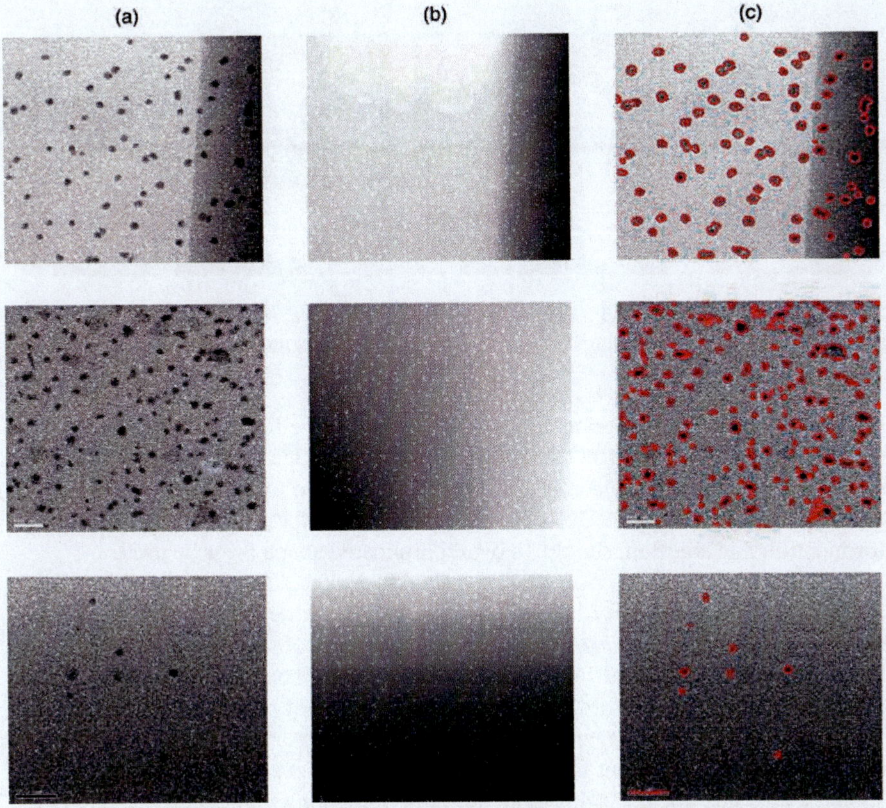

Fig. 3.7 Results of Vo and Park (2018) when the method is applied to the NANOPARTICLE dataset. **(a)** Input image. **(b)** Estimated background. **(c)** Estimated foreground (Reprinted with permission from Vo and Park (2018))

3.4 Foreground Segmentation

After the image binarization is performed, a set of the foreground regions are identified. Some of them may represent the image regions of individual materials, and some others contain multiple materials overlapping with each other. For the latter case, the foreground regions need to be further partitioned to the regions of individual materials. This step is referred to as the *foreground segmentation*. Foreground segmentation is rather different from the foreground-background separation because there is no enough contrast in terms of image intensity within foreground regions; see the example in Fig. 3.7a. The lack of enough contrast makes the intensity information as well as its gradient (i.e., edge information) hard to be exploited for segmentation.

Due to the lack of interior contrast, foreground segmentation has been heavily relying on use of prior knowledge in the form of the shapes of individual foreground

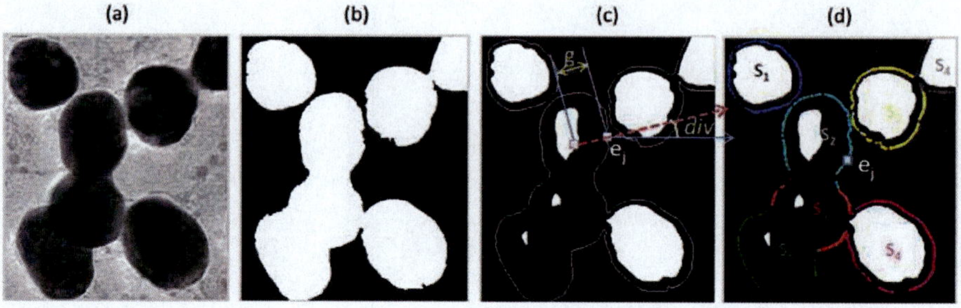

Fig. 3.8 Steps for foreground segmentation and inference. **(a)** Original image. **(b)** Binary silhouette. **(c)** Markers. **(d)** Contour evidences (Reprinted with permission from Park et al. 2013)

regions. The shape priors popularly used are spheres or ellipses for particle analysis (Zafari et al. 2015; Qian et al. 2016), or convex shapes (Park et al. 2013) for some broader classes of materials. There does not seem to exist shape priors more general than these classes. One reason may be due to the intractability of problem formulation and solution, should an overcomplicated shape prior be used.

Once assumed a shape prior, the first step of foreground segmentation is to extract individual foregrounds out of the overlapping agglomerate. This step is also known as the marker generation step because the outcomes of the step are the markers that identify the individual foregrounds; see Fig. 3.8c. The marker of a foreground region is a subset of the foreground region, and an ideal marker should satisfy the following three conditions: (a) there is only one marker for each foreground region, (b) the markers of different foreground regions do not overlap, and (c) the markers should locate near the centroids of the foreground regions for better representing the location of the foregrounds.

After the markers are identified, the foreground regions corresponding to the markers are then estimated by either a region growing approach, which expands the marker back to the corresponding foreground region (Schmitt and Hasse 2009), or a contour evidence association, which collects the contour evidences and relates them to each of the markers (Park et al. 2013); the contour evidence association is illustrated in Fig. 3.8d. In Sect. 3.4.1, we present several approaches for marker generation, whereas in Sect. 3.4.2, we explain how the region growing approach and contour evidence association approach work.

3.4.1 Marker Generation

When foreground materials are spherical or ellipsoidal, the centroids of the spheres or ellipses are estimated and then treated as the markers. Several local filtering methods can fulfill this task, such as the iterative voting method (Parvin et al. 2007) and its variant (Qian et al. 2016). In the original iterative voting approach, Parvin et al. (2007) use Canny's edge detector (Canny 1986) to locate the potential edge pixels on the spherical or elliptical contours of the foreground materials and then use the edge pixels to estimate the centroid location. The potential risk in this approach is the inaccuracy of Canny's edge detectors when it is applied to low contrast images, which can in turn cause inaccurate estimation of the markers. Figure 3.9a shows the outcome of applying the iterative voting method to an image containing 20 particles. The method has three misses and two false detections. Qian et al. (2016) modify the algorithm by controlling the influence of the edge pixels on marker estimation based on a confidence score of the edge pixel detection. The edge pixels detected with a strong confidence are more influential in marker estimation than the edge pixels detected with a lower confidence. The weighting scheme improves the robustness of marker estimation in the presence of inaccurate edge pixel detections. Figure 3.9b shows the outcome of the modified iterative voting approach, which comes with no false markers nor missed estimation of a marker.

When the shapes of foreground materials are non-spherical but still convex, the ultimate erosion method can be applied for obtaining the markers. The ultimate erosion is an iterative application of the morphological erosion operation, discussed in Sect. 2.6, on the image binarization outcome until the binary silhouette of the foregrounds is split into non-overlapping markers. Apparently, in this case, accurate image binarization is the essential prerequisite for the success of the ultimate erosion procedure.

To describe the idea of ultimate erosion, let us consider an image containing n foreground materials. Let C_i be a set of the image pixels in the interior or on the boundary of foreground i, for $i = 1, \ldots, n$. The set C_i should be a convex set if the shape of the foreground is convex. The union $C = \bigcup_{i=1}^{n} C_i$ should be the set of all foreground pixels in the image, which can be obtained from actual images through image binarization. The ultimate erosion applies iteratively a morphological erosion operation to C. Recall that the morphological erosion of C with respect to a structural element B is

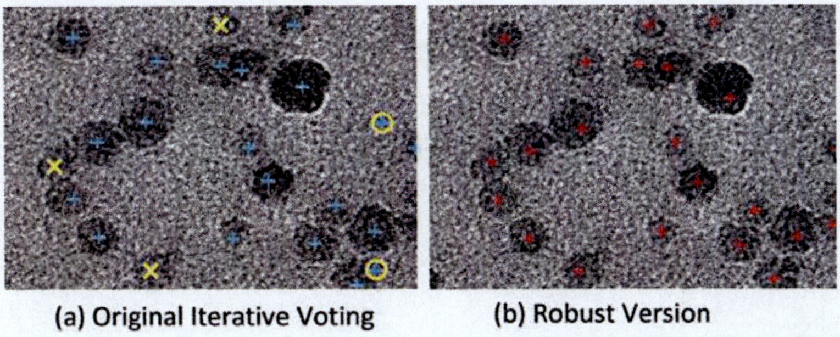

Fig. 3.9 Examples of applying the iterative voting methods: (**a**) The original approach by Parvin et al. (2007). The three misses are indicated by yellow X's and two false detections are marked by yellow circles. (**b**) The modified approach (Qian et al. 2016). (Reprinted with permission from Qian et al. 2016)

$$C \ominus B = \bigcap_{z \in B} C_{-z}.$$

Intuitively, if B is a closed ball in $\mathbb{R}^2$ centered at x with radius one, the result of the set operator is equivalent to peeling off C from its boundary by one pixel. Repeated applications of the operator may disconnect the junctions of overlapping objects. The question is when to stop the morphological erosion.

A popular choice is to keep applying the erosion process to each object just before it is completely removed. This is how the name, 'ultimate erosion' or shortly UE, comes from (Dougherty 1994, Chapter 2). Park et al. (2013) show that the UE is capable of identifying exactly one marker for each C_i in C when C_i's are convex and only mildly overlapping with each other. Assumption 3.1 formally states the mildly overlapping conditions. Figure 3.10 provides illustrative examples explaining Assumption 3.1.

Assumption 3.1 (Chained Cluster of Overlapping Objects Is Mildly Overlapping) The intersection of every three of the n convex sets composing C is at most one point and for every pair $i \neq j$, $C_i \backslash C_j$ is not empty, and it is connected. □

Ideally, UE is supposed to identify all markers of convex-shaped foregrounds from the binary image. The real images, however, contain noises, meaning that the corresponding binarization outcomes could be noisy as well. As a result, the binary silhouette of foregrounds may include image pixels falsely identified as foreground and C_i may not be perfectly convex. In other words, this noise effect could produce falsely identified markers which does not otherwise exist. Park et al. (2013) propose a noise-resistent morphological erosion process, named 'ultimate erosion for convex sets' (UECS), which finds an early stopping time for the iterative erosion process in order prevent the noisy binary image of a convex foreground from being oversegmented into multiple markers. Figure 3.11c, d present the markers produced by UE and UECS, respectively.

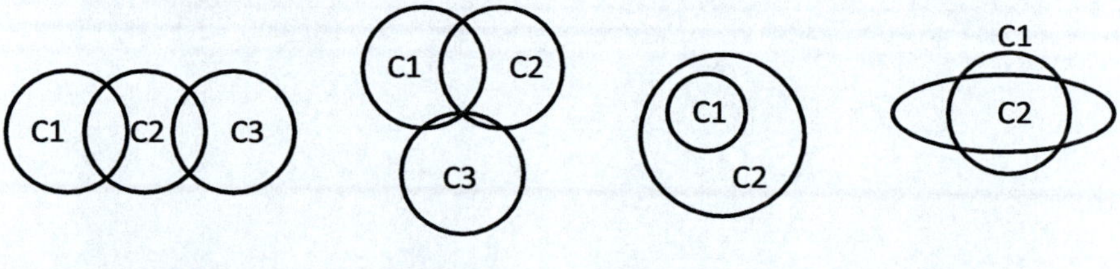

(a) Cases satisfying Assumption 1 C1 ∩ C2 ∩ C3 = empty (b) C1 ∩ C2 ∩ C3 = non-empty (c) C1 \ C2 is empty (d) C1 \ C2 is disconnected

Fig. 3.10 Intuitive examples explaining Assumption 3.1: **(a)** and **(b)** satisfy the assumption, whereas **(c)** and **(d)** violate the assumption. (Reprinted with permission from Park et al. 2013)

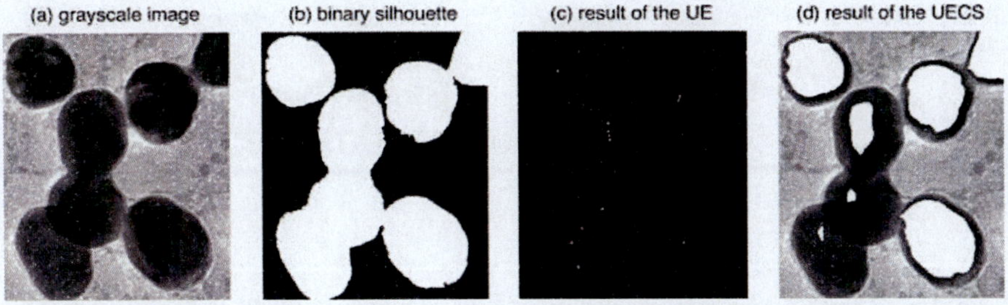

Fig. 3.11 Marker generation for convex-shape foregrounds: (**a**) the original grayscale image, (**b**) the binary silhouette of clustered convex shaped objects, (**c**) markers identified by ultimate erosion (UE) (**d**) markers identified by the ultimate erosion for convex sets (UECS). (Reprinted with permission from Park et al. 2013)

3.4.2 Initial Foreground Segmentation

Once a marker is identified for a foreground object, it is exploited for the purpose of estimating the region of the foreground object. The first step of the estimation is to collect some evidences for the regional boundary that divides the foreground region and the remainder of the image. The collected evidences are then used to estimate the regional boundary. There are two major approaches to collect the boundary evidences.

The first approach is a regional growing approach that expands the marker by an iterative application of morphological dilation to the marker. The expansion stops when the expanded marker collides with other expanded markers. The contour pixels of the expanded markers serve as the boundary evidences for the subsequent estimation of the regional boundary. Marker-controlled watershed methods in fact follow this approach (Dougherty 1994). A shortcoming of the uniform expansion of all markers is that some expanded markers become too large relative to the foreground region, whereas some other expanded markers are still too small. For example, in Fig. 3.12a, marker 4 expanded too much, while marker 3 expanded too little. This could happen particularly when the initial marker sizes differ significantly. Schmitt and Hasse (2009) and Qian et al. (2016) suggest a quick remedy to this issue. Their idea is to record the number of erosion steps used to generate the marker, and the number of the morphological dilation steps used to expand the marker is set proportionally to the number of erosion steps. This simple revision could improve the accuracy of collecting the boundary evidences; compare the marker expansions in the two subplots of Fig. 3.12.

The second approach of collection boundary evidence is to define the contour evidences explicitly, through an edge-to-marker association approach (Park et al. 2013). In such an approach, the edge pixels are first extracted from an input image and then associated with each individual marker, according to a relevance measure.

Suppose n markers identified by UECS are denoted by $\{T_1, T_2, \ldots, T_n\}$, where T_i is the marker of C_i. Here a marker is represented by a set of point coordinates in

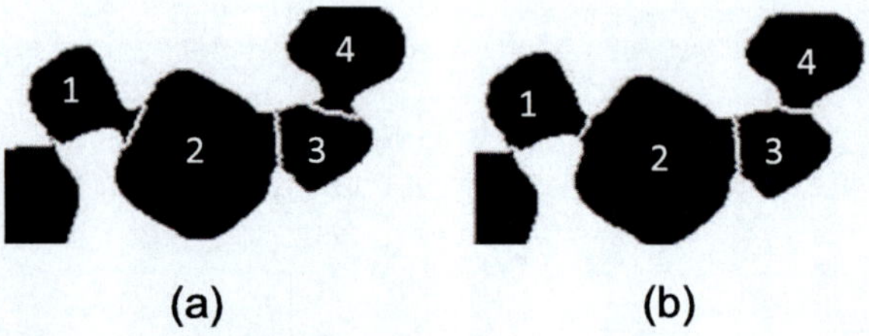

Fig. 3.12 Comparison of the two regional growing approaches: (**a**) uniform marker expansions (Dougherty 1994); and (**b**) adaptive marker expansion in Qian et al. (2016). (Reprinted with permission from Qian et al. 2016)

the marker. An edge detection algorithm such as Canny's method (Canny 1986) can be employed to find m edge pixel coordinates, denoted by $E = \{e_1, \ldots, e_m\}$. Park et al. (2013) define a compound measure to measure the relevance of e_j to T_i; let this relevance measure denoted by $\text{rel}(e_j, T_i)$.

One component composing the compound $\text{rel}(e_j, T_i)$ measure is a distance from e_j to T_i, the same as what is used in the marker-growing approach. The distance measure is defined so as to exclude the edge points that locate closely to an irrelevant marker by chance. The distance is defined with respect to C (the same C used in Sect. 3.4.1) as

$$g(e_j, T_i) = \min_{x \in T_i} g_j(x), \tag{3.5}$$

where $g_j(x)$ is the Euclidean distance $|e_j - x|$ if the line from e_j to x entirely resides within C and ∞ when any portion of the line is outside C. By the convexity of C_i, if e_j is an element of the C_i's contour, the line from $x \in T_i$ to e_j must be in C_i and also in C. Such treatment helps avoid over-emphasizing the relevance of e_j to markers that are irrelevant but close to e_j.

The other component in the compound measure is the divergence index of e_j from T_i, which compares the direction of intensity gradient at e_j with the direction of line from $x \in T_i$ to e_j. Technically, it is expressed as a cosine function:

$$div(e_j, T_i) = \min_{x \in T_i} \frac{\mathbf{g}(e_j) \cdot \mathbf{l}(x, e_j)}{\|\mathbf{g}(e_j)\| \cdot \|\mathbf{l}(x, e_j)\|},$$

where $\mathbf{g}(e_j)$ is the direction of intensity gradient at e_j and $\mathbf{l}(x, e_j)$ is the direction of line from $x \in T_i$ to e_j. The use of the divergence index is motivated by how electron micrography works. In a typical electron micrograph, the regions occupied by material foregrounds have lower image intensities than the background. For this reason, if e_j is a substance of the C_i's contour, the gradient at e_j diverges from T_i. Since C_i is convex, the gradient direction is very close to the vector direction from T_i to e_j, that is to say, the cosine of the angle between the two directions is close to being maximized. In Fig. 3.8c, the solid-line arrow outbound from e_j is the (image intensity) gradient vector at e_j, namely $\mathbf{g}(e_j)$, and the dotted-line arrow represents the straight line from T_i to e_j, namely $\mathbf{l}(x, e_j)$. The divergence index is the cosine of the angle between the two vectors.

Compounding $g(e_j, T_i)$ and $div(e_j, T_i)$, the relevance measure of e_j to T_i is defined as

$$\mathrm{rel}(e_j, T_i) = \frac{1}{1 + g(e_j, T_i)/nIter} + \frac{div(e_j, T_i) + 1}{2}, \tag{3.6}$$

where each of the two terms is normalized to (0, 1] and $nIter$ is the number of erosion iterations in UECS. If $i = \arg\max_k \mathrm{rel}(e_j, T_k)$, e_j becomes an element of the contour evidences for C_i.

3.4.3 Refine Foreground Segmentation with Shape Priors

By using shape priors, one can possibly refine foreground segmentation and enhance the quality of the outcomes. The refinement is done by fitting the contour evidences associated with the markers of foreground regions through a contour model regulated by shape priors. When a contour is assumed spherical or elliptical, a simple parametric model can be used to model the contour of each foreground region. Figure 3.13a presents an elliptic shape model (Qian et al. 2016) that can be parameterized using five parameters $[x_0, y_0, a_0, b_0, \theta_0]$, where x_0 and y_0 are the coordinates of the center, a_0 and b_0 are the lengths of the long and short axes, and θ_0 is the orientation of the particle. The major advantage of using such simple parametric models to fit the contour evidences is that the outcome is robust to background noises surrounding foreground regions. The fitted contour is less influenced by a few falsely identified contour evidences.

When it comes to the fitting of an elliptical shape, Qian et al. (2016) choose to use the second-moment fitting method (Taylor et al. 1983), which finds an ellipse that has the same mass center and same second moments as those of the detected particle region. This treatment uses all the pixels inside the contour of a detected particle, rather than rely on the detected contour of a particle. This choice, which avoids using the detected contour alone, is due to the contour's sensitivity to shape noises. In the presence of image noise, many detected contours can end up with an irregular shape; see the example (the gray region) in Fig. 3.13a. This second-moment method produces much more robust shape fitting outcomes, as evident by the comparison between Fig. 3.13b and c.

When the contour of a foreground region is not spherical or elliptical, a more general contour model needs to be used. Park et al. (2013) propose a B-spline based parametric curve for such representation. Park et al. (2013) further develop a mixture of the B-spline contour models for accommodating various shape possibilities. Estimating the contour and classifying it into one of the possible shapes are performed simultaneously through a mixture prior and the expectation-maximization procedure

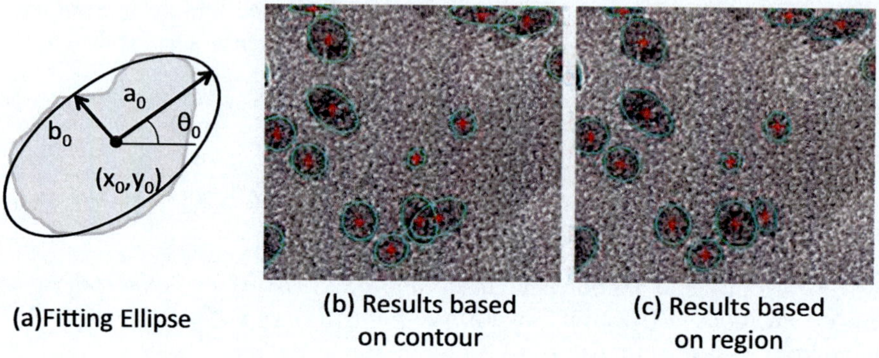

Fig. 3.13 Refine foreground segmentation with an elliptical shape prior: (**a**) Parametrization of an elliptical shape; (**b**) The fitting outcomes based on contour alone; (**c**) The fitting outcomes based on all the pixels in a detected particle region. (Reprinted with permission from Qian et al. 2016)

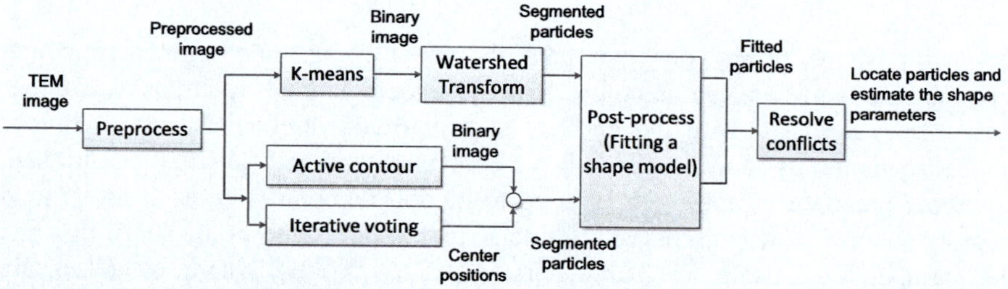

Fig. 3.14 The two pipelines of image segmentation for making use of the complementary image information. (Reprinted with permission from Qian et al. 2016)

(Dempster et al. 1977). Understanding this joint estimation/classification procedure requires the knowledge of morphology analysis, so its discussion is deferred to Chap. 4.

3.5 Ensemble Method for Segmenting Low Contrast Images

Material images can be noisy, producing very low contrast between foregrounds and background. For such low contrast images, separating a foreground region from the noisy background and the overlapping neighboring foregrounds could be challenging. For achieving a more robust image segmentation, one can use an ensemble method that runs complementary image segmentation approaches and combines their best outcomes. Qian et al. (2016) propose such an ensemble approach to identify nanoparticles in a low contract microscopic image. An overview of Qian et al. (2016)'s ensemble method is presented in Fig. 3.14.

Qian et al. (2016)'s ensemble approach runs two image segmentation processes in parallel. Each of the processes is composed of three stages—image binarization, initial foreground segmentation, and the refinement of the initial foreground segmentation with prior knowledge on foreground shapes. The first stage produces the binary image of nanoparticles possibly overlaping with each other, the second stage segments the binary image into the regions of individual nanoparticles, and the third stage fits an ellipse to each of the regions, assuming that particle shapes are all elliptical. The approach is specialized for identifying elliptically shaped nanoparticles.

The two image segmentation pipelines differ in the first two stages. The first pipeline uses a simple k-means clustering (Hartigan and Wong 1979) in the first stage (binarization), and then applies a watershed segmentation in the second stage. The second pipeline uses an active contour (Chan and Vese 2001) in the first stage to perform image binarization. In the second stage, it uses an iterative voting method (Parvin et al. 2007) to generate the markers of individual foregrounds, each of which is associated with the edge pixels of the binary image (resulting from the first stage) through the edge-to-marker association (Park et al. 2013). The edge pixels associated with each marker is fitted to an ellipse to estimate the elliptical foreground boundary.

Each of the two pipelines may produce different foreground segmentation outcomes. Let $I = \{I(i), i = 1, \cdots, N_I\}$ and $J = \{J(j), j = 1, \cdots, N_J\}$ denote the respective outcomes of the two processes, where $I(i)$ represents the ith nanoparticle region identified by the first pipeline, and $J(j)$ represents the ith nanoparticle region identified by the second pipeline. Please note that $N_I \neq N_J$, i.e., the numbers of the nanoparticles identified by the two pieplines may differ. Each of the nanoparticle regions is elliptical, and $I(i)$ is represented by the outline of the elliptical region, which is parameterized by the five parameters shown in Fig. 3.13a. Let $[x_0(I(i)), y_0(I(i)), a_0(I(i)), b_0(I(i)), \theta_0(I(i))]$ denote the five parameters for $I(i)$. The set of pixels within the ellipse is labeled as $P_{I(i)}$ and its cardinality $|P_{I(i)}|$ represents the area of the corresponding region. The notations for $J(j)$ are defined likewise.

To make good use of the different outcomes from the two pipelines, it is crucial to understand three possible scenarios of conflict/consensus outcomes. As illustrated in Fig. 3.15, a region $I(i)$ identified from the first process and $J(j)$ from the second process may have only a slight overlap or no overlap at all (Column (a), Fig. 3.15), it is unlikely that they are related to the common particle in the image. This scenario is referred to as an unrelated segmentation. When $I(i)$ and $J(j)$ virtually coincide with each other, manifesting in a heavy overlap between the detection regions, they point to the same underlying particle and are then referred to as a consensus segmentation. When $I(i)$ and $J(j)$ occupy the same region in the image, but the estimated particle regions have serious disagreement, either in number (one approach detects one particle, while the other detects two, for instance) or in key shape parameters (including the center location and size), these outcomes are referred to as a conflicting segmentation. The consensus segmentation and the conflicting segmentation are illustrated in Columns (b) and (c) of Fig. 3.15,

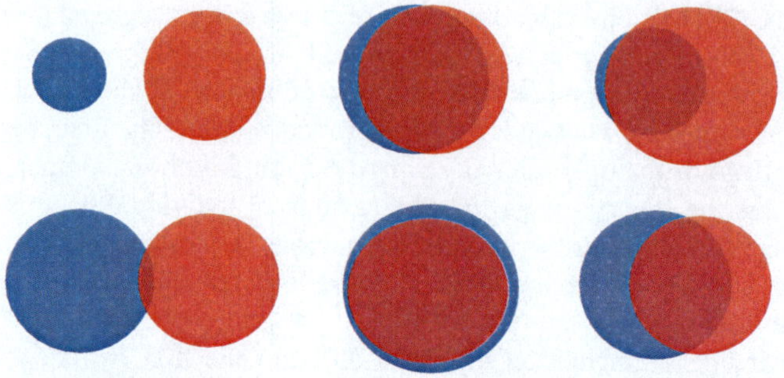

Fig. 3.15 Three possible relationships between $I(i)$ (blue) and $J(j)$ (red): (**a**) Two detection results are not related to the same particle. (**b**) Two results coincide with each other. (**c**) Two results are in conflict. (Reprinted with permission from Qian et al. 2016)

respectively. Determining whether $I(i)$ and $J(j)$ are unrelated, consensus, or conflicting depends on the degree of overlap between $I(i)$ and $J(j)$; more details are presented in Sect. 3.5.1.

The unrelated and consensus segmentations are relatively straightforward to deal with. When the two pipelines of processing reach a consensus, it enhances the credibility of the respective segmentations. It is safe to take the consensus outcomes and add them into the final detection results without further processing. One can compute the shape parameters of the final particle by averaging the corresponding parameters of $I(i)$ and $J(j)$. Then one can remove these particles from the sets of I and J, so that only the conflicting and unrelated detections are left to be resolved.

Denote the sets of the remaining particles as $\tilde{I} = \{\tilde{I}(1), \cdots, \tilde{I}(N_{\tilde{I}})\}$ and $\tilde{J} = \{\tilde{J}(1), \cdots, \tilde{J}(N_{\tilde{J}})\}$, where $N_{\tilde{I}}$ and $N_{\tilde{J}}$ are the numbers of particles in the two revised sets, respectively. To distinguish in between unrelated and conflicting ones, a conflict matrix $\mathbf{M} = (M_{ij})$ is defined, where $\mathbf{M}$ is an $N_{\tilde{I}} \times N_{\tilde{J}}$ binary matrix, and if $\tilde{I}(i)$ and $\tilde{J}(j)$ are conflicting, $M_{ij} = 1$; if they are unrelated, $M_{ij} = 0$. Figure 3.16 shows a simple example of conflicting detections and the corresponding conflict matrix. In Fig. 3.16, one can see that $\tilde{I}(1)$ is conflicting with $\tilde{J}(1)$, while $\tilde{I}(2)$ is conflicting with both $\tilde{J}(2)$ and $\tilde{J}(3)$; this is reflected in the 2×3 conflict matrix to the right.

For each pair of conflicting $\tilde{I}(i)$ and $\tilde{J}(j)$, the quality of the two conflicting segmentations, $\tilde{I}(i)$ and $\tilde{J}(j)$, needs to be assessed in order to make a better selection. Qian et al. (2016) define the segmentation quality measure that quantifies the fitness of a segmentation to the original image. The specific definition of the measure is provided in Sect. 3.5.2, but for now let us suppose that the quality measures are available and expressed in the form of a fitness score—Qian et al. (2016) denote the fitness score vector for the conflicting segmentations from the first pipeline by $\mathbf{s}_{\tilde{I}} = [s_{\tilde{I}(1)}, \cdots, s_{\tilde{I}(N_{\tilde{I}})}]^T$ and denote that from the second pipeline

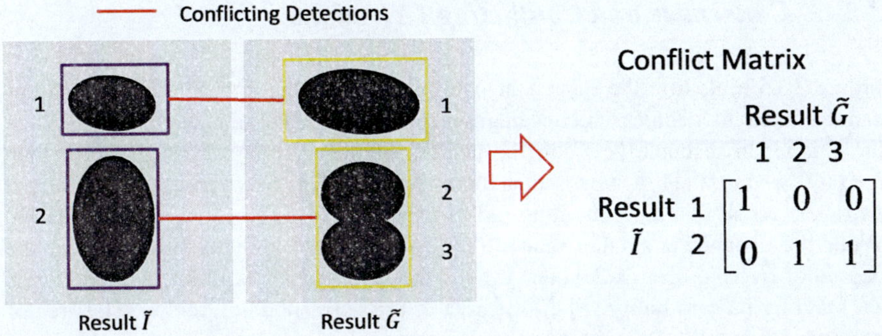

Fig. 3.16 An example of conflicting detections (left) and the corresponding conflict matrix (right). (Reprinted with permission from Qian et al. 2016)

by $\mathbf{s}_{\tilde{J}} = [s_{\tilde{J}(1)}, \cdots, s_{\tilde{J}(N_{\tilde{J}})}]^T$. With these quality measure notations defined, an optimization problem is formulated to optimize the selections for all conflicting cases. Let $b_{\tilde{I}(i)}$ and $b_{\tilde{J}(j)}$ be the binary decision variables to indicate the outcome of the conflict resolution: if $\tilde{I}(i)$ (or $\tilde{J}(j)$) is chosen as the final detection outcome, then $b_{\tilde{I}(i)}$ (or $b_{\tilde{J}(j)}$) will be set as one, otherwise it is set as zero. Aggregating all the decision variables associated with individual foreground detections, the decision vector for the first segmentation process is expressed as $\mathbf{b}_{\tilde{I}} = [b_{\tilde{I}(1)}, \cdots, b_{\tilde{I}(N_{\tilde{I}})}]^T$, and that for the second process is $\mathbf{b}_{\tilde{J}} = [b_{\tilde{J}(1)}, \cdots, b_{\tilde{J}(N_{\tilde{J}})}]^T$. A binary integer programming (BIP) problem to optimize the decision vectors is formulated as below:

$$\begin{aligned} &\max_{\mathbf{b}_{\tilde{I}}, \mathbf{b}_{\tilde{G}}} \quad \mathbf{s}_{\tilde{I}}^T \mathbf{b}_{\tilde{I}} + \mathbf{s}_{\tilde{G}}^T \mathbf{b}_{\tilde{G}}, \\ &\text{subject to} \quad \mathbf{b}_{\tilde{I}}^T \mathbf{M} \mathbf{b}_{\tilde{G}} = 0. \end{aligned} \tag{3.7}$$

The objective function is the summation of the fitness scores of the chosen segmentations. Qian et al. (2016) aim to obtain the highest total fitness score for an image. The constraint function is to ensure that only one of the conflicting segmentations will be chosen for each conflicting case. To see this, rewrite the constraint as

$$\sum_{i=1}^{N_{\tilde{I}}} \sum_{j=1}^{N_{\tilde{J}}} b_{\tilde{I}(i)} M_{ij} b_{\tilde{J}(j)} = 0, \tag{3.8}$$

meaning that if $\tilde{I}(i)$ and $\tilde{J}(j)$ are a pair of conflicting detections, namely $M_{ij} = 1$, then $b_{\tilde{I}(i)}$ and $b_{\tilde{J}(j)}$ cannot be one simultaneously. Section 3.5.3 describes the optimization algorithm that efficiently solves the problem.

3.5.1 Consensus and Conflicting Detections

Figure 3.15 hints that the degree of overlap between the two detection outcomes can be used to decide which category a pair of detections belongs to. When the Euclidean distance between the centers of the two detections is larger than $a_0(I(i)) + a_0(J(j))$, it means that there is no overlap between the two detected particles (recall that a_0 is the long radius of a particle). The pair is then unrelated. When the distance is smaller than $a_0(I(i)) + a_0(J(j))$, one needs to quantify the degree of overlap. Let us denote by $P_{I(i)}$ the set of pixels within the fitted ellipse $I(i)$ and by its cardinality $|P_{I(i)}|$ the area of the corresponding region. The area of overlap is then $|P_{I(i)} \cap P_{J(j)}|$. Qian et al. (2016) calculate the maximum overlapping ratio $r_{\max}$ and minimum overlapping ratio $r_{\min}$ as follows:

$$
\begin{aligned}
r_{\max}(I(i), J(j)) &= \max\left\{\frac{|P_{I(i)} \cap P_{J(j)}|}{|P_{I(i)}|}, \frac{|P_{I(i)} \cap P_{J(j)}|}{|P_{J(j)}|}\right\}, \\
r_{\min}(I(i), J(j)) &= \min\left\{\frac{|P_{I(i)} \cap P_{J(j)}|}{|P_{I(i)}|}, \frac{|P_{I(i)} \cap P_{J(j)}|}{|P_{J(j)}|}\right\}.
\end{aligned}
\tag{3.9}
$$

Qian et al. (2016) then set two thresholds, an upper ratio r_U and a lower ratio r_L, such that if $r_{\max}(I(i), J(j)) < r_L$, they deem the overlapping region small enough to declare $I(i)$ and $J(j)$ unrelated; if $r_{\min}(I(i), J(j)) < r_U$ and $r_{\max}(I(i), J(j)) > r_L$, they recommend that the two detection outcomes are considered related but different, namely that the two particles form a pair of conflicts; if $r_{\min}(I(i), J(j)) > r_U$, they consider this as a consensus detection.

3.5.2 Measure of Segmentation Quality

Calculating the fitness score for each particle is equivalent to evaluating the quality of the image segmentation, in which a regional part of TEM images is separated into the particle and its surrounding background area. Zhang et al. (2008) survey different evaluation methods for image segmentation quality when the ground truth is unknown. They point out a simple principle that is still widely used: the inter-region disparity should be large and the intra-region variability should be small. For instance, Fisker et al. (2000) maximize the difference in the average intensities between the foreground and its surrounding background for detecting a particle. To measure the inter-region disparity and the intra-region similarity, Qian et al. (2016) define a neighboring region Q for particles in $\tilde{I}$ and $\tilde{J}$. Consider a particle $\tilde{I}(i)$ (the same can be done to $\tilde{J}(j)$). Its foreground information is in $P_{\tilde{I}(i)}$ and the surrounding background information is in $Q_{\tilde{I}(i)}$ (illustrated in Fig. 3.17). In identifying $Q_{\tilde{I}(i)}$, Qian et al. (2016) double the size of $P_{\tilde{I}(i)}$, namely $|Q_{\tilde{I}(i)} \cup P_{\tilde{I}(i)}| = 2|P_{\tilde{I}(i)}|$, so that $|Q_{\tilde{I}(i)}| = |P_{\tilde{I}(i)}|$.

The measure of the inter-region disparity and that of the intra-region similarity are based on the sum of squares of pixel intensities. The sum of squares are

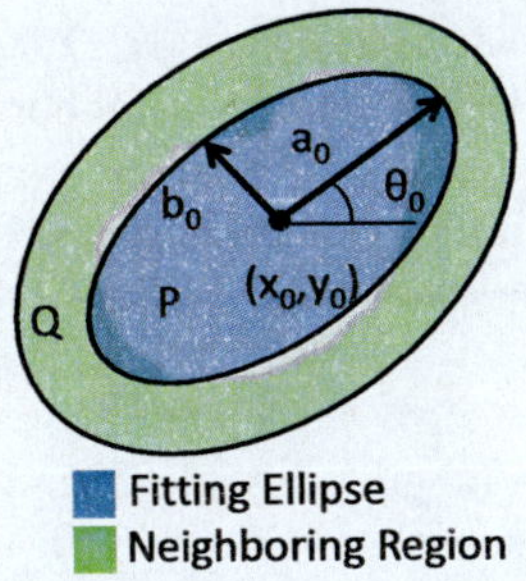

Fig. 3.17 The foreground region P (blue) and its neighboring region Q (green) for a detected particle. (Reprinted with permission from Qian et al. 2016)

proportional to the variance of the intensities within a region, so a large value indicates disparity while a small value indicates similarity. For a good segmentation, the sum of squares of the whole region should be much larger than that of separated background or foreground. For an arbitrary region A in the image, its sum of squares of the intensity, denoted by $SS(A)$, is calculated by:

$$SS(A) = \sum_{(x,y)\in A} [R(x, y) - \bar{R}(A)]^2, \tag{3.10}$$

where $\bar{R}(A)$ is the average intensity of all pixels inside A. Qian et al. (2016) then define the fitness score of $\tilde{I}(i)$ as:

$$\begin{aligned} s_{\tilde{I}(i)} = SS(P_{\tilde{I}(i)} \cup Q_{\tilde{I}(i)}) - [SS(P_{\tilde{I}(i)}) + SS(Q_{\tilde{I}(i)})] \\ -\lambda |P_{\tilde{I}(i)} \cup Q_{\tilde{I}(i)}|, \end{aligned} \tag{3.11}$$

where the first term $SS(P_{\tilde{I}(i)} \cup Q_{\tilde{I}(i)})$ measures the inter-region disparity, and the second term $[SS(P_{\tilde{I}(i)}) + SS(Q_{\tilde{I}(i)})]$ measures the intra-region similarity. The greater their difference, the stronger indication it is to think that $\tilde{I}(i)$ is part of the particle's foreground. The third term is a noise filter. Its inclusion forces the difference between the inter-region disparity and the intra-region similarity to be great enough so as to qualify $\tilde{I}(i)$ as a genuine particle, helping reduce the false detections in a noisy image. If $\tilde{I}(i)$ is a single unrelated particles, which means it has no conflicting detection in another set of results, it will be selected if and only if $s_{\tilde{I}(i)}$ is larger then 0.

In Eq. (3.11), the first term is the total sum of squares of the whole region and the second term is the within-group sum of squares. According to the property of variance (Mardia et al. 1980), their difference equals the between-group sum of squares, i.e.,

$$\begin{aligned} |P_{\tilde{I}(i)}|[\bar{R}(P_{\tilde{I}(i)}) - \bar{R}(P_{\tilde{I}(i)} \cup Q_{\tilde{I}(i)})]^2 \\ + |Q_{\tilde{I}(i)}|[\bar{R}(Q_{\tilde{I}(i)}) - \bar{R}(P_{\tilde{I}(i)} \cup Q_{\tilde{I}(i)})]^2, \end{aligned} \tag{3.12}$$

where $\bar{R}(P_{\tilde{I}(i)})$, $\bar{R}(Q_{\tilde{I}(i)})$ and $\bar{R}(P_{\tilde{I}(i)} \cup Q_{\tilde{I}(i)})$ are the average intensities of the foreground, its neighboring region, and the combined whole area, respectively. By the choice of neighboring region made above, namely $|Q_{\tilde{I}(i)}| = |P_{\tilde{I}(i)}|$ (they may not be exactly the same but the difference is negligible), it means:

$$\bar{R}(P_{\tilde{I}(i)} \cup Q_{\tilde{I}(i)}) = (\bar{R}(P_{\tilde{I}(i)}) + \bar{R}(Q_{\tilde{I}(i)}))/2. \tag{3.13}$$

Plugging Eqs. (3.12) and (3.13) into Eq. (3.11), Qian et al. (2016) express the fitness score as

$$s_{\tilde{I}(i)} = |P_{\tilde{I}(i)} \cup Q_{\tilde{I}(i)}| \left\{ \left(\frac{\bar{R}(P_{\tilde{I}(i)}) - \bar{R}(Q_{\tilde{I}(i)})}{2} \right)^2 - \lambda \right\}. \tag{3.14}$$

It is now clear how the third term in Eq. (3.11) works—if the intensity difference between the foreground and background is smaller than the threshold $2\sqrt{\lambda}$, then, the fitness score $s_{\tilde{I}(i)}$ turns negative, and consequently, $\tilde{I}(i)$ will not be chosen as a particle.

3.5.3 *Optimization Algorithm for Resolving Conflicting Segmentations*

To solve the optimization problem in Eq. (3.7) efficiently, one needs to address two more problems: (a) There are hundreds to thousands of particles in $\tilde{I}$ and $\tilde{J}$ in a TEM image. Solving the optimization in its current form is time consuming. (b) The constraint in Eq. (3.7) is not linear, which prevents a straightforward application of some existing efficient methods. It is necessary to decompose the original problem into smaller-sized subproblems, as well as to linearize the constraint.

The way to decompose the original optimization problem is to decompose the conflict matrix **M**. If **M** can be expressed in a block form with zero off-diagonal submatrices, then, each block submatrix can be used to form a separate BIP problem and be solved in parallel. A simple example is a two-block **M**, such as

$$\mathbf{M} = \begin{bmatrix} \mathbf{M}_1 & \mathbf{0} \\ \mathbf{0} & \mathbf{M}_2 \end{bmatrix}, \tag{3.15}$$

then Eq. (3.7) can be decomposed into two BIP problems:

$$\begin{matrix} \max_{\mathbf{b}_{\tilde{I}_1}, \mathbf{b}_{\tilde{J}_1}} \mathbf{s}_{\tilde{I}_1}^T \mathbf{b}_{\tilde{I}_1} + \mathbf{s}_{\tilde{J}_1}^T \mathbf{b}_{\tilde{J}_1} & \max_{\mathbf{b}_{\tilde{I}_2}, \mathbf{b}_{\tilde{J}_2}} \mathbf{s}_{\tilde{I}_2}^T \mathbf{b}_{\tilde{I}_2} + \mathbf{s}_{\tilde{J}_2}^T \mathbf{b}_{\tilde{J}_2} \\ \text{subject to } \mathbf{b}_{\tilde{I}_1}^T \mathbf{M}_1 \mathbf{b}_{\tilde{J}_1} = 0 & \text{subject to } \mathbf{b}_{\tilde{I}_2}^T \mathbf{M}_2 \mathbf{b}_{\tilde{J}_2} = 0, \end{matrix} \tag{3.16}$$

where $\mathbf{s}_{\tilde{I}} = [\mathbf{s}_{\tilde{I}_1}; \mathbf{s}_{\tilde{I}_2}]$ and $\mathbf{s}_{\tilde{J}} = [\mathbf{s}_{\tilde{J}_1}; \mathbf{s}_{\tilde{J}_2}]$. After solving those two subproblems, the minimizer of the original problem can be easily obtained by combining their individual solutions, namely $\mathbf{b}_{\tilde{I}} = [\mathbf{b}_{\tilde{I}_1}; \mathbf{b}_{\tilde{I}_2}]$ and $\mathbf{b}_{\tilde{J}} = [\mathbf{b}_{\tilde{J}_1}; \mathbf{b}_{\tilde{J}_2}]$.

The decomposition of the BIP can also be seen as a problem to find the connected independent subgraph. Qian et al. (2016) regard the $N_{\tilde{I}} + N_{\tilde{J}}$ particles in $\tilde{I}$ and $\tilde{J}$ as the nodes for building an undirected graph **J**. Then, Qian et al. (2016) connect two nodes if they form a pair of conflicting detection and obtain the corresponding adjacent matrix **W**. If one can find an independent connected subgraph containing, for example, $\tilde{I}(1)$, $\tilde{I}(2)$ and $\tilde{J}(1)$, $\tilde{J}(2)$, $\tilde{J}(3)$, that means there is no conflicting relationship between them and any other particles. So one can form a subproblem only concerning those five particles. The solution of that subproblem is the same as the corresponding part of the whole problem. To find all connected independent subgraphs in **J**, Qian et al. (2016) adopt the spectrum analysis method (Von Luxburg 2007).

Von Luxburg (2007) says that the number of independent connected subgraphs of **J** equals to the multiplicity of zero eigenvalue of its normalized graph Laplacian matrix:

$$\mathbf{L} = \mathbf{I} - \mathbf{D}^{-\frac{1}{2}}\mathbf{W}\mathbf{D}^{-\frac{1}{2}}, \tag{3.17}$$

where **W** is the adjacent matrix of the graph **J**, **I** is the identical matrix which has the same size of **W**, and **D** is the diagonal matrix of the row (or column) sum of **W**. Following the procedure in Von Luxburg (2007), Qian et al. (2016) suggested checking if the graph **J** is decomposable (i.e., checking the multiplicity of zero eigenvalue of **L**). If this multiplicity is $K > 1$, then **J** can be decomposed to a set of K independent connected subgraphs. Then one can break **M** into K block submatrices $\{\mathbf{M}_k\}_{i=1}^K$, and the fitness score vectors $\mathbf{s}_{\tilde{I}}$ and $\mathbf{s}_{\tilde{J}}$ into $\{\mathbf{s}_{\tilde{I}_k}\}_{i=1}^K$ and $\{\mathbf{s}_{\tilde{J}_k}\}_{i=1}^K$, respectively. As such, the original BIP can be decomposed to K smaller subproblems that can be solved in parallel. The kth subproblem is:

$$\begin{aligned} &\max_{\mathbf{b}_{\tilde{I}_k}, \mathbf{b}_{\tilde{J}_k}} \; \mathbf{s}_{\tilde{I}_k}^T \mathbf{b}_{\tilde{I}_k} + \mathbf{s}_{\tilde{J}_k}^T \mathbf{b}_{\tilde{J}_k}, \\ &\text{subject to} \quad \mathbf{b}_{\tilde{I}_k}^T \mathbf{M}_k \mathbf{b}_{\tilde{J}_k} = 0. \end{aligned} \tag{3.18}$$

Further, Qian et al. (2016) show that the constraint in Eq. (3.7) can be linearized. Because $\mathbf{b}_{\tilde{I}}$, $\mathbf{b}_{\tilde{J}}$ and **M** are binary vectors/matrix, the original constraint can be replaced by the following inequality:

$$\mathbf{M}^T \mathbf{b}_{\tilde{I}} + N_{\tilde{I}} \mathbf{b}_{\tilde{J}} \leq N_{\tilde{I}} \mathbf{1}_{N_{\tilde{J}}}, \tag{3.19}$$

where $\mathbf{1}_{N_{\tilde{J}}}$ represents an $N_{\tilde{J}} \times 1$ vector whose elements are all 1's.

Qian et al. (2016) show that the original constraint and Eq. (3.19) are equivalent. For the constraint in Eq. (3.7), it is obvious to see that the constraint is violated if and only if there exists any pair of i and j satisfying $M_{ij} = 1$, $b_{\tilde{I}(i)} = 1$ and $b_{\tilde{J}(j)} = 1$. Qian et al. (2016) proceed and demonstrate that Eq. (3.19) is violated under the same condition, as follows. Equation (3.19) produces $N_{\tilde{J}}$ linear inequalities. Let us consider the jth inequality:

$$\sum_{i=1}^{N_{\tilde{I}}} M_{ij} b_{\tilde{I}(i)} + N_{\tilde{I}} b_{\tilde{J}(j)} \leq N_{\tilde{I}}. \tag{3.20}$$

1. If $b_{\tilde{J}(j)} = 0$, because M_{ij} and $b_{\tilde{I}(i)}$ are both binary variables taking either zero or one, $\sum_{i=1}^{N_{\tilde{I}}} M_{ij} b_{\tilde{I}(i)} \leq N_{\tilde{I}}$ is always true. This suggests that regardless of the choice of $\mathbf{b}_{\tilde{I}}$, the constraint in (3.20) is satisfied.
2. If $b_{\tilde{J}(j)} = 1$, $N_{\tilde{I}} b_{\tilde{J}(j)}$ equals to $N_{\tilde{I}}$. If there exits any i satisfying $M_{ij} = 1$ and $b_{\tilde{I}(i)} = 1$, then $\sum_{i=1}^{N_{\tilde{I}}} M_{ij} b_{\tilde{I}(i)}$ is larger than zero, making the inequality untrue. In order for the inequality to hold, the first term must be 0, meaning when $b_{\tilde{J}(j)} = 1$, M_{ij} and $b_{\tilde{I}(i)}$ cannot be one at the same time.

The above argument extends to all j's.

As such, one can replace the original constraint with the inequality in Eq. (3.19), which is linear. As the objective function is also linear, one can use efficient linear binary programming methods to solve the optimization problem, such as a branch-and-bound algorithm (Narendra and Fukunaga 1977).

3.6 Case Study: Ensemble Method for Nanoparticle Detection

This section presents the application of the ensemble method described in the previous section for detecting nanoparticles in several noisy microscope images. The ensemble method has two sets of parameters to specify before it runs: (1) r_U and r_L that are used to classify the conflicting segmentations in Sect. 3.5.1, and (2) λ is used in Sect. 3.5.2 to measure the segmentation quality. Qian et al. (2016) empirically choose $r_U = 0.8$ and $r_L = 0.2$, and test many TEM images and find that these choices produce rather robust categorizations consistent with human interpretation. Qian et al. (2016) set the pixel intensity gap to be about one-tenth of the grayness levels from the brightest to the darkest in the TEM images, in order to differentiate a particle's foreground from its surrounding background. For noisy TEM images, this gap appears reasonable. Given that the TEM images in question have roughly 200 grayness levels, it suggests that the gap is going to be 20, and according to Eq. (3.14), this sets $\lambda = 100$.

This section includes two numerical studies. The first study in Sect. 3.6.1 is primarily designed to compare the ensemble approach with the two individual image segmentation processes included in the ensemble approach. The second study in

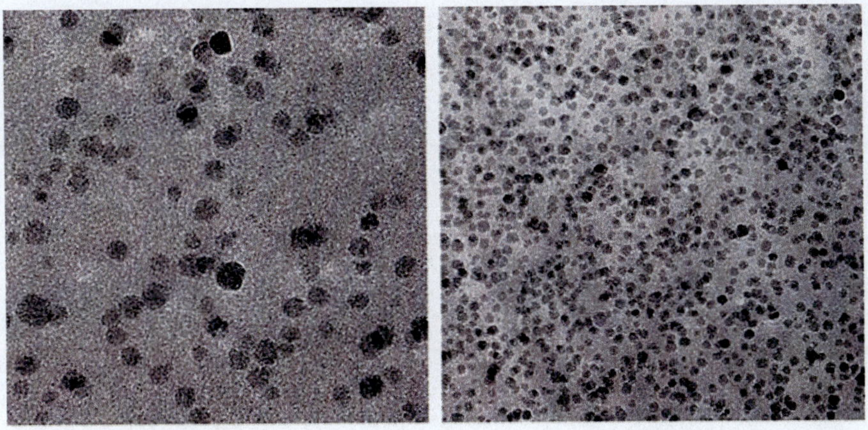

(a) (b)

Fig. 3.18 Two test TEM images of silica nanoparticles. (**a**) F3-2_7. (**b**) F10_8 (Reprinted with permission from Qian et al. 2016)

Sect. 3.6.2 provides the numerical performance of the ensemble approach for a comprehensive set of electron microscopic images having different resolutions.

3.6.1 Ensemble versus Individual Segmentation

Figure 3.18 presents two TEM images obtained under different instrumental resolutions and to be used in this study. The images are labeled as "F3-2_7" and "F10_8", respectively. In the images, the darker dots represent the nanoparticles, whereas the gray background represents the host material. Using the two images, Qian et al. (2016) compare the ensemble method with each of the individual image segmentation methods used in the ensemble method.

Figure 3.19 presents the image segmentation outcomes from the ensemble method. The images are color coded: green means a consensus segmentation, blue means that the outcome from the first segmentation process prevails, and yellow means that from the second segmentation process prevails. In the low resolution image ("F10_8"), there are 721 consensus segmentations, out of the 1, 100 particles finally detected. Among the 379 conflicting segmentations, 162, or 43%, final outcomes come from the first pipeline, whereas 217, or 57%, come from the second pipeline. The respective numbers for the medium resolution image ("F3-2_7") are: 103 total particles, 85 consensus segmentations, 18 conflicting segmentations. Among the 18 conflicting outcomes, 9, or 50%, are from the first pipeline, whereas the other 9, or 50%, from the second pipeline. One can observe that the ensemble method does improve upon the individual pipelines of processing. Qian et al. (2016) state that this is a key advantage of the ensemble approach, as it makes use of

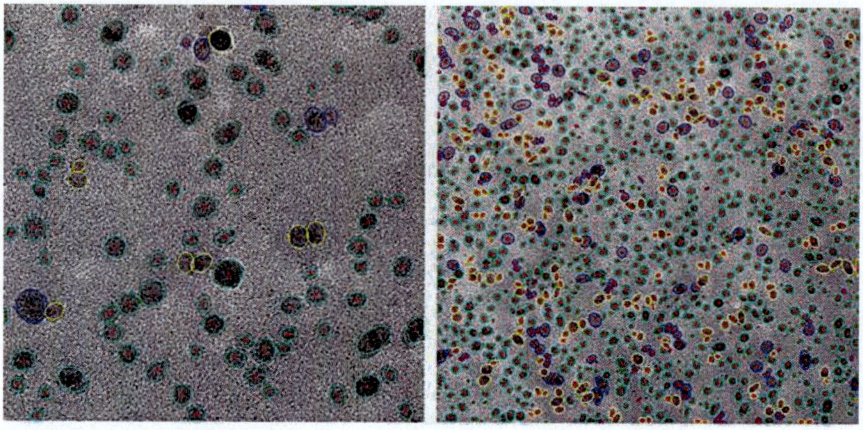

(a) (b)

Fig. 3.19 Comparison of individual processing pipelines. The left image corresponds to Fig. 3.18a (medium-resolution image), where the right image corresponds to Fig. 3.18b (low-resolution image). Green particles are those from the consensus detections; blue particles are from the first pipeline; yellow particles are from the second pipeline. (**a**) Medium resolution. (**b**) Low resolution (Reprinted with permission from Qian et al. 2016)

the image information fully and compensate for the limitations of the approaches emphasizing too much on one type of image information.

3.6.2 Numerical Performance of the Ensemble Segmentation

To quantify the performance of the ensemble image segmentation method, Qian et al. (2016) run the algorithm on 32 TEM images and report the number of particles they are able to identify. For the medium and high resolution images, Qian et al. (2016) are able to manually label the particles and treat the manual outcome as the ground truth. These detection results are included in Table 3.2.

In Table 3.2, for the individual pipeline of processing, Qian et al. (2016) report the numbers of the total particles identified as well as the number of the consensus segmentations and the conflicting segmentations resolved by the ensemble method. The percentages of the conflicting segmentations selected from each of the two individual pipelines are also shown in the table. For further comparison, Qian et al. (2016) define the dissimilarity between the outcomes and the ground truth as the average distance between the nearest centers of different point sets, and show the boxplots of the comparison results in Fig. 3.20. The smaller the dissimilarity, the better the outcome. Figures 3.21 and 3.22 present the processed outcomes of the images.

Table 3.2 Comparison of particle detections for medium and high resolution images. (*Source*: Qian et al. 2016)

TEM image	Ground truth	Seg. #1	Seg. #2	Consensus seg.	Selected from seg. #1		Selected from seg. #2		Ensemble method
					Number	Percentage	Number	Percentage	
Medium resolution images									
F3-2_6	103	99	97	85	9	56.3%	7	43.7%	101
F3-2_7	104	100	99	85	9	50%	9	50%	103
F3-2_8	100	99	98	73	10	35.7%	18	64.3%	101
F8-2_6	113	108	111	98	5	35.7%	9	64.3%	112
F8-2_7	134	126	131	119	1	7.7%	12	92.3%	132
F8-2_8	148	143	142	119	14	51.8%	13	48.2%	146
F10_10	214	201	195	114	64	66%	33	34%	211
F10_12	179	175	162	141	30	88.2%	4	11.8%	175
High resolution images									
F3-2_9	24	24	25	14	0	0%	10	100%	24
F3-2_10	26	26	24	11	4	28.6%	10	71.4%	25
F3-2_11	26	33	25	20	5	71.4%	2	28.6%	27
F8-2_10	42	44	37	17	12	50%	12	50%	41
F8-2_11	44	41	35	20	16	69.6%	7	30.4%	43
F10_13	37	41	36	23	4	33.3%	8	66.7%	35
F10_15	47	50	34	17	24	82.8%	5	17.2%	46
F10-2_17	25	31	25	19	4	66.7%	2	33.3%	25

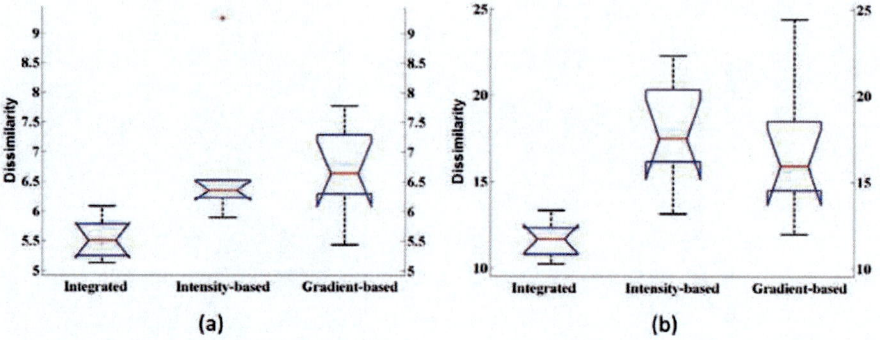

Fig. 3.20 The boxplot of the dissimilarity metric for (**a**) medium resolution images and (**b**) high resolution images. (Reprinted with permission from Qian et al. 2016)

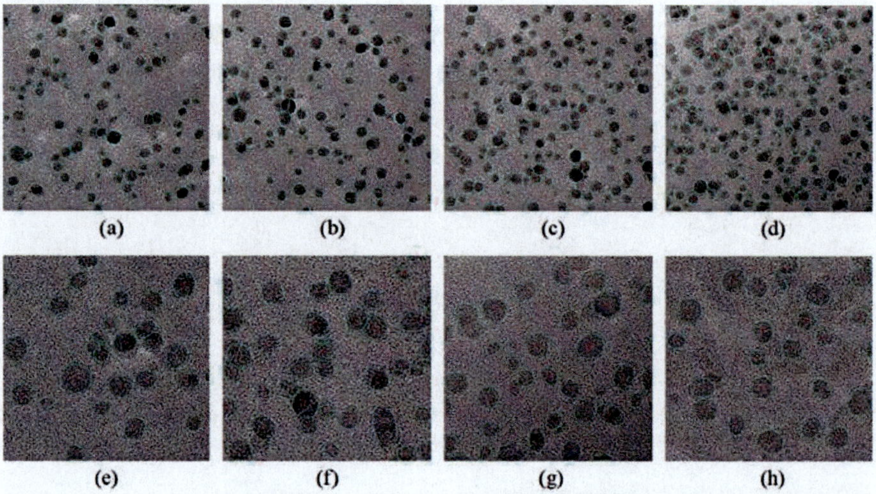

Fig. 3.21 The processed outcomes of medium-resolution (top row) images and high-resolution (bottom row) images. (**a**) F3-2_7. (**b**) F8-2_6. (**c**) F8-2_8. (**d**) F10_10. (**e**) F3-2_11. (**f**) F8-2_10. (**g**) F10_13. (**h**) F10-2_17 (Reprinted with permission from Qian et al. 2016)

The results presented in Table 3.2 and Fig. 3.20 demonstrate the effectiveness of the ensemble approach. Both of intensity-based and gradient-based processing contribute to the ensemble results but combining their strengths allows the ensemble method to achieve a high degree of accuracy across the samples. Qian et al. (2016) also conduct an analysis of variance (ANOVA) (Scheffe 1999) on the dissimilarity of three groups (ensemble approach, intensity-based only, and gradient-based only) for the medium- and high-resolution images. For the medium-resolution images, the p-value of an one-way ANOVA test is 0.0124 between the ensemble approach and the intensity-based approach and 0.0013 between the ensemble approach and

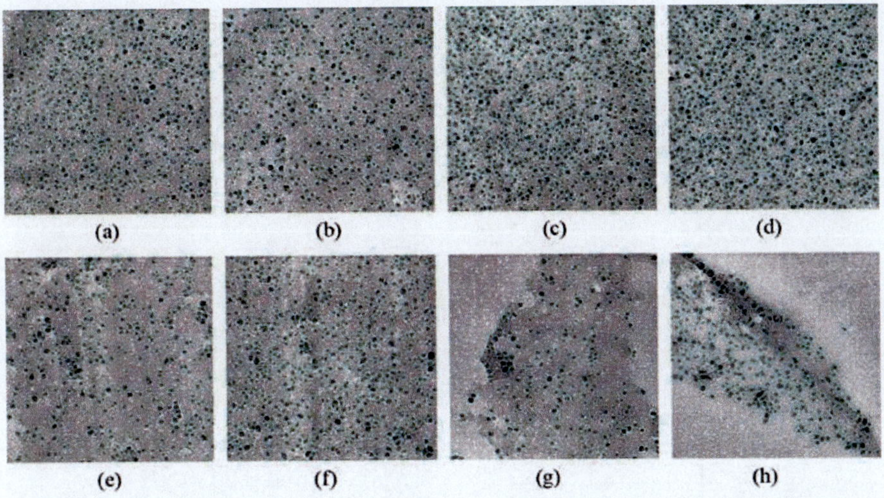

Fig. 3.22 The processed outcomes of low-resolution images (top row) and the images with uneven background (bottom row). (**a**) F8-2_4. (**b**) F8-2_5. (**c**) F10_8. (**d**) F10-2_3. (**e**) F3-2_4. (**f**) F3-2_15. (**g**) F8-2_15. (**h**) F10-2_13. (Reprinted with permission from Qian et al. 2016)

the gradient-based approach. For the high resolution images, the p-value is 0.0001 between the ensemble approach and the intensity-based approach and 0.0025 between the ensemble approach and the gradient-based approach.

For the low resolution images including those with uneven background, it is difficult to manually count and identify all the particles, as there are usually hundreds or even thousands of particles. What Qian et al. (2016) did is to present the processed outcomes of individual images in Fig. 3.22, so that people can visually sense how the method performs. Qian et al. (2016) present a table, similar to Table 3.2, but it does not have the ground truth column. For the intensity-based and gradient-based approaches, Qian et al. (2016) again report the numbers of particles it detects and the numbers of the conflicted outcomes resolved by the ensemble method. Combining both Table 3.3 and Fig. 3.22, Qian et al. (2016) state that the ensemble method presents an advantage in achieving robust detections when the image quality varies.

Tables 3.2 and 3.3 also suggest that the two pipelines of processing make similar contributions for the low, medium and high-resolution TEM images. But for those images with uneven background, more conflicting outcomes are selected from the intensity-based processing than from the gradient-based processing. Qian et al. (2016) believe that the unevenness in background intensity causes confusion in using the gradient information, making the intensity-based processing more accurate and the gradient-based processing less so.

Table 3.3 Comparison of particle detections for low resolution images and images with uneven backgrounds

TEM image	Seg. #1	Seg. #2	Consensus seg.	Selected from seg. #1		Selected from seg. #2		Ensemble method
				Number	Percentage	Number	Percentage	
Low resolution images								
F3-2_16	826	695	403	257	61.6%	160	38.4%	820
F8_8	1197	997	595	425	67.9%	201	32.1%	1221
F8-2_4	871	822	575	189	62.4%	141	37.6%	878
F8-2_5	633	678	510	65	40.4%	96	59.6%	671
F10_7	885	924	667	109	43.3%	143	56.7%	919
F10_8	1041	1077	721	162	42.7%	217	57.3%	1100
F10_9	1115	1153	730	211	46.6%	242	53.4%	1183
F10-2_3	1053	1096	763	153	43.5%	199	56.3%	1115
Uneven background images								
F3-2_4	502	487	294	133	60.7%	86	39.3%	513
F3-2_5	465	463	228	150	55.3%	121	44.7%	499
F3-2_15	815	712	466	222	64.5%	122	35.5%	810
F8_13	291	200	95	124	73.8%	44	26.2%	263
F8-2_15	327	309	159	133	65.3%	60	34.7%	332
F8-2_16	556	398	199	247	73.3%	90	26.7%	536
F10-2_12	480	187	102	303	95.3%	15	4.7%	420
F10-2_13	290	259	165	80	60.6%	52	39.4%	297

Source: Qian et al. (2016). With permission

References

Allili MS, Ziou D (2007) Globally adaptive region information for automatic color–texture image segmentation. Pattern Recognition Letters 28(15):1946–1956

Baxter WT, Grassucci RA, Gao H, Frank J (2009) Determination of signal-to-noise ratios and spectral SNRs in cryo-EM low-dose imaging of molecules. Journal of Structural Biology 166(2):126–132

Bernse J (1986) Dynamic thresholding of grey-level images. In: Proceedings of the 8th International Conference on Pattern Recognition, 1986, pp 1251–1255

Bradley D, Roth G (2007) Adaptive thresholding using the integral image. Journal of Graphics, GPU, and Game Tools 12(2):13–21

Canny J (1986) A computational approach to edge detection. IEEE Transactions on Pattern Analysis and Machine Intelligence 8(6):679–698

Caselles V, Kimmel R, Sapiro G (1997) Geodesic active contours. International Journal of Computer Vision 22(1):61–79

Chan TF, Vese LA (2001) Active contours without edges. IEEE Transactions on Image Processing 10(2):266–277

Chen Y, Thiruvenkadam S, Tagare HD, Huang F, Wilson D, Geiser EA (2001) On the incorporation of shape priors into geometric active contours. In: Proceedings IEEE Workshop on Variational and Level Set Methods in Computer Vision, Ieee, pp 145–152

Dempster AP, Laird NM, Rubin DB (1977) Maximum likelihood from incomplete data via the EM algorithm. Journal of the Royal Statistical Society: Series B (Methodological) 39(1):1–22

Dougherty E (1994) Digital Image Processing Methods. Crc

Felzenszwalb P, Huttenlocher D (2004) Efficient graph-based image segmentation. International Journal of Computer Vision 59(2):167–181

Fisker R, Carstensen JM, Hansen MF, Bødker F, Mørup S (2000) Estimation of nanoparticle size distributions by image analysis. Journal of Nanoparticle Research 2(3):267–277

Fujiyoshi Y (2013) Low dose techniques and cryo-electron microscopy. In: Electron Crystallography of Soluble and Membrane Proteins, Humana Press, Totowa, NJ., pp 103–118

Gatos B, Pratikakis I, Perantonis SJ (2006) Adaptive degraded document image binarization. Pattern Recognition 39(3):317–327

Hartigan JA, Wong MA (1979) A k-means clustering algorithm. Applied Statistics 28:100–108

Howe NR (2013) Document binarization with automatic parameter tuning. International Journal on Document Analysis and Recognition 16(3):247–258

Kass M, Witkin A, Terzopoulos D (1988) Snakes: Active contour models. International Journal of Computer Vision 1(4):321–331

Lu S, Su B, Tan CL (2010) Document image binarization using background estimation and stroke edges. International Journal on Document Analysis and Recognition 13(4):303–314

Mardia KV, Kent JT, Bibby JM (1980) Multivariate Analysis. Academic Press, San Diego, California, USA

Marturi N, Dembélé S, Piat N (2014) Scanning electron microscope image signal-to-noise ratio monitoring for micro-nanomanipulation. Scanning: The Journal of Scanning Microscopies 36(4):419–429

Narendra PM, Fukunaga K (1977) A branch and bound algorithm for feature subset selection. IEEE Transactions on Computers 100(9):917–922

Niblack W (1985) An Introduction to Digital Image Processing. Strandberg Publishing Company

Otsu N (1979) A threshold selection method from gray-level histograms. IEEE Transactions on Systems, Man, and Cybernetics 9(1):62–66

Park C, Huang JZ, Ji J, Ding Y (2013) Segmentation, inference and classification of partially overlapping nanoparticles. IEEE Transactions on Pattern Analysis and Machine Intelligence 35(3):669–681

Parker JR (1991) Gray level thresholding in badly illuminated images. IEEE Transactions on Pattern Analysis and Machine Intelligence 13(8):813–819

Parker JR, Jennings C, Salkauskas AG (1993) Thresholding using an illumination model. In: Document Analysis and Recognition, 1993., Proceedings of the Second International Conference on, Ieee, pp 270–273

Parvin B, Yang Q, Han J, Chang H, Rydberg B, Barcellos-Hoff MH (2007) Iterative voting for inference of structural saliency and characterization of subcellular events. IEEE Transactions on Image Processing 16(3):615–623

Phansalkar N, More S, Sabale A, Joshi M (2011) Adaptive local thresholding for detection of nuclei in diversity stained cytology images. In: Communications and Signal Processing (ICCSP), 2011 International Conference on, Ieee, pp 218–220

Qian Y, Huang JZ, Li X, Ding Y (2016) Robust nanoparticles detection from noisy background by fusing complementary image information. IEEE Transactions on Image Processing 25(12):5713–5726

Ramírez-Ortegón MA, Tapia E, Ramírez-Ramírez LL, Rojas R, Cuevas E (2010) Transition pixel: A concept for binarization based on edge detection and gray-intensity histograms. Pattern Recognition 43(4):1233–1243

Rousseeuw PJ, Leroy AM (2005) Robust Regression and Outlier Detection. John Wiley & Sons

Sauvola J, Pietikäinen M (2000) Adaptive document image binarization. Pattern Recognition 33(2):225–236

Scheffe H (1999) The Analysis of Variance. John Wiley & Sons, West Sussex, UK

Schmitt O, Hasse M (2009) Morphological multiscale decomposition of connected regions with emphasis on cell clusters. Computer Vision and Image Understanding 113(2):188–201

Sezgin M, Sankur B (2004) Survey over image thresholding techniques and quantitative performance evaluation. Journal of Electronic Imaging 13(1):146–168

Shi J, Malik J (2000) Normalized cuts and image segmentation. IEEE Transactions on Pattern Analysis and Machine Intelligence 22(8):888–905

Stathis P, Kavallieratou E, Papamarkos N (2008) An evaluation technique for binarization algorithms. Journal of Universal Computer Science 14(18):3011–3030

Su B, Lu S, Tan CL (2010) Binarization of historical document images using the local maximum and minimum. In: Proceedings of the 9th IAPR International Workshop on Document Analysis Systems, Acm, pp 159–166

Su B, Lu S, Tan CL (2013) Robust document image binarization technique for degraded document images. IEEE Transactions on Image Processing 22(4):1408–1417

Taylor WR, Thornton JM, Turnell WG (1983) An ellipsoidal approximation of protein shape. Journal of Molecular Graphics 1(2):30–38

Tian Y, Duan F, Zhou M, Wu Z (2013) Active contour model combining region and edge information. Machine Vision and Applications 24(1):47–61

Vo GD, Park C (2018) Robust regression for image binarization under heavy noise and nonuniform background. Pattern Recognition 81:224–239

Von Luxburg U (2007) A tutorial on spectral clustering. Statistics and Computing 17(4):395–416

Zafari S, Eerola T, Sampo J, Kälviäinen H, Haario H (2015) Segmentation of overlapping elliptical objects in silhouette images. IEEE Transactions on Image Processing 24(12):5942–5952

Zhang H, Fritts JE, Goldman SA (2008) Image segmentation evaluation: A survey of unsupervised methods. Computer Vision and Image Understanding 110(2):260–280

Zhang K, Zhang L, Song H, Zhou W (2010) Active contours with selective local or global segmentation: A new formulation and level set method. Image and Vision Computing 28(4):668–676

Chapter 4
Morphology Analysis

4.1 Basics of Shape Analysis

One of the first mathematical definitions of a shape can be found in the paper of David G. Kendall (Kendall 1984), the pioneer of the statistical shape theory. The definition reads (Mardia and Dryden 1989): "*Shape is all geometric information that remains when location, scale and rotation effects are filtered out from an object.*" This general definition leads further to many quantitative definitions of a shape. For a simple example, consider an ellipse in $\mathbb{R}^2$, which outline is described by the coordinates (x, y) satisfying the quadratic equation,

$$[x - x_o, y - y_o]\boldsymbol{A}[x - x_o, y - y_o]^T = 1, \tag{4.1}$$

where $(x_o, y_o) \in \mathbb{R}^2$ is the centroid location of the ellipse, and $\boldsymbol{A}$ is a 2×2 positive definite matrix. The centroid and the positive definite matrix represent this ellipse and its outline. After the location, orientation and scale effects are removed from the quantities, the remaining quantities would define the shape of the ellipse. First of all, the centroid location (x_o, y_o) determines the location of the ellipse and is thus removed. The positive definite matrix can be decomposed into the form,

$$\boldsymbol{A} = \boldsymbol{P}\boldsymbol{D}\boldsymbol{P}^T,$$

where $\boldsymbol{D}$ is a 2×2 diagonal matrix with positive diagonal elements, and $\boldsymbol{P}$ is a 2×2 rotation matrix in the form of

$$\boldsymbol{P} = \begin{bmatrix} \cos\theta & \sin\theta \\ -\sin\theta & \cos\theta \end{bmatrix}.$$

C. Park, Y. Ding, *Data Science for Nano Image Analysis*, International Series in Operations Research & Management Science 308,
https://doi.org/10.1007/978-3-030-72822-9_4

The rotation matrix determines the orientation of the ellipse, and the diagonal matrix $\boldsymbol{D}$ determines the major and minor axis lengths of the ellipse. Therefore, when we remove the location, rotation and scale effects from Eq. (4.1), the only degrees of freedom remaining is the ratio of the diagonal elements of $\boldsymbol{D}$,

$$\max(D_{11}, D_{22})/\min(D_{11}, D_{22}),$$

where D_{ij} is the (i, j)th element of $\boldsymbol{D}$. This ratio is known as the aspect ratio, which represents the shape of an ellipse. Statistical shape theory generalizes this idea to a broader class of shapes for presenting quantitative representations of shapes and their associated shape spaces. This section reviews the popular approaches.

4.1.1 *Landmark Representation*

One way to describe the shape of an outline is by locating a finite number of representative points on the outline, which are referred to as landmark points. The three vertices of a triangle are a good example of landmark points. A shape is then described by the coordinates of the landmark points (Mardia and Dryden 1989). The matrix of the landmark coordinates is referred to as a configuration matrix. For example, if a shape in an M-dim space of $\mathbb{R}^M$ is represented by K landmark points, the configuration matrix is a $K \times M$ matrix of the K coordinates, i.e.,

$$\tilde{\boldsymbol{X}} = \begin{bmatrix} \tilde{X}_{11} & \dots & \tilde{X}_{1M} \\ \tilde{X}_{21} & \dots & \tilde{X}_{2M} \\ \vdots & \ddots & \vdots \\ \tilde{X}_{K1} & \dots & \tilde{X}_{KM} \end{bmatrix}. \tag{4.2}$$

The configuration matrix determines a shape of an outline only up to a location, scale and rotation transformation of the outline. Apparently, the location, scale, rotation information should be removed from the configuration matrix, in order to attain location, scale, and rotation invariance, i.e., to produce the shape information. There are multiple approaches to do so. In this section we review three representing approaches, the Kendall's shape representation in Sect. 4.1.1.1, the Procrustes tangent coordinates in Sect. 4.1.1.2 and the Bookstein's shape representation in Sect. 4.1.1.3. In the last subsection, we briefly discuss the issues related.

4.1.1.1 Kendall's Shape Representation

David Kendall is one of the earliest pioneers in statistical shape analysis and landmark shape representation (Kendall 1984; Kendall et al. 2009). His work is represented by the Kendall's Shape Space.

We first introduce the Helmert submatrix, which is used to remove the location information for a configuration matrix. The Helmert submatrix $\boldsymbol{H}$ is the $(K-1) \times K$ matrix that is achieved by removing the first row of a full $K \times K$ Helmert matrix $\boldsymbol{H}_F$. A full $K \times K$ Helmert matrix is an orthogonal matrix with its first row elements all equal to $1/\sqrt{K}$, and the other rows are orthogonal to the first row. The specific form of the matrix is given in Lancaster (1965). The $(j+1)$th row of the matrix comes with the first j elements equal to $-1/\sqrt{j(j+1)}$, its $(j+1)$th element equal to $1/j\sqrt{j(j+1)}$, and the remaining elements equal to zero. If one multiplies a configuration matrix $\tilde{\boldsymbol{X}}$ by $\boldsymbol{H}_F$, the first row of the multiplication result represents the centroid location of the K landmark coordinates in $\tilde{\boldsymbol{X}}$, and the other rows does not depend on the centroid location due to the orthogonality of the Helmert matrix. In other words, the multiplication of the configuration matrix by the Helmert submatrix $\boldsymbol{H}$,

$$\boldsymbol{H}\tilde{\boldsymbol{X}},$$

does not depend on the location.

The scale information is further removed by normalizing the centered matrix $\boldsymbol{H}\tilde{\boldsymbol{X}}$:

$$\boldsymbol{X} = \boldsymbol{H}\tilde{\boldsymbol{X}}/||\boldsymbol{H}\tilde{\boldsymbol{X}}||_F,$$

where $||\cdot||_F$ is the Frobenius norm. The normalized matrix $\boldsymbol{X} \in \mathbb{R}^{(K-1)\times M}$ is referred to as a preshape because $\boldsymbol{X}$ is yet invariant to the rotation transformation. The space of the preshapes is

$$\mathcal{S}_M^K := \{\boldsymbol{X} \in \mathbb{R}^{(K-1)\times M} : ||\boldsymbol{X}||_F = 1\},$$

which is an unit sphere in $\mathbb{R}^{(K-1)\times M}$. Let $SO(M)$ be a group of the rotation matrices, defined by

$$SO(M) = \{\boldsymbol{O}_M \in \mathbb{R}^{M\times M} : \boldsymbol{O}_M\boldsymbol{O}_M^T = \boldsymbol{I}, \boldsymbol{O}_M^T\boldsymbol{O}_M = \boldsymbol{I}, det(\boldsymbol{O}_M) = 1\}.$$

We can then write a rotated version of a preshape $\boldsymbol{X} \in \mathcal{S}_M^K$ as

$$\boldsymbol{X}\boldsymbol{O}_M \text{ for } \boldsymbol{O}_M \in SO(M).$$

The shape of $\boldsymbol{X}$ is defined as an equivalent class,

$$[X] = \{\boldsymbol{X}\boldsymbol{O}_M : \boldsymbol{O}_M \in SO(M)\},$$

which is a collection of all different rotated versions of the preshape. The Kendall's shape space is the quotient space $\mathcal{S}_M^K/SO(M)$,

$$\Sigma_M^K = \{[X] : X \in \mathcal{S}_M^K\}.$$

A simple example of the Kendall's shape space is the space of triangles in $\mathbb{R}^2$. A triangle is represented with three landmark points locating at the three vertices of the triangle, and the corresponding configuration matrix is 3×2 matrix,

$$\tilde{X} = \begin{bmatrix} \tilde{X}_{11} & \tilde{X}_{12} \\ \tilde{X}_{21} & \tilde{X}_{22} \\ \tilde{X}_{31} & \tilde{X}_{32} \end{bmatrix}.$$

The Helmert submatrix for centering the configuration is

$$H = \begin{bmatrix} -1/\sqrt{2}, & 1/\sqrt{2}, & 0 \\ -1/\sqrt{6}, & -1/\sqrt{6}, & 2/\sqrt{6} \end{bmatrix},$$

and the corresponding centered configuration matrix is

$$H\tilde{X} = \begin{bmatrix} -\tilde{X}_{11}/\sqrt{2} + \tilde{X}_{21}/\sqrt{2}, & -\tilde{X}_{12}/\sqrt{2} + \tilde{X}_{22}/\sqrt{2} \\ -\tilde{X}_{11}/\sqrt{6} - \tilde{X}_{21}/\sqrt{6} + 2\tilde{X}_{31}/\sqrt{6}, & -\tilde{X}_{12}/\sqrt{6} - \tilde{X}_{22}/\sqrt{6} + 2\tilde{X}_{32}/\sqrt{6} \end{bmatrix}.$$

The size of the configuration is

$$||H\tilde{X}||_F^2 = \frac{4}{6}\sum_{k=1}^{3}\sum_{m=1}^{2} \tilde{X}_{km}^2,$$

and the preshape is the centered configuration matrix normalized by the size,

$$X = H\tilde{X}/||H\tilde{X}||_F.$$

The Kendall's shape space of triangles is defined by the group of triangle preshapes invariant to a rotation matrix. An element in the space is the shape of a triangle,

$$[X] = \{XO_2 : O_2 \in SO(2)\}.$$

Shape analysis on shapes in $\mathbb{R}^2$ has been popular due to many practical applications (Bookstein 1986; Goodall 1984), including morphology analysis of material images, because the vast majority of material images are available in two dimensions. For shapes in $\mathbb{R}^2$, it is convenient to convert the $K \times 2$ real configuration matrix in Eq. (4.2) to a $K \times 1$ vector of complex numbers,

$$\tilde{z} = \begin{bmatrix} \tilde{X}_{11} + i\tilde{X}_{12} \\ \tilde{X}_{21} + i\tilde{X}_{22} \\ \vdots \\ \tilde{X}_{K1} + i\tilde{X}_{K2} \end{bmatrix}. \tag{4.3}$$

The configuration is centered to

$$\boldsymbol{H}\tilde{z}.$$

The corresponding preshape is achieved as

$$z = \boldsymbol{H}\tilde{z}/||\boldsymbol{H}\tilde{z}||. \tag{4.4}$$

In this representation, a rotation operator by angle θ in the complex space is simply denoted by

$$\exp(i\theta)z.$$

The shape distance between two preshapes, z_1 and z_2, can be defined by

$$d(z_1, z_2) = \cos^{-1} |z_1^* z_2|,$$

which $*$ denotes the complex conjugate transpose. The distance is rotation-invariant in that $d(z_1, z_2) = d(\exp(i\theta_1)z_1, \exp(i\theta_2)z_2)$ for every θ_1 and $\theta_2 \in [0, 2\pi]$. Mardia and Dryden (1999) used the complex Watson distribution over the angular distance to define a probability distribution of preshapes deviating from a mean shape $\boldsymbol{\mu}$, which has the complex Watson density

$$p_w(z; \boldsymbol{\mu}, \kappa) = c_1(\kappa)^{-1} \exp\{\kappa |z^* \boldsymbol{\mu}|^2\}, \tag{4.5}$$

where κ is a concentration parameter and $c_1(\kappa)$ is a normalizing constant depending on κ. Kent (1994) used the complex Bingham distribution over z, which has a density of

$$p_b(z; \boldsymbol{A}) = c_2(\boldsymbol{A})^{-1} \exp\{z^* \boldsymbol{A} z\}, \tag{4.6}$$

where $\boldsymbol{A}$ is a $K \times K$ Hermitian matrix and $c_2(\boldsymbol{A})$ is a normalizing constant dependent on $\boldsymbol{A}$. Please note that $p_b(z_p; \boldsymbol{A})$ is invariant to a rotation in that $p_b(z; \boldsymbol{A}) = p_b(e^{i\theta} z; \boldsymbol{A})$, so it is valid for a shape distribution. The complex Watson distribution in Eq. (4.5) is a special case of the complex Bingham distribution when $\boldsymbol{A} = \kappa \boldsymbol{\mu} \boldsymbol{\mu}^*$. Many statistical inference problems have been formulated and solved with these statistical models. A very good review of the past works can be found in Dryden and Mardia (2016).

4.1.1.2 Procrustes Tangent Coordinates

Another approach to define a probability distribution of shapes with landmark data is to take a tangent space to a shape space and impose a probability distribution on the tangent space. Since the tangent space is Euclidean, a conventional multivariate distribution such as a multivariate normal distribution can be imposed on the tangent space.

We in the sequel describe the main idea for the 2-dim cases, i.e., for shapes in $\mathbb{R}^2$ and the corresponding shape space of Σ_2^K. Please note that the shape space Σ_2^K of landmark data is a manifold. The approach first takes the tangent space of the shape space at the 'center' of the space. Here, the notion of the 'center' can be defined statistically. Suppose that there are N outlines of objects and the corresponding preshapes in the form of complex space representation are

$$\{z_n \in \mathbb{C}^{K-1}; n = 1, \ldots, N\},$$

then, the 'center' of the shape space can be defined as the mean of the N shapes. The generalized Procrustes analysis is used to estimate the mean shape.

The generalized Procrustes analysis is an iterative method to align data matrix so that they are better conforming to each other. It was first introduced to analyze sensory profiling data (Gower 1975), and it has been a popular choice to align preshapes in terms of rotation, and/or scaling parameters and eventually take the mean shape of the aligned data. The iterative procedure is performed as follows.

Step 1. The initial estimate of the mean shape is randomly chosen among the N available preshapes.

Step 2. A partial Procrustes analysis is to optimize only the rotation parameter θ while aligning each of the N available configurations to the mean shape $\boldsymbol{\mu}$, i.e.,

$$\hat{z}_n := \exp(i\theta_n)z_n, \tag{4.7}$$

where

$$\theta_n = \underset{\theta \in [0,2\pi)}{\operatorname{argmin}} \, || \exp(i\theta)z_n - \boldsymbol{\mu}||^2.$$

The optimal solution for θ_n is given by $\theta_n = -\mathrm{Arg}(\boldsymbol{\mu}^* z_n)$ (Mardia and Dryden 1989). A full Procrustes analysis optimizes both of the scale and rotation parameters to align z_n to $\boldsymbol{\mu}$,

$$\hat{z}_n = s_n \exp(i\theta_n)z_n \tag{4.8}$$

with $(s_n, \theta_n) = \mathrm{argmin}_{s>0,\theta\in[0,2\pi)} ||s \exp(i\theta)z_n - \boldsymbol{\mu}||^2$.

Step 3. Update the mean shape with the arithmetic mean of the aligned configurations,

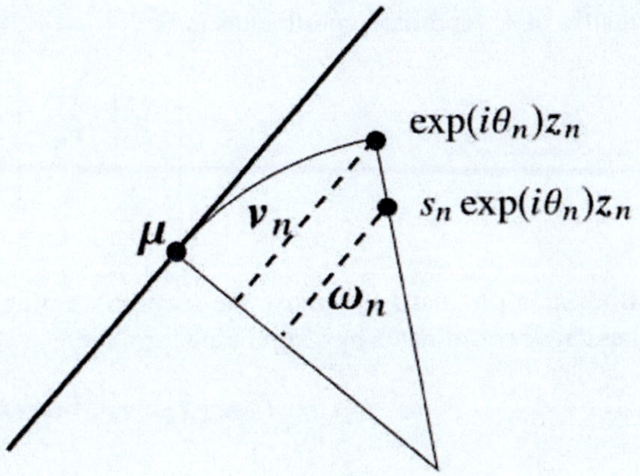

Fig. 4.1 Procrustes tangent coordinates of landmark data. v_n and ω_n denote the partial and full Procrustes tangent coordinates respectively

$$\mu = \frac{1}{N}\sum_{n=1}^{N} \tilde{z}_n.$$

Step 4. Repeat **Step 2** and **Step 3** until the value of μ converges.

After the convergence, the tangent space of the shape space Σ_2^K at μ is attained. Each preshape z_n is projected to the tangent space for obtaining the tangent coordinates. The partial Procrustes tangent coordinate for z_n is given by

$$v_n = \exp\{i\theta_n\}[I_{K-1} - \mu\mu^*]z_n, \tag{4.9}$$

where I_K represents a K-dimensional identity matrix, and the matrix $[I_{K-1} - \mu\mu^*]$ is a projection matrix into the space orthogonal to μ. The partial Procrustes tangent coordinate is basically the projection of the rotationally aligned preshape into the tangent space. Similarly, the full Procrustes tangent coordinate is given by

$$\omega_n = s_n \exp\{i\theta_n\}[I_{K-1} - \mu\mu^*]z_n. \tag{4.10}$$

Figure 4.1 compares the partial and full Procrustes tangent coordinates with respect to the tangent space at μ.

The tangent space is a $(2K - 3)$-dim hyperplane embedded in a $2(K - 1)$-dim Euclidean space. A multivariate normal distribution with a zero mean vector and a covariance matrix can be defined on the tangent space for serving as the probability distribution of the tangent coordinates.

4.1.1.3 Bookstein's Shape Coordinates

Bookstein (1986) studied shapes in $\mathbb{R}^2$ and proposed a very practical shape representation, based on landmark coordinates. Let $\tilde{X}$ denote the configuration matrix of K landmark coordinates in $\mathbb{R}^2$,

$$\tilde{\boldsymbol{X}} = \begin{bmatrix} \tilde{X}_{11} & \tilde{X}_{12} \\ \tilde{X}_{21} & \tilde{X}_{22} \\ \vdots & \vdots \\ \tilde{X}_{K1} & \tilde{X}_{K2} \end{bmatrix}. \tag{4.11}$$

Bookstein proposed to remove the rotation, scaling and location information from landmark coordinates by a rigid transformation,

$$\tilde{\boldsymbol{U}} = (\tilde{\boldsymbol{X}} - \mathbf{1}_K \boldsymbol{c}^T)(s\boldsymbol{O}), \tag{4.12}$$

where $\mathbf{1}_K$ represents a K-dimensional column vector of ones, $s \in \mathbb{R}^+$ is the scaling parameter, $\boldsymbol{O}$ is a K-dimensional orthogonal matrix representing the orientation parameter, and $\boldsymbol{c} \in \mathbb{R}^2$ is the location parameter of the rigid body transformation. The transformation parameters are chosen such that the first and second landmark coordinates are transformed to $(0, 0)$ and $(1, 0)$ respectively, for which the following two conditions should satisfy

$$s([\tilde{X}_{11}, \tilde{X}_{12}] - \boldsymbol{c}^T)\boldsymbol{O} = (0, 0)$$
$$s([\tilde{X}_{21}, \tilde{X}_{22}] - \boldsymbol{c}^T)\boldsymbol{O} = (1, 0)$$

The first condition gives $\boldsymbol{c} = [\tilde{X}_{11}, \tilde{X}_{12}]^T$, and the second condition gives

$$s\boldsymbol{O} = \frac{1}{(\tilde{X}_{21} - \tilde{X}_{11})^2 + (\tilde{X}_{22} - \tilde{X}_{12})^2} \begin{bmatrix} \tilde{X}_{21} - \tilde{X}_{11} & \tilde{X}_{12} - \tilde{X}_{22} \\ \tilde{X}_{22} - \tilde{X}_{12} & \tilde{X}_{21} - \tilde{X}_{11} \end{bmatrix}.$$

The first two rows of $\tilde{\boldsymbol{U}}$ are always $[0, 0]$ and $[1, 0]$. After the first two rows are removed, the remaining $K - 2$ rows are referred to as the Bookstein coordinates, which is denoted as $\boldsymbol{U}$.

There are a few approaches conducting shape analysis using the Bookstein coordinates. The first approach is to use a standard multivariate analysis with the Bookstein coordinates. For example, we can impose a multivariate normal distribution to describe the probability distribution of the Bookstein coordinates. The distribution parameters are estimated using the maximum likelihood estimates of the mean and covariance vectors. Another approach is to use an offset normal distribution (Dryden and Mardia 1991; Kume and Welling 2010). In the offset normal approach, the configuration matrix is assumed to be a multivariate normal random variable,

$$\text{vec}(\tilde{\boldsymbol{X}}) \sim \mathcal{N}(\tilde{\boldsymbol{\mu}}, \tilde{\boldsymbol{\Sigma}}),$$

and the probability distribution of the Bookstein coordinates $\boldsymbol{U}$ is induced from the multivariate normal distribution through the change of variables. To describe the idea, let $\boldsymbol{X}_p$ represents the $(K-2) \times K$ matrix that is achieved by removing the first two rows from $\tilde{\boldsymbol{X}} - \mathbf{1}_K \boldsymbol{c}^T$,

$$\boldsymbol{X}_p = \boldsymbol{L}(\tilde{\boldsymbol{X}} - \mathbf{1}_K \boldsymbol{c}^T),$$

where $\boldsymbol{L} = [\mathbf{0}_{K-2}, \mathbf{0}_{K-2}, \boldsymbol{I}_{K-2}]$. The probability distribution for $\text{vec}(\boldsymbol{X}_p)$ can be expressed as

$$\text{vec}(\boldsymbol{X}_p) \sim \mathcal{N}(\boldsymbol{\mu}, \boldsymbol{\Sigma}),$$

where $\boldsymbol{\mu} = \boldsymbol{L}\tilde{\boldsymbol{\mu}}$, and $\boldsymbol{\Sigma} = (\boldsymbol{I}_2 \otimes \boldsymbol{L})\tilde{\boldsymbol{\Sigma}}(\boldsymbol{I}_2 \otimes \boldsymbol{L}^T)$.

From Eq. (4.12), we can obtain

$$\begin{aligned} \boldsymbol{U} &= \boldsymbol{X}_p(s\boldsymbol{O}) \\ &= \boldsymbol{X}_p \frac{1}{(\tilde{X}_{21} - \tilde{X}_{11})^2 + (\tilde{X}_{22} - \tilde{X}_{12})^2} \begin{bmatrix} \tilde{X}_{21} - \tilde{X}_{11} & \tilde{X}_{12} - \tilde{X}_{22} \\ \tilde{X}_{22} - \tilde{X}_{12} & \tilde{X}_{21} - \tilde{X}_{11} \end{bmatrix}. \end{aligned}$$

Its inverse is

$$\boldsymbol{X}_p = \boldsymbol{U} \begin{bmatrix} \tilde{X}_{21} - \tilde{X}_{11} & \tilde{X}_{22} - \tilde{X}_{12} \\ \tilde{X}_{12} - \tilde{X}_{22} & \tilde{X}_{21} - \tilde{X}_{11} \end{bmatrix}.$$

The vectorization version can be written as

$$\text{vec}(\boldsymbol{X}_p) = \boldsymbol{W}_{\boldsymbol{u}}\boldsymbol{h},$$

where $\boldsymbol{h} = (\tilde{X}_{21} - \tilde{X}_{11}, \tilde{X}_{22} - \tilde{X}_{12})^T$, and

$$\boldsymbol{W}_{\boldsymbol{u}} = (\boldsymbol{I}_2 \otimes \boldsymbol{U}) \begin{bmatrix} 1 & 0 \\ 0 & -1 \\ 0 & 1 \\ 1 & 0 \end{bmatrix}.$$

Let $\boldsymbol{u} = \text{vec}(\boldsymbol{U})$. From the density of $\text{vec}(\boldsymbol{X}_p)$, the joint probability density of $\boldsymbol{u}$ and $\boldsymbol{h}$ can be induced,

$$f(\boldsymbol{u}, \boldsymbol{h}) = \frac{1}{(2\pi)^{K-1}|\boldsymbol{\Sigma}|^{1/2}} \exp\{-\frac{1}{2}(\boldsymbol{W}_{\boldsymbol{u}}\boldsymbol{h}-\boldsymbol{\mu})^T \boldsymbol{\Sigma}^{-1}(\boldsymbol{W}_{\boldsymbol{u}}\boldsymbol{h}-\boldsymbol{\mu})\}|J(\boldsymbol{X}_p \to (\boldsymbol{u}, \boldsymbol{h}))|,$$

where $|J(\boldsymbol{X}_p \to (\boldsymbol{u}, \boldsymbol{h}))|$ is the Jacobian of the transformation from $\boldsymbol{X}_p$ to $(\boldsymbol{u}, \boldsymbol{h})$). We can integrate out $\boldsymbol{h}$ to get the marginal distribution of $\boldsymbol{u}$. The marginal distribution is referred to as the Mardia–Dryden offset normal distribution.

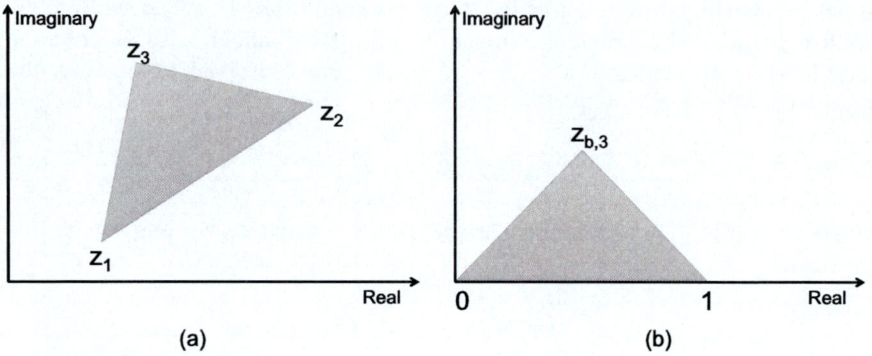

Fig. 4.2 Bookstein shape coordinates of a triangle. (**a**) Landmark representation **z**. (**b**) Bookstein coordinates of (**a**)

The Bookstein coordinates can also be expressed using the complex notation of landmark coordinates as described in Eq. (4.3). Let $\tilde{z}$ denote a $K \times 1$ complex vector of K landmark coordinates in $\mathbb{R}^2$, and its kth element is denoted by $\tilde{z}_k$, which represents the k landmark coordinates. In order to remove the effect of the rotation, scale and translation from the landmark coordinates, Bookstein proposed to transform $\tilde{z}$ by a rigid transformation such that the first landmark coordinate is placed at zero and the second landmark coordinate is placed at one. A rigid transformation in the complex space is described as

$$z_b = s(\tilde{z} - z_c)\exp\{i\theta\}, \tag{4.13}$$

where $s > 0$ is the scaling factor, $\theta \in [0, 2\pi)$ is the rotation angle, and $z_c \in C$ is the translation. With specific choices of the transformation parameters,

$$s = \frac{1}{|\tilde{z}_2 - \tilde{z}_1|},$$

$$\exp\{i\theta\} = \frac{|\tilde{z}_2 - \tilde{z}_1|}{\tilde{z}_2 - \tilde{z}_1}, \text{ and}$$

$$z_c = \tilde{z}_1,$$

the first element of z_b becomes zero, and the second element of z_b becomes one $(1 + 0i)$. Since the first two elements of z_b are constant after the transformation, the remaining $K - 2$ elements of z_b can be obtained, which is the Bookstein's coordinates in a complex form. For example, a triangle in $\mathbb{R}^2$ can be represented by three landmarks placed on the three vertices, as illustrated in Fig. 4.2, and its Bookstein shape coordinates consists of two real numbers, zero and one, and a complex number, $z_{b,3}$.

4.1.1.4 Related Issues

One major issue with this landmark representation is the tedium to identify landmark coordinates on a large number of outlines or contours for statistical analysis, although there are automated landmark detection algorithms available for some biological objects (Vandaele et al. 2018) and a human face (Burgos-Artizzu et al. 2013). Another issue is that there may not be clear landmark points for some shapes, e.g., nanoparticles which shapes are closer to spheres. For such shapes, the landmark-based representation is not so easy to apply.

4.1.2 Parametric Curve Representation

Another way to describe the shape of an outline is to use a parametric curve of the outline in $\mathbb{R}^2$. The outline in $\mathbb{R}^2$ is a closed curve, which has the same topology as an unit circle, and the closed parametric curve in $\mathbb{R}^2$ is defined as a continuous map from an unit circle $\mathbb{S}^1$ to $\mathbb{R}^2$,

$$t \in \mathbb{S}^1 \mapsto (x(t), y(t)) \in \mathbb{R}^2,$$

or equivalently, a continuous map $z(t)$ from a unit circle $\mathbb{S}^1$ to a complex coordinate in $\mathbb{C}$, namely

$$t \in \mathbb{S}^1 \mapsto z(t) = x(t) + iy(t) \in \mathbb{C}. \tag{4.14}$$

The parametric curve is affected by the location, orientation and scale of the outline, whose effect needs to be removed from the parametric curve to define the quotient feature or the shape feature of interest. Several shape spaces have been defined, based on different representations of a parametric curve and different shape features of the representations.

The continuous map, $z(t)$, is popularly represented by basis expansions with predetermined bases in the form of

$$z(t) = \sum_{l=1}^{L} \alpha_l \phi_l(t), \tag{4.15}$$

where ϕ_l's are the complex-valued basis functions defined on $\mathbb{S}^1$ and $\alpha_l \in \mathbb{C}$ are the unknown parameters. Popular choices of the basis functions are Fourier basis (Zahn and Roskies 1972; Staib and Duncan 1992), wavelet basis (Bengtsson and Eklundh 1991; Tieng and Boles 1997), and spherical harmonics (Brechbühler et al. 1995). Once given a choice of the basis functions, the associated parameters, $\{\alpha_l, l = 1, .., L\}$, define the outline. The shape of the outline is defined after removing the location, orientation and scale effect from the parameters. We will discuss a little more of this approach in Sect. 4.1.2.1

The aforementioned curve representations have not considered the well-known reparametrization issue with a parametric curve, that is, a parametric curve $z(t)$ and a choice of its reparametrization $z \circ \gamma(t)$ represents the same outline for every reparametrization function γ in $\Gamma = \{\gamma : \mathbb{S}^1 \rightarrow \mathbb{S}^1 : \gamma(0) = 0, \gamma(2\pi) = 2\pi, \gamma \text{ is a diffeomorphism}\}$, where the $\circ$ operator is the function composition operator. Therefore, the proper shape representation of a curve $z(t)$ also needs to be invariant to the reparametrization of the curve. There have been an increasing interest to tackle this issue in the shape space of continuous curves. Younes (1998) studied the space of geometric curves in $\mathbb{R}^2$ and defined the distance between two curves as the deformation energy from one to another. Klassen et al. (2004) studied the shape space of closed curves and defined the geodesic distance between two closed curves. This work was also among the first attempts to devise a numerical algorithm to compute the geodesic distance for closed curves. Joshi et al. (2007) and Srivastava et al. (2010) proposed a new representation of open curves and closed curves, the square-root velocity function (SRVF) representation, and defined a metric space of curves with a Riemannian metric. Based on the metric, a mean and covariance of shapes, as well as the principal component analysis in the new space, were defined. Kurtek et al. (2012) defined two statistical distributions on the new representation, a wrapped normal distribution and truncated normal distribution. We discuss in Sect. 4.1.2.2 the more recent works of Srivastava et al. (2010) and Kurtek et al. (2012).

4.1.2.1 Fourier Shape Descriptor

The Fourier shape descriptor uses the Fourier basis functions to represent a closed curve,

$$z(t) = \sum_{l=1}^{L} \alpha_l \exp\{2\pi i(l-1)t\},$$

where $\exp\{2\pi i(l-1)t\}$ is the lth Fourier basis function, and α_l is the corresponding Fourier coefficient. The closed curve is parameterized by a set of the Fourier coefficient values. The centroid of the closed curve is defined by

$$c_z = \int_{\mathbb{S}^1} z(t)dt,$$

and the integration turns out to be the first Fourier coefficient, α_1. After removing the centroid from $z(t)$, the location-invariant feature of the closed curve is represented by the closed curve centered around the origin,

$$z_c(t) = z(t) - c_z = \sum_{l=2}^{L} \alpha_l \exp\{2\pi i(l-1)t\}.$$

One can judge the scale effect and rotation effect on the curve. Rotating $z_c(t)$ with angle $\theta \in [0, 2\pi]$ and scaling it by $s \in \mathbb{R}^+$ leads to

$$\begin{aligned} \tilde{z}_c^{\theta,s}(t) &= s z_c(t) \exp\{i\theta\} \\ &= \sum_{l=2}^{L} \tilde{\alpha}_l^{\theta,s} \exp\{2\pi i(l-1)t\}. \end{aligned}$$

with $\tilde{\alpha}_l^{\theta,s} = s\alpha_l \exp\{i\theta\}$ as the new Fourier coefficients. The Fourier descriptor is the features of the Fourier coefficients invariant to θ and s. A popular choice is

$$\left\{ \frac{|\tilde{\alpha}_l^{\theta,s}|}{|\tilde{\alpha}_2^{\theta,s}|} : l = 3, \ldots, L \right\}.$$

With this descriptor, a shape is represented by the $L-2$ real values. As such, the corresponding shape space is $\mathbb{R}^{L-2}$. The metric that measures the difference between two shapes is the Euclidean distance in $\mathbb{R}^{L-2}$.

4.1.2.2 Square-Root Velocity Function (SRVF) Representation

Consider a closed curve $\beta : \mathbb{S}^1 \to \mathbb{R}^2$,

$$t \in \mathbb{S}^1 \mapsto \beta(t) = (x(t), y(t)) \in \mathbb{R}^2.$$

To study the shape of β, Srivastava et al. (2010) introduced the square-root velocity function (SRVF) of β,

$$q(t) = \frac{\dot{\beta}(t)}{\sqrt{||\dot{\beta}(t)||}},$$

where $||\cdot||$ is the $\mathbb{L}^2$-norm, and $\dot{\beta}(t)$ denotes the first order derivative of β with respect to t. The $q(t)$ is invariant to the translation and rescale of β, because the derivative is invariant to the translation and normalized. Given $q \in \mathbb{L}^2(\mathbb{S}^1, \mathbb{R}^2)$, β can be determined uniquely up to a translation and scale transformation by the equation

$$\beta(t) = \int_0^t ||q(s)||q(s)ds = \int_0^t \dot{\beta}(s)ds = \beta(t) - \beta(0),$$

which is just the constant translation of $\beta(t)$.

The SRVF satisfies some conditions. Since β is a closed curve, $q \in \mathbb{L}^2(\mathbb{S}^1, \mathbb{R}^2)$ also satisfies

$$\int_{\mathbb{S}^1} ||q(t)||q(t)dt = \beta(2\pi) - \beta(0) = \mathbf{0}.$$

In addition, it has an unit length,

$$\int_{\mathbb{S}^1} ||q(t)||^2 dt = \int_{\mathbb{S}^1} ||\dot{\beta}(t)||dt = 1.$$

Therefore, the SRVF of closed curves belongs to

$$C^c := \left\{ q \in \mathbb{L}^2(\mathbb{S}^1, \mathbb{R}^2) : \int_{\mathbb{S}^1} ||q(t)||^2 dt = 1, \int_{\mathbb{S}^1} ||q(t)||q(t)dt = \mathbf{0} \right\},$$

which in turn belongs to the unit hypersphere in $\mathbb{L}^2(\mathbb{S}^1, \mathbb{R}^2)$. This space is a Riemannian manifold (Srivastava et al. 2010), and its Riemannian metric measuring the distance between two preshapes, $q_0, q_1 \in C^c$, is defined by the length of the geodesic path connecting the two preshapes, i.e.,

$$d_c(q_0, q_1) := \inf\{\alpha : [0, 1] \to C^c : \alpha(0) = q_0, \alpha(1) = q_1\}.$$

This Riemannian manifold forms a preshape space of closed curves, because the SRVF of a closed curve is yet invariant to the rotation $O \in SO(2)$ and the reparametrization $\gamma \in \Gamma$, where Γ is the set of all reparametrizations,

$$\Gamma = \{\gamma : \mathbb{S}^1 \to \mathbb{S}^1 : \gamma(0) = 0, \gamma(2\pi) = 2\pi, \gamma \text{ is a diffeomorphism}\}.$$

For the SRVF $q \in C^c$ of β and $\gamma \in \Gamma$, the SRVF of $\beta \circ \gamma$ is $(q \circ \gamma)\sqrt{\dot{\gamma}}$. The shape of a closed curve β is defined by the orbit of SRVFs,

$$[q] = \{O(q \circ \gamma)\sqrt{\dot{\gamma}} : (O, \gamma) \in SO(2) \times \Gamma\}.$$

The shape space of closed curves is defined as the quotient space, $C^c \backslash SO(2) \times \Gamma$,

$$S = \{[q] : q \in C^c\}.$$

The metric on the quotient space is induced by the Riemannian metric of C^c,

$$d_s([q_0], [q_1]) = \inf_{(O,\gamma) \in SO(2) \times \Gamma} d_c(q_0, O(q_1 \circ \gamma)\sqrt{\dot{\gamma}}).$$

Kurtek et al. (2012) developed statistical analysis on the shape space, defining the notations of mean and covariance and shape distributions. To describe the idea, let $\{\beta_n; n = \ldots, N\}$ denote a collection of closed curves, and their SRVFs are denoted by $\{q_n : n = 1, \ldots, N\}$. The mean shape of the closed curves is defined using the sample Karcher mean,

$$\bar{q} = \arg\min_{[q]\in\mathcal{S}} \sum_{n=1}^{N} d_s([q], [q_n])^2. \tag{4.16}$$

The sample covariance is defined on the tangent space of the Riemannian manifold $\mathcal{S}$ at the sample mean $\bar{q}$, which is denoted by $\mathcal{T}_{\bar{q}}(\mathcal{S})$. The projection of a point $p \in \mathcal{S}$ is referred to as an inverse exponential map, which is defined as

$$\exp_{\bar{q}}^{-1}([q]) = \frac{\theta}{\sin(\theta)}(p - \bar{q}\cos(\theta)),$$

where $\theta = \arccos(\int_{\mathbb{S}^1} p(t)\bar{q}(t)dt$. When each closed curve β_n is observed at m sampled points, the sampled tangent space feature is a $2m$ dimensional vector with its jth element equal to the value of the inverse exponential map at the jth sample point. Let $\boldsymbol{v}_n$ denote the sampled tangent space feature for β_n and let $\boldsymbol{V}$ denote a $2m \times N$ matrix of all the sampled tangent space features for all N observations. The Karcher covariance matrix is defined by $\boldsymbol{K} = \boldsymbol{V}\boldsymbol{V}^T$, and the corresponding principal component analysis can be defined as the following: Given a singular value decomposition $\boldsymbol{M} = \boldsymbol{U}\boldsymbol{D}\boldsymbol{V}^T$, the principal coefficients for β_n are

$$\boldsymbol{U}^T\boldsymbol{v}_n$$

Two shape distributions are defined on the shape space $\mathcal{S}$, a wrapped Gaussian distribution and a truncated Gaussian distribution.

The wrapped Gaussian approach is to first assume a multivariate Gaussian distribution on the tangent space feature $\boldsymbol{v}_n$,

$$\boldsymbol{v}_n \sim \mathcal{N}(\boldsymbol{0}, \boldsymbol{K}). \tag{4.17}$$

Since the exponential map projecting $\boldsymbol{v}_n$ back to the corresponding shape $[q_n]$ is defined as

$$\exp_{\bar{q}}(\boldsymbol{v}_n) := \cos(||\boldsymbol{v}_n||)\bar{q} + \frac{\sin(||\boldsymbol{v}_n||)}{||\boldsymbol{v}_n||}\boldsymbol{v}_n.$$

By the change of variance, the probability distribution of shape $[q_n]$ can be induced from the multivariate normal distribution. The resulting probability distribution is a wrapped Gaussian distribution.

The truncated Gaussian distribution can be similarly induced by assuming the truncated normal distribution of the tangent space feature instead of the multivariate Gaussian distribution.

4.2 Shape Analysis of Nanoparticles

Shape analysis in nano images has been mostly performed manually for a small number of nanomaterials selected and cropped from material images. Its automated analysis was first discussed for nanoparticle shape analysis in the Park's PhD dissertation (Park 2011) and his other papers (Park et al. 2012, 2013), which entail some applications of statistical shape theory for star-shaped nanoparticles and convex-shaped nanoparticles.

4.2.1 Shape Analysis for Star-Shaped Nanoparticles

As illustrated in Fig. 4.3, most of the nanoparticle shapes reported in literature are as simple as ellipse, triangle, rectangle and star, all of which belong to a class of star shapes. Park et al. (2012) contemplated the question of how to classify the outlines of the nanoparticles, extracted from microscope images, into one of the star shapes. In Park et al. (2012), every particle outline extracted from a digital microscope image is first represented in the form of a parametric curve, $z(t)$. Then, the parametric curve is transformed into the centroid distance function as shown in Fig. 4.4,

$$r(t) = |z(t) - c_z|, t \in \mathbb{S}^1.$$

The centroid distance representation is invariant to the location of the outline. A preshape is defined on the centroid distance representation after further removing the scale effect through normalization:

$$r_p(t) = \frac{r(t)}{\int_{\mathbb{S}^1} r(t)dt}.$$

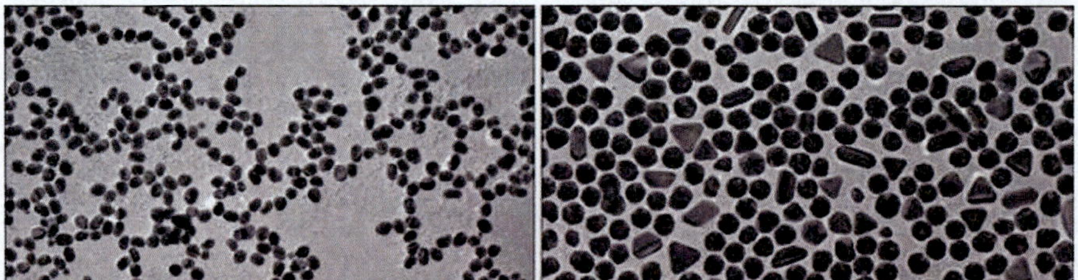

Fig. 4.3 Examples of TEM images. **(a)** spherical and elliptical particles, **(b)** a mixture of spherical, triangular and rod particles. (Reprinted with permission from Park et al. 2012)

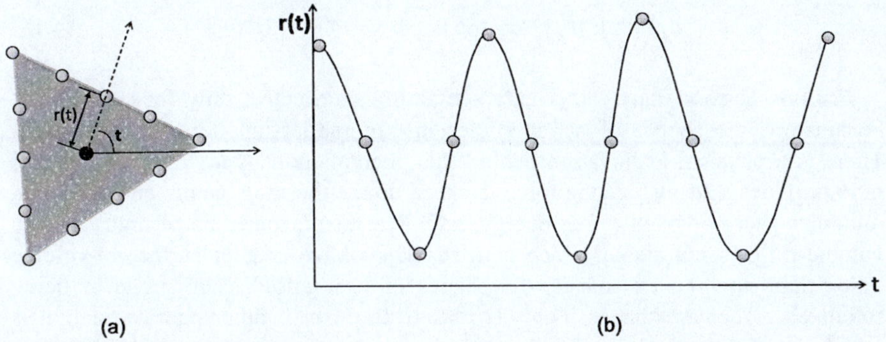

Fig. 4.4 Representation of nanoparticle outlines. **(a)** Outline of a triangular particle: the black dot is the gravity center of a triangular nanoparticle, and the gray points are pixels sampled from the outline of the nanoparticle, **(b)** centroid distance representation of **(a)**. (Reprinted with permission from Park et al. 2012)

Note that $\int_{\mathbb{S}^1} r(t)dt$ is the average centroid distance, and using the average centroid distance is a popular way to measure the size of a shape (called centroid size) for the purpose of scaling (Dryden and Mardia 1998, pages 23–24). The r_p can be considered as a rescale version of $r(t)$ with the inverse of the average distance as the scaling factor.

The space of the preshapes is a space of constrained functions, i.e.,

$$\mathcal{R}_p := \left\{ r_p : \mathbb{S}^1 \mapsto \mathbb{R}^+; \int_{\mathbb{S}^1} r(t)dt = 1 \right\}.$$

Please note that a preshape is not yet rotation invariant, so different elements in the preshape space may come from the outlines of the same shape with different orientations. To define a shape from the preshape, we first define a rotation operator on the preshape. Let $\tilde{z}(t)$ denote $z(t)$ rotated around its centroid c_z by θ. According to Park et al. (2012), the preshape of $\tilde{z}(t)$ is equivalent to a circular shifting of $r_p(t)$ by θ, such as

$$\tilde{r}_p^{\theta}(t) = r_p \circ \gamma_\theta(t),$$

where $\gamma_\theta(t) := (t + \theta)$ modulo 2π denotes the circular shift by θ. A shape of a preshape $z(t)$ is defined by a collection of different rotations of the preshape,

$$[r_p] = \{r_p \circ \gamma_\theta(t) \in \mathcal{R}_p; \theta \in [0, 2\pi]\}.$$

A collection of all possible $[r_p]$'s is referred to as the shape space, denoted by $\mathcal{R}_s$. The shape space is not Euclidean, but instead belongs to a curved manifold with a Riemannian metric. Let $[r_p]$ and $[r_p']$ be two elements of the shape space. The distance between the two elements is defined by a rotation-invariant metric,

$$d_p\Big([r_p], [r'_p]\Big) = \min_{\theta} ||r_p \circ \gamma_\theta(t) - r'_p(t)||_2. \quad (4.18)$$

For non-convex shapes, this parametrization is effective only for star shapes. Fortunately, the shapes of nanoparticles in our applications are mostly convex. There is a physical explanation behind this phenomenon. A high surface-volume ratio provides a strong driving force to speed up the thermodynamic processes that minimize thermodynamic free energy, and as a result, materials of high surface-volume ratio is not stable. Since convex shapes have smaller surface-to-volume ratios than non-convex shapes, the shapes of nanoparticles are prone to being stabilized to convex shapes. For this reason, the use of other parametrization is appropriate for the analysis of nanoparticles.

4.2.1.1 Embedding of the Shape Manifold to Euclidean Space

It is difficult to deal with the shape representation $[r_p]$. First of all, it is high dimensional. Theoretically, $r_p(t)$ has values at infinitely many t's, so it is infinitely dimensional. In practice, microscope images are in discretized digital forms, so that we only have the values of $r(t)$ at a finite number of discrete locations, $\{t_1, \ldots, t_n\}$. But the value of n is usually very large. Understandably, dealing with such high dimensional space and a moderate number of data samples is challenging. In addition, the shape representation belongs to a Riemannian manifold. Working with the manifold space for subsequent analysis such as classification and clustering is complicated, because any conventional approaches based on the Euclidean space cannot be applied. Because of the two complexities, Park et al. (2012) suggested embedding the manifold data to a low-dimensional Euclidean space that approximates the manifold.

The basic idea of the embedding is to find a nonlinear map $\boldsymbol{\phi}$ that transforms the shape manifold to a q-dim Euclidean space such that the metric in the Euclidean space approximates the metric in the original manifold space, as illustrated in Fig. 4.5. Let d_p the geodesic distance defined in the original shape space, and let $\boldsymbol{\phi}$ denote the map function. The choice of the map should minimize the distortion of the metric by the map,

$$d_p([r_p], [r'_p]) \approx (\boldsymbol{\phi}([r_p]) - \boldsymbol{\phi}([r'_p]))^T(\boldsymbol{\phi}([r_p]) - \boldsymbol{\phi}([r'_p])).$$

and make the approximated distance in the q-dim Euclidean space between two curves as close to that in the original space as possible.

The map can be achieved by the Isomap algorithm (Tenenbaum et al. 2000) given a set of preshapes $\{r_p^{(i)} : i = 1, ..., m\}$. In the algorithm, for every pair of two preshapes $r_p^{(i)}$ and $r_p^{(j)}$, the Riemannian metric d_p between them is evaluated, which is denoted by g_{ij}. So the problem can be rephrased as—given the distances $\{g_{ij}; i \neq j\}$, how to find the low dimensional features $\boldsymbol{\phi}_1, \ldots, \boldsymbol{\phi}_m$ of curves $r_p^{(1)}, \ldots, r_p^{(m)}$

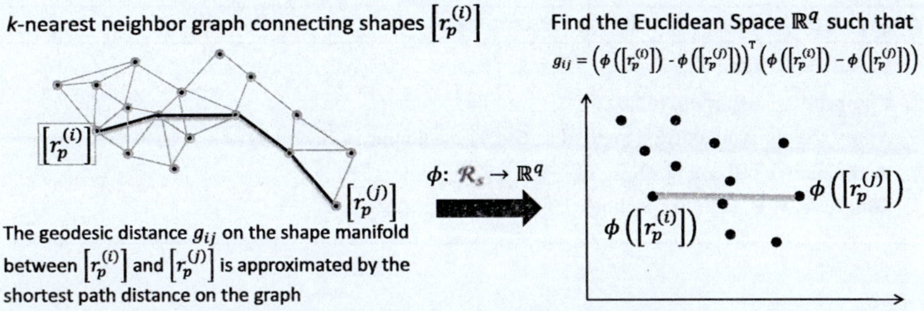

Fig. 4.5 Basic idea of feature extraction: $\boldsymbol{\phi}(\cdot)$ maps the parametric curve $[r_p]$ from the shape manifold onto a low-dimensional Euclidean space $\mathcal{R}^q$, such that the Euclidean distance between the transformed curves $\boldsymbol{\phi}([r_p])$ and $\boldsymbol{\phi}([r'_p])$ is "close" to the geodesic distance $d_p([r_p], [r'_p])$ defined on the shape manifold. The low dimensional features in $\boldsymbol{\phi}([r_p])$ is called the q features of the original parametric curve. (Reprinted with permission from Park et al. 2012)

such that the Euclidean distances between the features are close to the corresponding g_{ij}'s between the curves, that is,

$$g_{ij} \approx (\boldsymbol{\phi}_i - \boldsymbol{\phi}_j)^T (\boldsymbol{\phi}_i - \boldsymbol{\phi}_j). \tag{4.19}$$

Denote by $\boldsymbol{G}$ the dissimilarity matrix whose (i, j)-entry is g_{ij} and denote the doubly centered distance matrix

$$\tilde{\boldsymbol{G}}^2 = -\frac{1}{2}\boldsymbol{H}\boldsymbol{G}^2\boldsymbol{H}^T, \tag{4.20}$$

where $\boldsymbol{G}$ is the matrix whose elements are the squares of the elements of $\boldsymbol{G}$, and $\boldsymbol{H}$ is the $m \times m$ centering matrix with $\delta_{ij} - \frac{1}{m}$ as its (i, j)th element. If the distance matrix $\tilde{\boldsymbol{G}}^2$ is positive semi-definite, the classical multidimensional scaling (MDS) gives an explicit solution to the embedding problem. Consider the eigen-decomposition of $\tilde{\boldsymbol{G}}^2$,

$$\tilde{\boldsymbol{G}}^2 = \boldsymbol{V}\boldsymbol{\Lambda}\boldsymbol{V}^T,$$

where $\boldsymbol{V}$ is the $m \times m$ matrix of the m eigenvectors of $\tilde{\boldsymbol{G}}^2$ and $\boldsymbol{\Lambda}$ is the diagonal matrix of the corresponding eigenvalues. The low-dimensional Euclidean features of $\boldsymbol{\phi}_i$'s are achieved by taking

$$\boldsymbol{\Phi} = \boldsymbol{V}_q\boldsymbol{\Lambda}_q^{1/2},$$

where $\boldsymbol{V}_q$ is the $m \times q$ matrix of the first q leading eigenvectors and $\boldsymbol{\Lambda}_q^{1/2}$ is the diagonal matrix of the square root of the corresponding eigenvalues. Note that the

Algorithm 1 Feature extraction by Isomap

1. **Compute** $\boldsymbol{G}$ with $(\boldsymbol{G})_{ij} = d_p([\boldsymbol{r}_p^{(i)}], [\boldsymbol{r}_p^{(j)}])$.
2. **Compute** $\tilde{\boldsymbol{G}}^2$ using Eq. (4.20).
3. **Apply** the constant shifting method to $\tilde{\boldsymbol{G}}^2$. The outcome is $\tilde{\boldsymbol{G}}_c^2$
4. **Perform** eigen-decomposition: $\tilde{\boldsymbol{G}}_c^2 = \boldsymbol{V}\boldsymbol{\Lambda}\boldsymbol{V}^T$.
5. **Obtain** $\boldsymbol{\Phi} = \boldsymbol{V}_q\boldsymbol{\Lambda}_q^{1/2}$ for a choice of q.

ith row vector of the matrix $\boldsymbol{\Phi}$ corresponds to $\boldsymbol{\phi}_i$. When $\tilde{\boldsymbol{G}}^2$ is not positive definite, one can apply the constant shifting method (Choi and Choi 2007) to make the matrix positive definite, and the above procedure can then be applied. The steps for the nonlinear embedding is summarized in Algorithm 1.

4.2.1.2 Semi-Supervised Clustering of Shapes

Park et al. (2012) also discussed several possibilities to group nanoparticle shapes by their similarities. One possibility is to perform the clustering analysis on the low-dimensional features achieved by the manifold embedding. However, generic clustering methods tend to overly partition a dataset into too many groups than what is needed in nanomaterial research. Park et al. (2012) adopted a semi-supervised learning approach in which domain experts determine the number of shape groups and manually pick a small number of nanoparticles from each shape group. Those few chosen sample particle shapes serve as labeled data, and the existing semi-supervised learning (Zhu et al. 2003) was employed to classify the remaining unlabeled nanoparticle shapes into the predetermined shape groups. The details of the approach is summarized in the sequel.

Suppose that there are m particle outlines and the corresponding shape features, $\{\boldsymbol{\phi}_i; i = 1, ..., m\}$, which are achieved by the method described in Sect. 4.2.1.1. The l shape features among them are manually classified into K shape groups. The shape group belonging of the ith shape feature is represented by $k_i \in \{0, \ldots, K-1\}$, for $i = 1, \ldots, l$. Those l cases are referred to as labeled data. The remaining $m - l$ shape features are unlabeled, and they are expected to be labeled using a semi-supervised learning approach. Specifically, the semi-supervised learning proposed by Zhu et al. (2003) is applied. In this approach, labeled and unlabeled data are represented as vertices in a connected graph, where each edge is assigned a weight that measures the similarity between the two data points connected by the edge; the method produces a label function that is smooth on the graph and correctly matches the manually labeled data. While the method in Zhu et al. (2003) was originally designed for binary classification, Park et al. (2012) extended it for multiclass classification.

First, construct a connected graph $G = (V, E)$, where each node $v \in V$ corresponds to an element in $\{\boldsymbol{\phi}_i; i = 1, ..., m\}$, and E is the collection of edges. The edge connecting $\boldsymbol{\phi}_i$ and $\boldsymbol{\phi}_j$ is weighted by the scaled Euclidean distance between the two shape features

$$w_{ij} = \exp\left\{-(\boldsymbol{\phi}_i - \boldsymbol{\phi}_j)^T \boldsymbol{R}^{-1}(\boldsymbol{\phi}_i - \boldsymbol{\phi}_j)\right\},$$

where the scale matrix $\boldsymbol{R}$ is a $q \times q$ diagonal matrix with σ_d as its dth diagonal element. The choice of the scale matrix will be discussed later. Note that the scaled Euclidean distance in the feature space corresponds to the geodesic distance in the original (curved) shape space by the definition of the feature mapping. Therefore, the weights, w_{ij}, reflect well the similarities between parametric curves in the original space.

Second, estimate a vector-valued label function $\boldsymbol{y} = (y_0, \ldots, y_{K-1})$ on the graph G, where its kth element $y_k : V \mapsto \{0, 1\}$ is an indicator function that determines whether $v \in V$ belongs to the k shape group; $y_k(\boldsymbol{\phi}_i) = 1$ if $\boldsymbol{\phi}_i$ belongs to the kth shape group, and $y_k(\boldsymbol{\phi}_i) = 0$ otherwise. The label function outcomes are required to match the labeled data,

$$y_k(\boldsymbol{\phi}_i) = \delta_{k_i,k}, \qquad \text{for } k = 0, \ldots, K-1, \tag{4.21}$$

where δ is the Kronecker delta. For unlabeled data, it is assigned the label $k_i = \arg\max_{k=0,\ldots,K-1} y_k(\boldsymbol{\phi}_i)$ for $i = l+1, \ldots, m$.

The assigned labels are expected to be similar for similar shape features. That is, it is desirable to choose the label function $\boldsymbol{y}$ such that unlabeled points have the same labels as their neighboring points in the graph. These considerations motivate us to obtain the label function by minimizing with respect to $\boldsymbol{y}$ the following loss function

$$C(\boldsymbol{y}) = \frac{1}{2}\sum_{i,j} w_{ij} \|\boldsymbol{y}(\boldsymbol{\phi}_i) - \boldsymbol{y}(\boldsymbol{\phi}_j)\|^2 = \frac{1}{2}\sum_{k}\sum_{i,j} w_{ij}\{y_k(\boldsymbol{\phi}_i) - y_k(\boldsymbol{\phi}_j)\}^2,$$

subject to the constraints that $y_k(\boldsymbol{\phi}_i) = \delta_{k_i,k}$, $i = 1, \ldots, l$.

Let $\boldsymbol{W}$ denote the matrix whose (i, j)-th entry is w_{ij} and let $\boldsymbol{y}_k = (y_k(\boldsymbol{\phi}_1), \ldots, y_k(\boldsymbol{\phi}_M))^T$. The loss function can be rewritten as a quadratic form

$$C(\boldsymbol{y}) = \frac{1}{2}\sum_{k=0}^{K-1} \boldsymbol{y}_k^T \boldsymbol{\Delta} \boldsymbol{y}_k, \tag{4.22}$$

where $\boldsymbol{\Delta} = \boldsymbol{D} - \boldsymbol{W}$, and $\boldsymbol{D}$ is the $m \times m$ diagonal matrix whose i-th diagonal entry is $w_i = \sum_{j=1}^{M} w_{ij}$. To present the solution of this minimization problem, we write the matrix $\boldsymbol{W}$ and $\boldsymbol{D}$ and the vector $\boldsymbol{y}_k$ in block forms, corresponding to labeled parts and unlabeled parts, respectively:

$$\mathbf{W} = \begin{bmatrix} \mathbf{W}_{ll} & \mathbf{W}_{lu} \\ \mathbf{W}_{ul} & \mathbf{W}_{uu} \end{bmatrix}, \; \mathbf{D} = \begin{bmatrix} \mathbf{D}_{ll} & \mathbf{O} \\ \mathbf{O} & \mathbf{D}_{uu} \end{bmatrix}, \; y_k = \begin{bmatrix} \mathbf{y}_k^{(l)} \\ \mathbf{y}_k^{(u)} \end{bmatrix}, \tag{4.23}$$

where $\mathbf{O}$ denotes a matrix of zeros whose dimensions can be determined from the context. Note that the values of $\mathbf{y}_k^{(l)}$ are fixed by the constraints $y_k(\boldsymbol{\phi}_i) = \delta_{k_i,k}$, $i = 1, \ldots, l$. Ignoring the fixed term in the loss function, we have the loss function re-expressed as

$$\frac{1}{2}(\mathbf{y}_k^{(u)})^T(\mathbf{D}_{uu} - \mathbf{W}_{uu})\mathbf{y}_k^{(u)} - (\mathbf{y}_k^{(l)})^T \mathbf{W}_{ul} \mathbf{y}_k^{(u)}.$$

We get the following closed-from expression for minimizing the loss function with respect to $\mathbf{y}_k^{(u)}$,

$$\mathbf{y}_k^{(u)} = (\mathbf{D}_{uu} - \mathbf{W}_{uu})^{-1} \mathbf{W}_{ul}^T \mathbf{y}_k^{(l)}. \tag{4.24}$$

Given the label function in Eq. (4.24), we assign the label $k^* = \arg\max_k h_k(\boldsymbol{\phi}_i)$ to the unlabeled feature $\boldsymbol{\phi}_i$ for $i = l + 1, \ldots, m$.

Regarding the choice of a suitable scale $\boldsymbol{R}$ in the weighting function, Park et al. (2012) proposed a selection criterion based on Shannon's entropy, which is defined by

$$H(\boldsymbol{R}) := \sum_{i=l+1}^{m} H_i(\boldsymbol{R}), \tag{4.25}$$

where

$$H_i(\boldsymbol{R}) = H\left[\frac{y_0(\boldsymbol{\phi}_i)}{\sum_k y_k(\boldsymbol{\phi}_i)}, \frac{y_1(\boldsymbol{\phi}_i)}{\sum_k y_k(\boldsymbol{\phi}_i)}, \ldots, \frac{y_{K-1}(\boldsymbol{\phi}_i)}{\sum_k y_k(\boldsymbol{\phi}_i)}\right]$$

is the entropy associated with the i-th unlabeled case. Among the candidate values of $\boldsymbol{R}$, one that minimizes the criterion can be chosen. A smaller value of the entropy H_i implies that the labeled function values are more concentrated, so the confidence of labeling the ith case to the concentrated one is higher. The criterion basically chooses the scale parameter so that the confidence of the labeling would be highest.

In the application to the real TEM image shown on the right panel of Fig. 4.3, domain experts are asked to determine the number of shape groups and manually pick ten particles from each shape group to form the labeled data. According to the experts, it is sufficient to distinguish a nanoparticle into one of the four shapes: triangles, rectangles, circles, and rods. These four shapes were assigned labels 0–3 respectively. Figure 4.6 presents the results of applying the semi-supervised learning procedure described in this section.

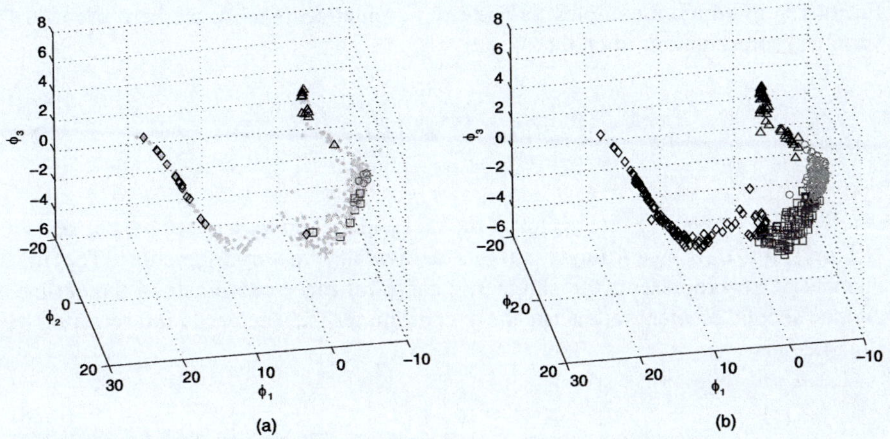

Fig. 4.6 Results of the semi-supervised learning. Each marker represents a low-dimensional embedding of a parametric curve. The triangular, rectangular, circular and diamond-shaped markers represent, respectively, triangular, rectangular, circular and rod-shaped particles. The left panel shows a mix of a few manually assigned shape labels and the unassigned shapes marked using the small dots. **(a)** Initial assignment of labels. **(b)** After the semi supervised learning (Reprinted with permission from Park et al. 2012)

4.2.2 *Shape Analysis for a Broader Class of Nanoparticles*

This section introduces a nanoparticle shape analysis for more complex than star shapes in $\mathbb{R}^2$, based on the statistical models and methodology described in Park et al. (2013). The method makes use of the outline-based shape representation. In particular, the outline of an object is represented by a general parametric curve like the one discussed in Sect. 4.1.2. A nonparametric shape distribution was defined on the normalized outline after removing the scale, rotation and translation effect from it. This subsection discusses the shape representation, shape parameter estimation, shape classification, and shape inference issues.

4.2.2.1 Shape Representation

Recall in Eq. (4.15) that the outline of an object is represented by a parametric curve with spline basis, i.e.,

$$z(t) = \sum_{l=1}^{L} \alpha_l \phi_l(t), \tag{4.26}$$

where $\phi_l(t)$ is the lth periodic B-spline bending function of degree d and $\alpha_l \in \mathbb{C}$ is the corresponding complex coefficient in the B-spline function. Different from

Eq. (4.15) in which a complex variable representation is used, we here use a real variable representation, such that

$$z(t) = \sum_{l=1}^{L} \boldsymbol{\alpha}^{(l)} \phi_l(t), \tag{4.27}$$

where $\boldsymbol{\alpha}^{(l)}$ is a bivariate real vector of the real and imaginary parts of α_l, i.e., $\boldsymbol{\alpha}^{(l)} = (\text{Re}(\alpha_l)), \text{Im}(\alpha_l))$. Accordingly, $z(t)$ is a bivariate vectorial function. The first element of $z(t)$ represents the parametric curve for the x-coordinate of the outline, and the second element represents the y-coordinate. The vectorial representation of Eq. (4.27) is

$$z(t) = (\boldsymbol{\phi}_t^T \otimes \boldsymbol{I}_2)\boldsymbol{\alpha}, \tag{4.28}$$

where $\boldsymbol{\alpha} = (\boldsymbol{\alpha}^{(l)}, \ldots, \boldsymbol{\alpha}^{(L)})^T$ is a $2L$ vector of all the spline coefficients, and $\boldsymbol{\phi}_t = (\phi_1(t), \ldots, \phi_L(t))^T$.

The coefficient vector $\boldsymbol{\alpha}$ is affected by shape information. It also varies with other uninterested factors such as scaling, shifting and rotation of shapes. Park et al. (2013) specifically related these factors to the outline and its shape. Given the scale parameter $s \in \mathbb{R}_+$, rotation angle $\theta \in [0, 2\pi]$ and translation $c \in \mathbb{R}^2$, the coefficient vector is

$$\boldsymbol{\alpha}^{(l)} = \frac{1}{s} \boldsymbol{R}_\theta \tilde{\boldsymbol{\alpha}}^{(l)} + \boldsymbol{c}, \tag{4.29}$$

where $\tilde{\boldsymbol{\alpha}}^{(l)}$ is the shape feature independent of the pose parameters, $\boldsymbol{c}$ is a two-dimensional column vector of horizontal and vertical translation, and $\boldsymbol{R}_\theta$ is a transformation matrix for a rotation by θ in counter-clock wise. The model for the whole feature $\boldsymbol{\alpha}$ is

$$\boldsymbol{\alpha} = \frac{1}{s} \boldsymbol{Q}_\theta \tilde{\boldsymbol{\alpha}} + \boldsymbol{H}\boldsymbol{c}, \tag{4.30}$$

where $\boldsymbol{Q}_\theta = \boldsymbol{I}_L \otimes \boldsymbol{R}_\theta$ is a Kronecker product of the $L \times L$ identity matrix with $\boldsymbol{R}_\theta$, and $\boldsymbol{H} = \boldsymbol{1}_L \otimes \boldsymbol{I}_2$ is a Kronecker product of the $L \times 1$ vector of ones with a 2×2 identity matrix. Therefore, different versions of the coefficient vectors can be achieved with the same normalized feature $\tilde{\boldsymbol{\alpha}}$ and different pose parameters such as s, θ and c.

Park et al. (2013) posed a mixture of multivariate normal distributions for the shape feature $\tilde{\boldsymbol{\alpha}}$ to represent the shape distribution of different groups of nanoparticle shapes. In the model, for the kth shape group, the normalized feature follow a multivariate normal distribution,

$$\tilde{\boldsymbol{\alpha}} | k \sim \mathcal{N}(\boldsymbol{\mu}_k, \boldsymbol{\Sigma}_k),$$

where $\boldsymbol{\mu}_k$ and $\boldsymbol{\Sigma}_k$ are, respectively, the mean vector and covariance matrix of the multivariate normal distribution. When the number of the shape groups is K, the unconditional distribution of $\tilde{\boldsymbol{\alpha}}$ follows a mixture of K multivariate normal distributions,

$$\tilde{\boldsymbol{\alpha}} \sim \sum_{k=1}^{K} w_k \mathcal{N}(\boldsymbol{\mu}_k, \boldsymbol{\Sigma}_k), \tag{4.31}$$

where $w_k \in [0, 1]$ is the mixing proportion parameter that represents the mixing proportion of the kth shape group in the mixture, and the parameters should satisfy $\sum_{k=1}^{K} w_k = 1$. This mixture distribution represent the shape distribution of nanoparticles.

4.2.2.2 Parameter Estimation

Park et al. (2013) described how the shape distribution in Eq. (4.31) can be estimated using the data of nanoparticle outlines extracted from a microscope image. Suppose that the outlines of N nanoparticles are extracted. The outline of the nth nanoparticle is represented in the vector form of Eq. (4.28). For digital images, the outline is only observed at a finite number of locations, $t \in \{t_{nj}; j = 1, \ldots, m_n\}$, and the observation may be corrupted by noises. Therefore, the observation model is

$$\tilde{\mathfrak{z}}_n(t_{nj}) = (\boldsymbol{\phi}_{t_{nj}}^T \otimes \boldsymbol{I}_2)\boldsymbol{\alpha}_n + \boldsymbol{\epsilon}_{nj}, \tag{4.32}$$

where $\boldsymbol{\epsilon}_{nj} \sim \mathcal{N}(\mathbf{0}, \sigma^2 \boldsymbol{I}_2)$ represents the random observation noises. Let $\tilde{\mathfrak{z}}_n$ denote a $2m_n \times 1$ vector with the two dimensional vector $\tilde{\mathfrak{z}}_n(t_{nj})$ as its j submatrix. Collectively, all of the n outline observations are denoted by $\mathcal{D} = \{\tilde{\mathfrak{z}}_n; n = 1, \ldots, N\}$.

The data can be linked to the parameters of the shape distribution in Eq. (4.31) as follows. From Eq. (4.32), the conditional probability of $\tilde{\mathfrak{z}}_n$, conditioned on $\boldsymbol{\alpha}_n$, is given by

$$\tilde{\mathfrak{z}}_n \sim \mathcal{N}(\boldsymbol{\Phi}_n \boldsymbol{\alpha}_n, \sigma^2 \boldsymbol{I}_{2m_n}), \tag{4.33}$$

where $\boldsymbol{\Phi}_n$ is a $2m_n \times 2L$ matrix formed by binding $\boldsymbol{\phi}_{t_{nj}}^T \otimes \boldsymbol{I}_2$'s in rowwise. The probability distribution of $\boldsymbol{\alpha}_n$ depends on the hidden shape group index and pose parameters for the nth outline. Let $k_n \in \{1, \ldots, K\}$ denote the index of the shape group that the nth outline belongs to. Let s_{nk}, θ_{nk}, and $\boldsymbol{c}_{nk}$ denote the pose parameters for the nth outline when it belongs to the kth shape group. Please note that the pose parameters might depend on the shape group index, so Park et al. (2013) denote them separately for each shape group. All of the unknown model parameters are collectively denoted as $\Theta = \{\sigma^2, s_{nk}, \theta_{nk}, \boldsymbol{c}_{nk} : n = 1, \ldots, N, k = 1, \ldots, K\} \cup \{w_k, \boldsymbol{\mu}_k, \boldsymbol{\Sigma}_k; k = 1, \ldots, K\}$. When k_n and the parameters Θ are given,

$$\boldsymbol{\alpha}_n | k_n = k, \Theta \sim \mathcal{N}\left(\boldsymbol{A}_{nk}\boldsymbol{\mu}_k + \boldsymbol{H}\boldsymbol{c}_{nk}, \boldsymbol{A}_{nk}\boldsymbol{\Sigma}_k\boldsymbol{A}_{nk}^T\right),$$

where $\boldsymbol{A}_{nk} = \frac{1}{s_{nk}}\boldsymbol{Q}_{\theta_{nk}}$, and $\boldsymbol{H} = \mathbf{1}_L \otimes \boldsymbol{I}_2$ is a Kronecker product of the $L \times 1$ vector of ones with a 2×2 identity matrix. Accordingly, the observation vector $\tilde{z}_n$, conditioned on $k_n = k$, follows

$$\tilde{z}_n | k_n = k, \Theta \sim \mathcal{N}\left(\boldsymbol{\Phi}_n\boldsymbol{A}_{nk}\boldsymbol{\mu}_k + \boldsymbol{\Phi}_n\boldsymbol{H}\boldsymbol{c}_{nk}, \boldsymbol{\Phi}_n\boldsymbol{A}_{nk}\boldsymbol{\Sigma}_k\boldsymbol{A}_{nk}^T\boldsymbol{\Phi}_n^T + \sigma^2\boldsymbol{I}_{2m_n}\right).$$

Park et al. (2013) denote its density function by $f_{n,k}(\tilde{z}_n; \Theta)$. The complete log likelihood of Θ is

$$g(\Theta | \{k_n : n = 1, .., N\}, \mathcal{D}) = \sum_{n=1}^{N}\sum_{k=1}^{K}\delta(k_n = k)\log(w_k f_{n,k}(\tilde{z}_n; \Theta)),$$

where δ is the Dirac delta function.

Park et al. (2013) proposed the Expectation Maximization (EM) algorithm to estimate the parameters Θ given data $\mathcal{D}$. In the EM algorithm, the hidden variables, k_n and $\boldsymbol{\alpha}$, are optimized jointly with the parameters Θ, through multiple iterations of the E-step and M-step. In the E-step of the algorithm, the posterior distributions of k_n and $\boldsymbol{\alpha}$ are obtained while Θ is fixed to their estimate from the previous M-step. Let $\hat{\Theta}$ denotes the estimates of Θ from the previous M-step. The hat notation of each element in Θ denotes the estimate from the previous M-step. The posterior distribution of k_n is given by

$$p(k_n = k | \mathcal{D}, \hat{\Theta}) = \frac{\hat{w}_k f_{n,k}(\boldsymbol{y}_n; \hat{\Theta})}{\sum_{k'=1}^{K}\hat{w}_{k'} f_{n,k'}(\boldsymbol{y}_n; \hat{\Theta})}. \tag{4.34}$$

The posterior distribution of $\boldsymbol{\alpha}$, conditioned $k_n = k$, is obtained as

$$\begin{aligned} &\boldsymbol{\alpha}_n | k_n = k, \mathcal{D}, \hat{\Theta} \sim \mathcal{N}(\boldsymbol{m}_{nk}, \boldsymbol{S}_{nk}) \\ &\quad \boldsymbol{m}_{nk} = \hat{\boldsymbol{A}}_{nk}\hat{\boldsymbol{\Sigma}}_k\hat{\boldsymbol{A}}_{nk}^t\boldsymbol{\Phi}_n^t\left(\hat{\sigma}^2\boldsymbol{I}_{2M} + \boldsymbol{\Phi}_n\hat{\boldsymbol{A}}_{nk}\hat{\boldsymbol{\Sigma}}_k\hat{\boldsymbol{A}}_{nk}^t\boldsymbol{\Phi}_n^T\right)^{-1} \\ &\quad\quad \left(\tilde{z}_n - \boldsymbol{\Phi}_n\hat{\boldsymbol{A}}_{nk}\boldsymbol{\mu}_k - \boldsymbol{H}\hat{\boldsymbol{c}}_{nk}\right) + \hat{\boldsymbol{A}}_{nk}\hat{\boldsymbol{\mu}}_k + \boldsymbol{H}\hat{\boldsymbol{c}}_{nk} \\ &\quad \boldsymbol{S}_{nk} = (\hat{\boldsymbol{A}}_{nk}^{-T}\hat{\boldsymbol{\Sigma}}_k^{-1}\hat{\boldsymbol{A}}_{nk}^{-T} + \hat{\sigma}^{-2}\boldsymbol{\Phi}_n^t\boldsymbol{\Phi}_n)^{-1}, \end{aligned} \tag{4.35}$$

where each of the hat matrices denotes the corresponding matrix evaluated with the estimate $\hat{\Theta}$. Let $q_{nk}(\boldsymbol{\alpha}_n; \hat{\Theta})$ denote the conditional density function of $\boldsymbol{\alpha}_n$. The expectation of the complete log likelihood g with respect to the posterior distributions is taken as

$$\tilde{g}(\Theta) = E[g(\Theta|\{k_n : n = 1, .., N\}, \mathcal{D})]$$
$$= \sum_{k=1}^{K} p(k_n = k|\mathcal{D}, \hat{\Theta}) \int g(\Theta|\{k_n : n = 1, .., N\}, \mathcal{D}) q_{nk}(\boldsymbol{\alpha}_n; \hat{\Theta}) d\boldsymbol{\alpha}_n.$$

In the M-step, the expectation $\tilde{g}(\Theta)$ is maximized with respect to Θ for finding the maximum likelihood estimates of the parameters in Θ. One computational issue is that one cannot derive the closed form expressions of the local optima for $\boldsymbol{\mu}_k$, $\boldsymbol{\Sigma}_k$, θ_{nk}, s_{nk} and c_{nk}, since their first order necessary conditions are entangled with one another in complicated forms. One could instead perform the M-step iteratively by the Newton Raphson, which would incur another expensive iterations for each M-step. There are two other possible options to proceed the M-step without iterations; the first one is to perform only one iteration of the Newton Raphson rather than to fully maximize it for every M-step, resulting in a GEM algorithm (Dempster et al. 1977), and another one is to use the ECM algorithm (Meng and Rubin 1993). The first option does not in general converge appropriately. Park et al. (2013) suggested the ECM algorithm. The ECM iteration converges. At the convergence, the outcomes of the ECM algorithm can be used for different purposes of shape analysis. The direct outcome is the shape classification that classifies the particle outlines by their shape groups, which we will describe in Sect. 4.2.2.3. In addition, one can estimate the outline of $z_n(t)$ from its noisy observation $\boldsymbol{y}_n$. The ECM is capable of estimating the complete outline $z_n(t)$ when the noisy observation $\boldsymbol{y}_n$ only contains a part of the complete outline due to the outline occlusion, because the ECM can utilize the outlines classified into the same group to fill in the missing portion. Additional detail of how to use the ECM for the contour estimation is given in Sect. 4.2.2.4.

4.2.2.3 Shape Classification

At the ECM convergence, the posterior probability of Eq. (4.34) can be calculated, and the shape group of the nth particle outline can be assigned to the k value that achieves the highest posterior probability. To numerically illustrate the results of the algorithm, we chose four micrographs. The four figures have various types of particle shapes. The classification results are presented by labeling the nanoparticles with one character symbol representing shape classes: 't' = triangle, 'b' = rectangle, 'c' = circle and 'r' = rod. Please refer to Fig. 4.7.

We compare the automated classification outcomes with what human would classify the shapes. In the top-left figure, the result was accurate except for two miss-classifications—the automated shape classification classified a triangle as a circle and classified a circle as a triangle. Such miss-classification are also observed in a few other cases at the bottom-left figure and the bottom-right figure. The circle-to-triangle misclassification is mostly caused by insufficient contour evidences. The other type of miss-classification was caused by a faulty data parametrization for

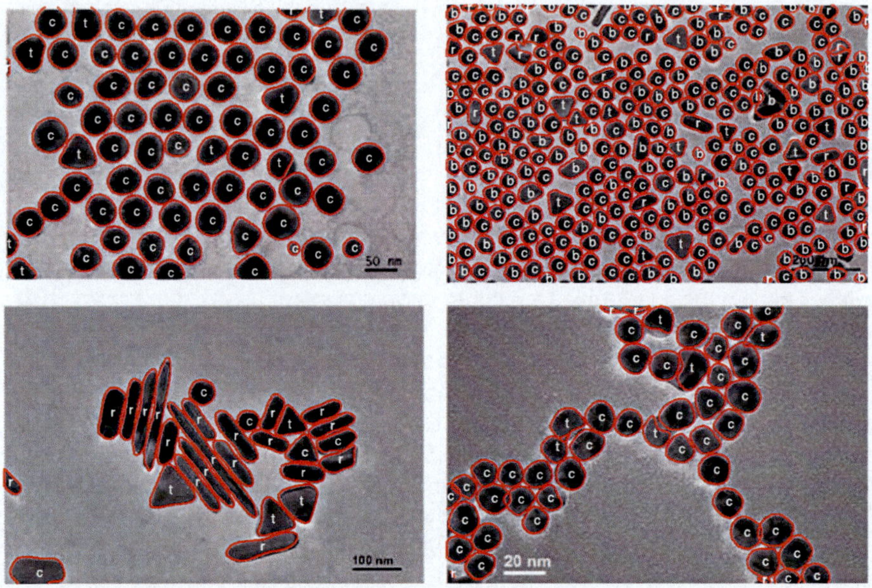

Fig. 4.7 Shape classification. Each particle's shape is labeled as: t = triangle, b = rectangle, c = circle and r = rod. (Reprinted with permission from Park et al. 2013)

the spline curves in the ECM. Looking for a more capable data parametrization is certainly desirable but does not appear to be a simple task.

4.2.2.4 Shape Inference

After the convergence, we have the estimates of the parameters in Θ, denoted by $\hat{\Theta}$. To estimate $z_n(t)$, given the shape classification outcomes, we can achieve the posterior expectation of $z_n(t)$, which will be the estimate of the complete contour. Note that

$$z_n(t) = \boldsymbol{\phi}_t^T \boldsymbol{\alpha}_n.$$

The posterior expectation of $z_n(t)$ would be $\boldsymbol{\phi}_t^T \hat{\boldsymbol{\alpha}}$ with $\hat{\boldsymbol{\alpha}}$ as the posterior expectation of $\boldsymbol{\alpha}$. From Eqs. (4.34) and (4.35), the posterior distribution of $\boldsymbol{\alpha}$ can be achieved as follows:

$$\boldsymbol{\alpha}_n | \mathcal{D}, \hat{\Theta} \sim \sum_{k=1}^{K} \beta_{nk} \mathcal{N}(\boldsymbol{\alpha}_n; \boldsymbol{m}_{nk}, \boldsymbol{S}_{nk}).$$

The posterior expectation $\hat{\boldsymbol{\alpha}}_n$ is

$$\hat{\alpha}_n = \left(\sum_{k=1}^{K} \hat{w}_k f_{n,k}(\tilde{z}_n; \hat{\Theta}) S_{nk}^{-1} \right)^{-1} \sum_{k=1}^{K} \hat{w}_k f_{n,k}(\tilde{z}_n; \hat{\Theta}) S_{nk}^{-1} m_{nk}.$$

4.2.3 Numerical Examples: Image Segmentation to Nanoparticle Shape Inference

This section highlights the major challenges in performing the shape inference with real microscope images and describes the step-by-step to overcome the challenges, starting with the image segmentation step for getting noisy outline data.

The major challenge in estimating the shape outlines is the overlaps among nanoparticles in microscope images. When particles overlap, their respective outlines are overlapped and mixed, so identifying the outlines are difficult. In addition, some portions of the nanoparticle outlines can be occluded due to the overlaps. As the degree of the particle overlaps increases, larger portions are occluded, and the shape inference task becomes more difficult.

We described in Fig. 3.1 the three major steps to estimate the outlines of nanoparticles under such difficulties. The first step is the image binarization to get the binary image containing all nanoparticles. The second step is the foreground segmentation to further segment the binary image into the foreground regions of individual nanoparticles. The boundaries of the foreground regions naturally become the noisy outline data for the respective nanoparticles. The second step comes in two sub-steps: foreground marker identification in Sect. 3.4.1 and initial foreground segmentation in Sect. 3.4.2. The noisy outline data contain only a portion of the outlines for those overlapping particles. The last step is to perform the shape inference as described in Sect. 4.2.2.4.

To illustrate how the three-step procedure is performed, we chose twelve different real micrographs obtained from the synthesis processes of gold nanoparticles. In order to see the change in the quality of nanoparticle shape inference for different degrees of overlaps among nanoparticles, we categorized the micrographs into three groups according to the degrees of overlap; low, medium and high. The micrographs of 'low' overlapping degree have slightly touching among particles. In the cases of the 'medium' degree, most nanoparticles overlaps and the overlapping structures conform with Assumption 3.1. The 'high' degrees of overlap are when nanoparticles overlap more severely so that the overlapping clearly violates Assumption 3.1.

For the first step of the procedure, we applied the alternative sequence filtering (Gonzalez and Woods 2002) for image denoising, followed by the Otsu's optimum global thresholding (Otsu 1979) for image binarization. For the second step, we used the UECS that was described in Sect. 3.4.2 for marker identification and then used the contour association approach in Sect. 3.4.2 to collect the outline data.

The results of the shape inference for the twelve micrographs are presented in Figs. 4.8, 4.9, 4.10, and 4.11. Each figure has four columns. The first column is the original micrograph. The second column is the marker identification result after

Fig. 4.8 Results from low-degree overlapping cases. (**a**) original images, (**b**) markers from UECS, (**c**) outline data collected by the contour evidence association approach introduced in Sect. 3.4.2 , (**d**) final contours by applying the ECM method. (Reprinted with permission from Park et al. 2013)

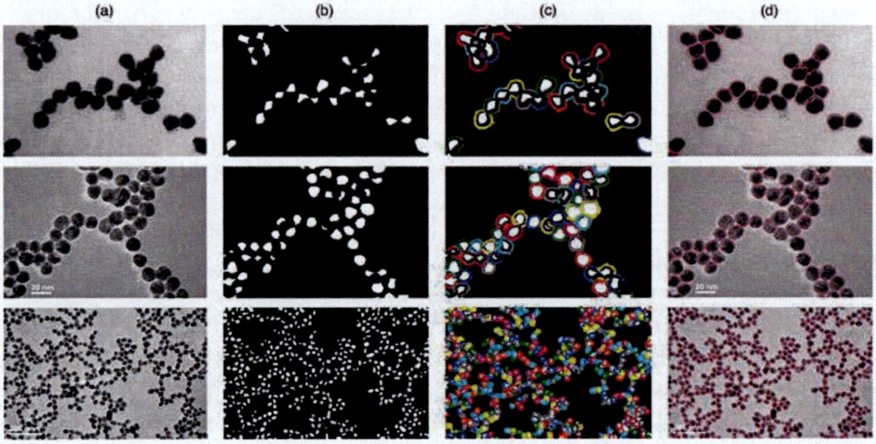

Fig. 4.9 Results from medium-degree overlapping cases. (**a**) original images, (**b**) markers from UECS, (**c**) outline data collected by the contour evidence association approach introduced in Sect. 3.4.2, (**d**) final contours by applying the ECM method. (Reprinted with permission from Park et al. 2013)

UECS, and the third column is the outline data collected by the contour association approach. Contour evidences (pixels at the boundaries of particles) have first been extracted by Canny's edge detection method (Canny 1986), and then, they were associated with the markers by using the contour association approach in Sect. 3.4.2. After the association, the algorithm filters out some noise edge outliers based on the mean and standard deviation of g defined in Eq. (3.5). In the third column of

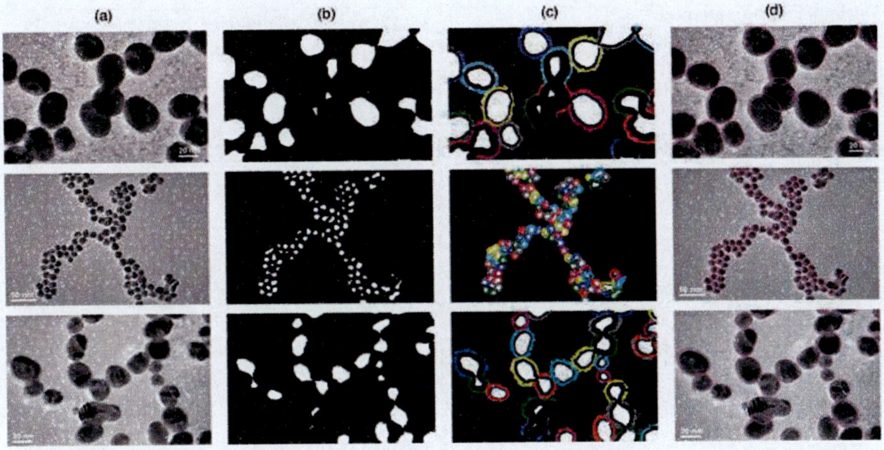

Fig. 4.10 Results from medium-degree overlapping cases. (**a**) original images, (**b**) markers from UECS, (**c**) outline data collected by the contour evidence association approach introduced in Sect. 3.4.2, (**d**) final contours by applying the ECM method. (Reprinted with permission from Park et al. 2013)

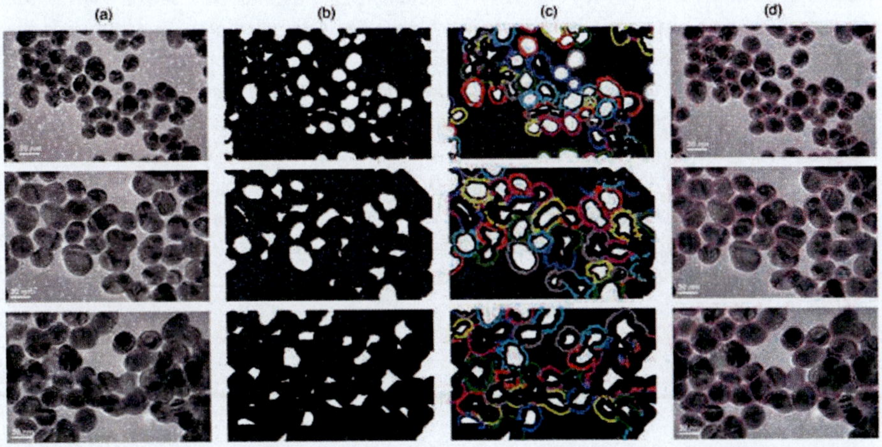

Fig. 4.11 Results from high-degree overlapping cases. (**a**) original images, (**b**) markers from UECS, (**c**) outline data collected by the contour evidence association approach introduced in Sect. 3.4.2, (**d**) final contours by applying the ECM method. (Reprinted with permission from Park et al. 2013)

the figures, the association to different markers is illustrated by different colors of the outline evidences. The last column shows the final result from the shape inference method in Sect. 4.2.2.4. The accuracy of the shape inference method is quantitatively evaluated. For each of the twelve micrographs, we manually counted the total number of nanoparticles and the number of the particles whose outlines are correctly inferred; these numbers are tabulated in Table 4.1.

Table 4.1 Comparison of performances on nanoparticle shape inference. (*Source*: Park et al. 2013)

Samples	Degree of overlap	Total # of particles	# of the correctly recognized ones
Fig. 4.8, row 1	Low	76	76
Fig. 4.8, row 2	Low	307	299
Fig. 4.8, row 3	Low	28	28
Fig. 4.9, row 1	Medium	28	26
Fig. 4.9, row 2	Medium	52	48
Fig. 4.9, row 3	Medium	459	437
Fig. 4.10, row 1	Medium	19	17
Fig. 4.10, row 2	Medium	108	103
Fig. 4.10, row 3	Medium	29	25
Fig. 4.11, row 1	High	63	54
Fig. 4.11, row 2	High	44	34
Fig. 4.11, row 3	High	45	33

4.3 Beyond Shape Analysis: Topological Data Analysis

The shape analysis is very useful in analyzing material structures when the structures do not have any interior holes or interior structures so that they can be characterized by their exterior outlines. For some material applications, however, topology that characterizes the topological structure of materials includes exterior and interior structures. Their topological relations are also important characteristics. For example, the topology of two-particle aggregates could be of great interest (Sikaroudi et al. 2018). For those applications, a point-set based shape representation and related metric space are more useful, which will be discussed in Sect. 10.7. Reviewing the topological data analysis in detail is beyond the scope of this book. For readers interested in this topic, please refer to the review paper by Wasserman (2018).

References

Bengtsson A, Eklundh JO (1991) Shape representation by multiscale contour approximation. IEEE Transactions on Pattern Analysis & Machine Intelligence 13(1):85–93

Bookstein FL (1986) Size and shape spaces for landmark data in two dimensions. Statistical Science 1(2):181–222

Brechbühler C, Gerig G, Kübler O (1995) Parametrization of closed surfaces for 3-d shape description. Computer Vision and Image Understanding 61(2):154–170

Burgos-Artizzu XP, Perona P, Dollár P (2013) Robust face landmark estimation under occlusion. In: Proceedings of the 2013 International Conference on Computer Vision, Ieee, pp 1513–1520

Canny J (1986) A computational approach to edge detection. IEEE Transactions on Pattern Analysis and Machine Intelligence 8(6):679–698

Choi H, Choi S (2007) Robust kernel isomap. Pattern Recognition 40(3):853–862

Dempster A, Laird N, Rubin D (1977) Maximum likelihood from incomplete data via the EM algorithm. Journal of the Royal Statistical Society Series B (Methodological) 39(1):1–38

Dryden IL, Mardia KV (1991) General shape distributions in a plane. Advances in Applied Probability 23:259–276

Dryden IL, Mardia KV (1998) Statistical Shape Analysis. John Wiley & Sons, West Sussex, UK

Dryden IL, Mardia KV (2016) Statistical Shape Analysis: With Applications in R, 2nd Edition. John Wiley and Sons Ltd., West Sussex, UK

Gonzalez RC, Woods RE (2002) Digital Image Processing, 3rd ed. Prentice Hall, Upper Saddle River, NJ

Goodall CR (1984) The Statistical Analysis of Growth in Two Dimensions. PhD thesis, Department of Statistics, Harvard University

Gower JC (1975) Generalized Procrustes analysis. Psychometrika 40(1):33–51

Joshi SH, Klassen E, Srivastava A, Jermyn I (2007) A novel representation for Riemannian analysis of elastic curves in Rn. In: Proceedings of the 2007 IEEE Conference on Computer Vision and Pattern Recognition, Ieee, pp 1–7

Kendall DG (1984) Shape manifolds, Procrustean metrics, and complex projective spaces. Bulletin of the London Mathematical Society 16(2):81–121

Kendall DG, Barden D, Carne TK, Le H (2009) Shape and Shape Theory. John Wiley & Sons, West Sussex, UK

Kent JT (1994) The complex Bingham distribution and shape analysis. Journal of the Royal Statistical Society: Series B (Methodological) 56(2):285–299

Klassen E, Srivastava A, Mio M, Joshi SH (2004) Analysis of planar shapes using geodesic paths on shape spaces. IEEE Transactions on Pattern Analysis and Machine Intelligence 26(3):372–383

Kume A, Welling M (2010) Maximum likelihood estimation for the offset-normal shape distributions using EM. Journal of Computational and Graphical Statistics 19(3):702–723

Kurtek S, Srivastava A, Klassen E, Ding Z (2012) Statistical modeling of curves using shapes and related features. Journal of the American Statistical Association 107(499):1152–1165

Lancaster H (1965) The helmert matrices. The American Mathematical Monthly 72(1):4–12

Mardia KV, Dryden IL (1989) The statistical analysis of shape data. Biometrika 76(2):271–281

Mardia KV, Dryden IL (1999) The complex Watson distribution and shape analysis. Journal of the Royal Statistical Society: Series B (Statistical Methodology) 61(4):913–926

Meng X, Rubin D (1993) Maximum likelihood estimation via the ECM algorithm: A general framework. Biometrika 80(2):267

Otsu N (1979) A threshold selection method from gray-level histograms. IEEE Transactions on Systems, Man, and Cybernetics 9(1):62–66

Park C (2011) Automated Morphology Analysis of Nanoparticles. PhD thesis, Texas A&M University

Park C, Huang JZ, Huitink D, Kundu S, Mallick BK, Liang H, Ding Y (2012) A multistage, semi-automated procedure for analyzing the morphology of nanoparticles. IIE Transactions 44(7):507–522

Park C, Huang JZ, Ji J, Ding Y (2013) Segmentation, inference and classification of partially overlapping nanoparticles. IEEE Transactions on Pattern Analysis and Machine Intelligence 35(3):669–681

Sikaroudi AE, Welch DA, Woehl TJ, Faller R, Evans JE, Browning ND, Park C (2018) Directional statistics of preferential orientations of two shapes in their aggregate and its application to nanoparticle aggregation. Technometrics 60(3):332–344

Srivastava A, Klassen E, Joshi SH, Jermyn IH (2010) Shape analysis of elastic curves in Euclidean spaces. IEEE Transactions on Pattern Analysis and Machine Intelligence 33(7):1415–1428

Staib LH, Duncan JS (1992) Boundary finding with parametrically deformable models. IEEE Transactions on Pattern Analysis & Machine Intelligence (11):1061–1075

Tenenbaum JB, Silva VD, Langford JC (2000) A global geometric framework for nonlinear dimensionality reduction. Science 290(5500):2319–2323

Tieng QM, Boles W (1997) Recognition of 2d object contours using the wavelet transform zero-crossing representation. IEEE Transactions on Pattern Analysis and Machine Intelligence 19(8):910–916

Vandaele R, Aceto J, Muller M, Péronnet F, Debat V, Wang CW, Huang CT, Jodogne S, Martinive P, Geurts P, Marée R (2018) Landmark detection in 2d bioimages for geometric morphometrics: A multi-resolution tree-based approach. Scientific Reports 8(1):538.1–538.13

Wasserman L (2018) Topological data analysis. Annual Review of Statistics and Its Application 5:501–532

Younes L (1998) Computable elastic distances between shapes. SIAM Journal on Applied Mathematics 58(2):565–586

Zahn CT, Roskies RZ (1972) Fourier descriptors for plane closed curves. IEEE Transactions on Computers 100(3):269–281

Zhu X, Ghahramani Z, Lafferty J (2003) Semi-supervised learning using Gaussian fields and harmonic functions. In: Proceedings of the 20th International Conference on Machine Learning, Washington, DC, pp 912–919

Chapter 5
Location and Dispersion Analysis

5.1 Basics of Mixing State Analysis

Material scientists tend to describe the quality of a mixing state using two phrases: "dispersion" and "distribution" (Manas-Zloczower 1997). *Dispersion* refers to the ability to break down agglomerates into small pieces, while *distribution* refers to the ability to make nanoparticles or other additive agents, large or small, to locate uniformly throughout a host material; please see an illustration in Fig. 5.1 for the meaning of the two terms. We also refer to the two effects as size (dispersion) and location (distribution) effects. Manas-Zloczower (1997) and Hui et al. (2008) state that the ideal mixing state is good dispersion and good distribution. To accomplish that goal, material scientists employ different equipment or mechanisms (Ray and Okamoto 2003; Zou et al. 2008) to break down large-sized nanoparticle agglomerates into small pieces (ideally, individual particles) and try to make the small pieces distribute uniformly throughout the host material.

Please note that in the right-bottom panel of Fig. 5.1, a good distribution/good dispersion is shown as particles on a perfectly aligned lattice. This is not the case in practice, where a good distribution is manifested as a randomly uniform pattern. The uniform *probability* distribution, known in the statistical literature as the complete spatial randomness (CSR) (Diggle 2013), rather than the perfectly aligned array, provides the baseline or benchmarking state to which an actual mixing state is to be compared with.

Quantification of nanoparticle mixing provides an objective assessment and can even serve as a surrogate for material properties. Doing so becomes a prerequisite to ensuring good quality of nanocomposites in its manufacturing process. Image processing plays an essential role in this analysis. It functions as a preprocessing action because the quantification process starts with taking image measurements of a nanocomposite sample through the use of an electron microscope. All images used

C. Park, Y. Ding, *Data Science for Nano Image Analysis*, International Series in Operations Research & Management Science 308,
https://doi.org/10.1007/978-3-030-72822-9_5

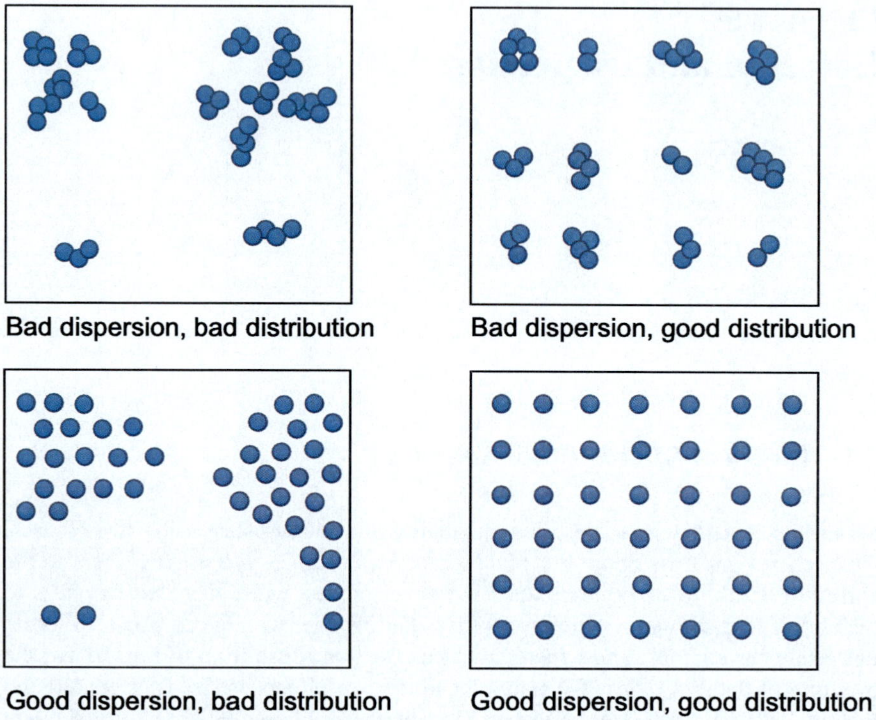

Fig. 5.1 Illustration of distribution and dispersion of particles in host materials. (Reprinted with permission from Manas-Zloczower 1997)

in this chapter are taken by TEMs, but the methods discussed herein apply to SEM images or optical images as well.

Figure 5.2 shows a TEM image of a nanocomposite sample. The dark image areas indicate the presence of nanoparticles or nanoparticle agglomerates. The TEM images are then processed by the image processing tool `ImageJ` (Ferreira and Rasband 2011). More sophisticated methods presented in the earlier chapters (Park et al. 2012, 2013) can be used for better results. The image processing tool produces the contours of the nanoparticles and nanoparticle agglomerates, which can be used to calculate their sizes and centroid locations for the subsequent quantification analysis of the mixing state.

Despite the desire of engineers to break down nanoparticle agglomerates into small pieces of similar sizes, the agglomerates are difficult to avoid in reality, as evident in the TEM image in Fig. 5.2. The resulting material inevitably contains particle agglomerates of various sizes and of irregular shapes. Obviously some physical processes disperse the particles better than other processes do, and it is indeed of great interest to material scientists to find out which process, or process setting, does the job the best.

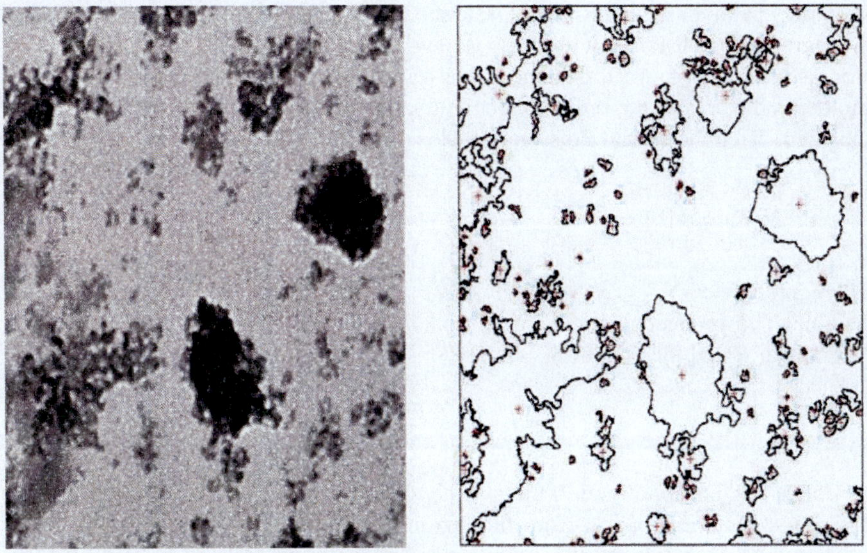

Fig. 5.2 TEM image of a nanocomposite sample. Left: the original TEM image; right: the extracted agglomerate boundaries and centroids. Centroids are indicated by the '+' signs. (Reprinted with permission from Dong et al. 2017)

5.2 Quadrat Method

As mentioned in Sect. 5.1, the location and dispersion analysis is related to how nanoparticles are distributed. The simplest mathematical treatment for analyzing the distribution may be to regard each nanoparticle as a dimensionless point and investigate the distribution of the points representing particles. Chapters 2 and 3 of Smith (2016) explain the basic ideas, models and procedures behind the spatial point process and the quadrat method. This section provides a brief review.

Imagine that the area of the host material is partitioned into many cells, and n particles are distributed over the cells. The uniform probability distribution implies that the particles have equal probabilities to show up in any of the cells. For each particle, one can define a set of the Bernoulli random variables $C_i(S_j)$, where $C_i(S_j) = 1$ indicates that particle i shows up in cell S_j, and $C_i(S_j) = 0$ implies otherwise. The number of particles in cell S_j can be counted as

$$N(S_j) = \sum_{i=1}^{n} C_i(S_j). \tag{5.1}$$

If there is a region consisting of a number of cells, such as $R = \{S_1, S_2, \ldots, S_m\}$, then the Bernoulli random variable, $C(\cdot)$, and the count function, $N(\cdot)$, can be similarly applied to this bigger region R.

In the spatial data analysis, the basic model of the complete spatial randomness is deemed a spatial Poisson process (Bailey and Gatrell 1995). The spatial Poisson process uses two parameters—one is known as the intensity and denoted by λ and another is the area of the region under study, denoted by A, or $A(R)$ if the region is specified. Then the spatial Poisson model is

$$\Pr[N(R) = k|\lambda] = \frac{[\lambda A(R)]^k}{k!} \exp\{-\lambda A(R)\}, \quad k = 0, 1, 2, \ldots, \tag{5.2}$$

where $N(R) = \sum_{j=1}^{m} N(S_j)$ and $A(R) = \sum_{j=1}^{m} A(S_j)$ if S_j's are mutually disjoint. The intensity parameter, λ, can be estimated by using the number of particles in R and the area of R, namely

$$\hat{\lambda} = \frac{n}{A}. \tag{5.3}$$

Under this baseline model, the statistical test is to see if the observed particle dispersion pattern deviates from the spatial Poisson distribution. Assume that a rectangular region R is partitioned into m equal-sized cells; the example in Fig. 5.3 has $m = 20$. The basic idea of the quadrat method is based on counting the number of particles in each cell and see if they are the same across the cells, subject to a random variation. Under the CSR assumption, i.e., under the null hypothesis, the expected number of particles in each cell should be the same. Following this intuition, it is apparent that the example in Fig. 5.3 is not spatially random.

Suppose that the whole area has the intensity of λ, which is estimated by using Eq. (5.3). Then, under the null hypothesis that the particles are distributed completely randomly, the number of particles per cell must be an independent sample from the Poisson distribution as specified in Eq. (5.2) with λ and the area size of A/m. Because λ and A/m are the same for all cells, the number of particles per cell follows an identical distribution.

The test for the CSR hypothesis can then be done by using a χ^2 goodness-of-fit test. If the observed number of particles for j-th cell is n_j, then the χ^2 statistic is

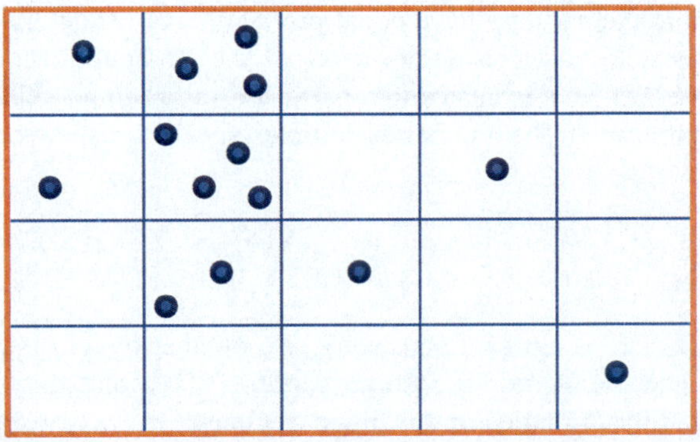

Fig. 5.3 Partition of an area into equal sized cells

$$\chi^2 = \sum_{j=1}^{m} \frac{(n_j - \frac{n}{m})^2}{n/m} = \sum_{j=1}^{m} \frac{(n_j - \bar{n})^2}{\bar{n}}. \tag{5.4}$$

The meaning of n/m is obvious. It is the expected number of particles per cell. Under the null hypothesis, all cells are expected to have the same number of particles, so that the expected number of particles per cell is simply the sample average, $\bar{n}$. So, the second expression in Eq. (5.4) follows naturally.

The χ^2 is to compare the difference between the observed number of particles per cell from its expected number. A large deviation indicates the violation of the CSR assumption. The χ^2 statistic follows asymptotically a chi-square distribution with $m - 1$ degrees of freedom. A statistical test can be readily carried out.

Consider the example in Fig. 5.3. If we number the cells as 1–5 for the first row, 6–10 for the second row and so on, then we have $N(S_1) = N(S_6) = N(S_9) = N(S_{13}) = N(S_{20}) = 1$, $N(S_{12}) = 2$, $N(S_2) = 3$, $N(S_7) = 4$, and $N(S)$ for the rest of cells is zero. The total number of particles is $n = 14$. The average number of particle per cell is $n/m = 0.7$. Using Eq. (5.4), we compute $\chi^2 = 34.6$, which is rather large. This test statistic corresponds to a p-value of 0.016, which, based on the conventional cut-off of 0.05, indicates a significant deviation from the completely random pattern.

Because the sample variance of the particle number across cells is

$$s^2 = \frac{1}{m-1} \sum_{j=1}^{m} (n_j - \bar{n})^2, \tag{5.5}$$

one can rewrite the expression in Eq. (5.4) as

$$\chi^2 = (m-1)\frac{s^2}{\bar{n}}. \tag{5.6}$$

Note that for a Poisson distribution, its variance is the same as its mean. What this means is that under the null hypothesis, the ratio, $s^2/\bar{n}$, is about one. This variance-to-mean ratio is known the *index of dispersion* (ID). When $s^2/\bar{n} < 1$, it means that the variation among cell-based counts are smaller than expected, suggesting possible manipulation or intervention (to make the particles more evenly distributed) instead of being entirely random, whereas when $s^2/\bar{n} > 1$, it means too much variation among cell-based counts, suggesting possible clustering of the particles. For the example in Fig. 5.3, $s^2/\bar{n} = 1.82$, which is greater than one, indicating particle clustering. This is consistent with the intuition based on visual inspection.

The quadrat method and the associated χ^2 test are not very powerful. Look at another example in Fig. 5.4. Although a visual inspection still leaves us with the feeling of nonuniform distribution or non-random pattern, the χ^2 statistic is now 18, translating to a p-value of 0.52, which is too large to reject the null hypothesis. The variance-to-mean ratio now is 0.95, rather close to one. The drawbacks of the

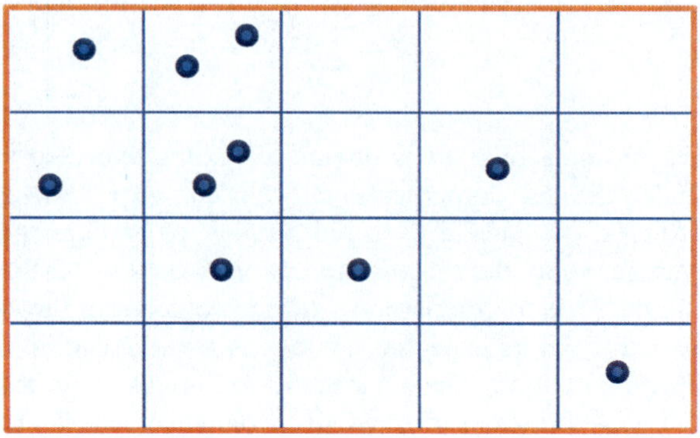

Fig. 5.4 Another example of particle distribution

quadrat method motivated researchers to develop a new school of methods, which are based on measuring the distance between particles.

5.3 Distance Methods

A more popular way to characterize the homogeneity of spatial point distribution is the distance-based approach. The basic idea is to calculate the pairwise Euclidean distances between the centroid locations of particles or particle agglomerates and use them as the inputs for calculating a scalar metric. The distance-based approach, in their default settings, assumes that each centroid is a dimensionless point. This assumption, once made, in effect neglects the sizes of particles or particle agglomerates.

One of the most popular distance methods in the spatial data analysis is Ripley's K function (Ripley 1976). There are a few other letter-named functions and some variants to K function that we will describe in the following subsections.

5.3.1 *The K Function and L Function*

The K function is defined as:

$$K(r) = \frac{\mathbb{E}(\text{number of extra points within distance } r \text{ of a randomly chosen point})}{\lambda}. \tag{5.7}$$

where $\mathbb{E}$ is the expectation operator. Recall that λ is the intensity of the point process, i.e., the number of points per unit area. $K(r)$ has a close form under many spatial point models. The most commonly used ideal model is the CSR model or the homogeneous spatial Poisson process; under this ideal model, $K_{\text{null}}(r) = \pi r^2$.

When it is unclear that the underlying spatial point process follows the CSR model, $K(r)$ can be estimated from an image realizing the spatial point process. First, the number of the points in the image is counted and denoted by n, and the area of the image is again denoted by A. The intensity λ can then be estimated by $\hat{\lambda} = \frac{n}{A}$, as in Eq. (5.3). We want to stress here that this $\hat{\lambda}$ is a *local* normalizing parameter as the microscopic images capturing nanoscale characteristics are always localized in the host material.

The K function can be estimated with $\hat{\lambda}$ and the pairwise distances of the n points. Let $d_{\ell_1\ell_2}$ denote the Euclidean distance between point ℓ_1 and ℓ_2, $\forall \ell_1, \ell_2 \in \{1, \ldots, n\}$. For the TEM images in Fig. 5.2, $d_{\ell_1\ell_2}$ is computed using the centroid coordinates of particles or particle agglomerates, ℓ_1 and ℓ_2, which are available to us after the image processing step. For point ℓ_1, the number of the other points within distance r from point ℓ_1 can be counted. The counts can be achieved for $\ell_1 = 1, \ldots, n$, and the counts can be averaged to estimate the K function,

$$\widehat{K}(r) = \frac{1}{A\hat{\lambda}^2} \sum_{\ell_1=1}^{n} \sum_{\ell_2 \neq \ell_1}^{n} \mathbb{1}\{d_{\ell_1\ell_2} < r\}, \tag{5.8}$$

where $\mathbb{1}\{x\}$ is the indicator function, namely $\mathbb{1}\{x\} = 1$ when condition x is satisfied or $\mathbb{1}\{x\} = 0$ otherwise. A larger $\widehat{K}$, as compared with the K value under the CSR assumption, implies clustering. The larger the $\widehat{K}$ value, the severer the clustering.

The above expression may run into problems when point ℓ_1 is very close to the boundary of an image, and consequently, part of the circle of radius r with the point may be outside the image area. One commonly used correction is Ripley's isotropic correction (Ripley 1991), which is to assign a value between zero and one to the circles that have part of their circumference outside the image area. So the corrected Ripley K function reads,

$$\widehat{K}(r) = \frac{1}{A\hat{\lambda}^2} \sum_{\ell_1=1}^{n} \sum_{\ell_2 \neq \ell_1}^{n} w_{\ell_1\ell_2} \mathbb{1}\{d_{\ell_1\ell_2} < r\}, \tag{5.9}$$

where $w_{\ell_1\ell_2}$ is the circumference proportion within the image area of a circle that has a radius of $d_{\ell_1\ell_2}$ and centers at point ℓ_1.

The K function under CSR is a quadratic curve. Besag (1977) proposes to transform the K function to a straight line through

$$L(r) = \sqrt{\frac{K(r)}{\pi}}, \tag{5.10}$$

which is known as the L function. The function is often plotted as $L(r) - r$, namely,

$$L(r) = \sqrt{\frac{K(r)}{\pi}} - r.$$

Marcon and Puech (2003) interprets the L function in the following way—"$L(r) = l$ *means that as many neighbors are found around reference points up to distance r as would be expected at distance $r + l$ if neighbor points are well distributed independently of reference points.*"

We keep Eq. (5.10) as the formal definition of the L function because we want to be consistent with how it is defined with the corresponding function in the R package.

5.3.2 The K_{mm} Function

Researchers notice that the treatment of volumic nanoparticles as volume-less points can cause inaccurate quantification of homogeneity. Penttinen et al. (1992) introduce a K_{mm} function to address this issue. It is an extension of the K function by adding a weight to each point ℓ_1 or ℓ_2. Let us first re-write the K function in the following way:

$$K(r) = \frac{\mathbb{E}\left(\sum_{\ell_1=1}^{n}\sum_{\ell_2\neq\ell_1}^{n} \mathbb{1}\{d_{\ell_1\ell_2} < r\}\right)}{A\lambda^2}. \tag{5.11}$$

The K_{mm} function is a weighted version of Eq. (5.11), i.e.,

$$K_{mm}(r) = \frac{\mathbb{E}\left(\sum_{\ell_1=1}^{n}\sum_{\ell_2\neq\ell_1}^{n} \mu_{\ell_1}\mu_{\ell_2}\mathbb{1}\{d_{\ell_1\ell_2} < r\}\right)}{A\lambda^2\mu^2}, \tag{5.12}$$

where μ_{ℓ_1} and μ_{ℓ_2} are the weights associated with the ℓ_1-th and ℓ_2-th points, respectively, and μ is the mean of all weights. When it is applies to nanomaterial applications, one can use the particle size as the weights, μ_{ℓ_1} and μ_{ℓ_2}. The estimate of the $K_{mm}(r)$ can be corrected as in Eq. (5.9),

$$\widehat{K}_{mm}(r) = \frac{\sum_{\ell_1=1}^{n}\sum_{\ell_2\neq\ell_1}^{n} \mu_{\ell_1}\mu_{\ell_2}w_{\ell_1\ell_2}\mathbb{1}\{d_{\ell_1\ell_2} < r\}}{A\hat{\lambda}^2\hat{\mu}^2}, \tag{5.13}$$

where $\hat{\mu}$ is the sample average of the weights. Apparently, when the weights of each point are the same, i.e. $\mu_{\ell_1} = \mu_{\ell_2} = \mu$, the K_{mm} is the same as Ripley's K function. The value of K_{mm} under null hypothesis is obtained when both weights and locations are completely randomly assigned.

The discussion above implies that the K function may still work effectively, if the nanoparticles, despite not dimensionless, are of a uniform shape (e.g., all round) and size (the same radius) on a given material sample as well as across different samples, because $\mu_{\ell_1}\mu_{\ell_2}$ would be uniform for that case. It is indeed reported so in the study by Li et al. (2014), who show that when the particle size is homogenous

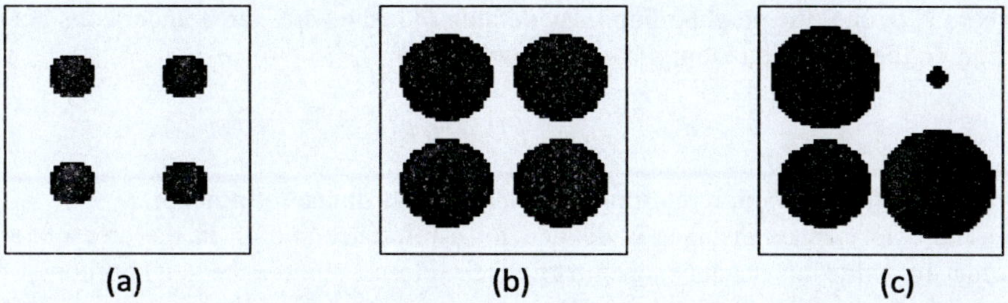

Fig. 5.5 Three simulated images. Window size is normalized to be [0, 1] × [0, 1]. The centers of the agglomerates in all images are {(0.25, 0.25), (0.75, 0.25), (0.25, 0.75), (0.75, 0.75)}. (**a**) The radius of each agglomerate is 0.1; (**b**) The radius of each agglomerate is 0.2; (**c**) The radii of the agglomerates, from top-left to bottom-right, are 0.23, 0.04, 0.2, and 0.25, respectively. (Reprinted with permission from Dong et al. 2017)

and the shape of particles are spherical, the K function can be a reasonable metric to quantify the nanoparticle mixing state in a nanocomposite sample.

But when the particles are not uniform in size and shape, the analysis and conclusion could be complex. Figure 5.5 presents a number of simulated images with different agglomerate types. Comparing this figure and Figs. 5.1 and 5.5a resembles the "good dispersion, good distribution" graph (bottom-right in Figs. 5.1) and 5.5b resembles the "bad dispersion, good distribution" graph (top-right in Fig. 5.1), whereas Fig. 5.5c resembles the "bad dispersion, bad distribution" graph (top-left in Fig. 5.1). When applying the original K function to the three images, it returns exactly the same value as the centroids of the agglomerates are the same. This is expected, as the K function ignores the dispersion (size) aspect altogether. When applying the K_{mm} function for which one uses the agglomerate size as the weights, it does distinguish Fig. 5.5a from Fig. 5.5c, or 5.5b from Fig. 5.5c, but it still cannot distinguish Fig. 5.5a from Fig. 5.5b.

5.3.3 The *F* Function and *G* Function

In addition to the K and L functions (and their variants), the F and G functions are two other commonly used functions in the spatial point pattern analysis.

In order to explain these two functions, we first introduce a few distance notations. The pairwise distance is already introduced above as $d_{\ell_1\ell_2}$. Two other distances to be used are the nearest neighborhood distance of a point and the empty space distance.

The nearest neighborhood distance of ℓ_i is denoted by $d_{\mathbb{N}}(\ell_i)$ and defined as

$$d_{\mathbb{N}}(\ell_i) = \min_{j \neq i} d(\ell_i, \ell_j) \quad \forall \ell_j \in \mathbb{N}, \tag{5.14}$$

where $\mathbb{N}$ defines the neighborhood. By default, $\mathbb{N}$ is the whole area under consideration. In the above equation, $d(\ell_i, \ell_j)$ is defined as

$$d(\ell_i, \ell_j) = \|\psi(\ell_i) - \psi(\ell_j)\|,$$

where $\psi(\ell_i)$ is a function returning the location coordinates of point ℓ_i.

The empty space distance is defined for a reference site, u, in the area under study. To define this distance, one first finds the closest point in the neighborhood of u and then uses their pairwise distance as the radius to draw a circle around u. This circle contains the so-called empty space for u because there is no point inside the circle.

Let us denote a collection of spatial points by $\mathbb{L} = \{\ell_1, \ldots, \ell_n\}$. The empty space distance is denoted by $d_{\text{ESD}}(\cdot)$ and defined as

$$d_{\text{ESD}}(u, \mathbb{L}) = \min_i \|\psi(u) - \psi(\ell_i)\| \quad \forall \ell_i \in \mathbb{L}. \tag{5.15}$$

The F function is introduced to describe probabilistically the size of the empty spaces over the study area and is in fact referred to as the empty space function. It is defined as the cumulative distribution function of the empty space distance, i.e.,

$$F(r) = \Pr[d_{\text{ESD}}(u, \mathbb{L}) \le r]. \tag{5.16}$$

Under the null hypothesis, i.e., the CSR assumption, $F_{\text{null}}(r) = 1 - \exp(-\lambda\pi r^2)$.

Here u is an arbitrary location in the study area. To empirically estimate the F function, one can create a grid of locations, $\{u_1, u_2, \ldots, u_m\}$. Together with the observed empty space distances, one can estimate the F function by

$$\widehat{F}(r) = \frac{1}{m}\sum_{j=1}^{m} w(u_j) \cdot \mathbb{1}\{d_{\text{ESD}}(u_j, \mathbb{L}) \le r\}, \tag{5.17}$$

where $w(\cdot)$ is an edge correction term, similar to $w_{\ell_1\ell_2}$ in Eq. (5.9) but could have used a different mechanism for correction.

To use the F function, one can plug $\hat{\lambda}$ into $F_{\text{null}}(r)$ and compare the corresponding value of $\widehat{F}(r)$ with $F_{\text{null}}(r)$. If $\widehat{F}(r) < F_{\text{null}}(r)$, it means that the observed empty space at distance threshold r occurs less likely than what the spatial Poisson model anticipates, leading to the suspicion of point clustering.

The G function has a strong relationship with the F function. The G function is the cumulative distribution function of the nearest neighbor distance for a typical point in the area. It is defined as

$$G(r) = \Pr[d_{\text{ESD}}(\ell, \mathbb{L}\backslash\{\ell\}) \le r | \ell \in \mathbb{L}], \tag{5.18}$$

where $d_{\text{ESD}}(\ell, \mathbb{L}\backslash\{\ell\})$ is the shortest distance from an arbitrary point ℓ in $\mathbb{L}$ to the rest of the points in the collection, excluding ℓ itself. Under the null hypothesis, the G function has the same expression as $F_{\text{null}}(\cdot)$, i.e., $G_{\text{null}}(r) = 1 - \exp(-\lambda\pi r^2)$.

Empirically, the G function can be estimated by

$$\widehat{G}(r) = \frac{1}{n}\sum_{i=1}^{n} w(\ell_i) \cdot \mathbb{1}\{d_{\mathbb{L}}(\ell_i) \leq r\}, \tag{5.19}$$

where $w(\cdot)$, again, is an edge correction term and the subscript "$\mathbb{L}$" means that the neighborhood $\mathbb{N}$ is now the set of $\mathbb{L}$.

The interpretation of the G function is the reverse of that of the F function. If $\widehat{G}(r) > G_{\text{null}}(r)$, it means that the observed nearest neighborhood distance at distance threshold r occurs more often than what the spatial Poisson model anticipates, leading to the suspicion of point clustering.

5.3.4 Additional Notes

The package `spatstat` contains the `R` functions to compute the three distance metrics and the four letter-named functions. The spatial points information should be stored in a point pattern object, say `X`. Then `pairdist(X)`, `nndist(X)`, `distmap(X)` compute, respectively, the pairwise distance, the nearest neighborhood distance, and the empty space distance. The `Fest`, `Gest`, `Lest` and `Kest` compute the F, G, L, and K functions, respectively. For a comprehensive instruction of using the `spatstat` package, please refer to Baddeley and Turner (2005) and Baddeley (2008).

One more note is about the difference between the quadrat method and the distance methods. They characterize different aspects of spatial point patterns—the first-order property versus the second-order property. The first order property measures the distribution of points in a study region and can be characterized by the intensity. The quadrat method looks at the first-order effect of the underlying spatial process. It is a local intensity-based method, and uses the intensities associated with local cells and the difference among them to signal the homogeneity in point dispersion or the lack thereof. The distance-based methods, like the K function, look at the second-order effect, which measure more directly the tendency of points to appear either clustered or randomly scattered. It is therefore not surprising that the distance-based methods are used more commonly for characterizing the dispersion and distribution of spatial point patterns.

5.4 A Revised *K* Function

In the spatial statistical literature, a point in the spatial point process is deemed dimensionless; this is true for both the quadrat method and the distances. The coordinates of the points observed in the underlying process are used as inputs to the subsequent analysis without the need to account for the size associated with the points (as the points do not have a size in the first place). When we discuss the K_{mm} function in Sect. 5.3.2, we mention that researchers have noticed the need to associate some properties with a point, because having that association could very well change the perception of homogeneity—this is the principal reason behind the proposal of the K_{mm} function. We also illustrate in Fig. 5.5 that while the K_{mm} function may serve some applications, it appears to fall short on some other applications.

In the research originally reported in Dong et al. (2017), while using the K function as the base method, Dong et al. (2017) propose two additional actions to account for the size (dispersion) effect of particle agglomerates. The first is to discretize the particle agglomerates into small enough, fine-scale particles and extract the centroid locations of the small particles for use in the K function. It turns out that this discretization action alone is insufficient to solve the problem. An adjustment is needed on the normalizing parameter in the K function. Dong et al. (2017) refer to the revised K function as the $\widetilde{K}$ function.

5.4.1 Discretiztaion

In revising the K function to account for the size effect of particle agglomerates, the first line of thought is to discretize the agglomerates into much smaller disjoint quadrats. Then each quadrat is treated as a new particle with its own centroid. By this action, the existence of large agglomerates is translated into a large number of small quadrats that are closely clustered together. Presumably, Ripley's K function, once applied to the small-sized quadrats, can reflect the closeness among the quadrats, so as to distinguish the particle mixing states with and without aggregation. When the quadrat sizes are small enough, then the remaining size effect, albeit not perfectly dimensionless, would hopefully no longer be significant.

Dong et al. (2017) show that if using smaller enough quadrats, the discretization under a discretized version of CSR provides good enough approximation of the K function under the CSR assumption.

Suppose that a TEM image is made up of $U \times U$ pixels. Dong et al. (2017) discretize the image by quadrats of size $k \times k$ pixels, producing $m = (U/k)^2$ number of quadrats in the image. For simplicity, let U be a multiple of k to avoid using quadrats of a different size near the boundary.

Image processing steps identify the particle agglomerate from a TEM image, so that one can assign a value of one for the particle pixels and a value of zero for each of the m pixels. Dong et al. (2017) label a quadrat of size $k \times k$ to one (particle quadrat) if at least $\lfloor \frac{k \times k}{2} \rfloor + 1$ pixels are of one and label the quadrat to zero (background quadrat) otherwise. Let q denote the fraction of the particle quadrats. Under the discretized version of CSR, different quadrats are independent from one another, each quadrat has a probability of q for being a particle and a probability of $1 - q$ for being the background. After the discretization, the centroid coordinates of the particle quadrats are fed into the calculation of the K function.

Let K_{discrete} be the theoretical value of the K function in the discretized image. Following the definition of the K function, Dong et al. (2017) derive

$$K_{\text{discrete}}(r) = \frac{(Q(\frac{Ur}{k}) - 1)\, q}{mq} = \frac{Q(\frac{Ur}{k}) - 1}{m}, \tag{5.20}$$

where $Q(h)$ is the number of quadrats inside the boundary of a circle of radius h.

To show the connection between $K_{\text{discrete}}(r)$ and the original $K(r)$, the key is to understand the function $Q(\cdot)$, which is known as the Gauss circle problem (Hardy 1999), and its solution is given as

$$Q(h) = \pi h^2 + \epsilon(h), \tag{5.21}$$

where $\|\epsilon(h)\| \leq a \cdot h$ and a is a constant.

Recall that under CSR, $K(r) = \pi r^2$. Dong et al. (2017) show that, under CSR,

$$\begin{aligned}
\|K_{\text{discrete}}(r) - K(r)\| &= \left\| \frac{\pi\left(\frac{U}{k} \cdot r\right)^2 + \epsilon\left(\frac{U}{k} \cdot r\right) - 1}{\left(\frac{U}{k}\right)^2} - \pi r^2 \right\| \\
&= \left\| \frac{\epsilon\left(\frac{U}{k} \cdot r\right) - 1}{\left(\frac{U}{k}\right)^2} \right\| \leq \left\| \frac{\epsilon\left(\frac{U}{k} \cdot r\right)}{\left(\frac{U}{k}\right)^2} \right\| + \left(\frac{k}{U}\right)^2 \\
&\leq \frac{a \cdot \frac{U}{k} \cdot r}{\left(\frac{U}{k}\right)^2} + \left(\frac{k}{U}\right)^2 = \frac{a \cdot k}{U} r + \left(\frac{k}{U}\right)^2.
\end{aligned}$$

For relatively small k's, $K_{\text{discrete}}(r)$ is very close to $K(r)$ under CSR. Figure 5.6 presents a numerical example in which the image size is normalized to $[0, 1] \times [0, 1]$ and the image is made up of 1024×1024 pixels (i.e., $U = 1024$). Dong et al. (2017) set k to be 512, 128, and 16, respectively, for examining the difference between $K(r)$ and the corresponding $K_{\text{discrete}}(r)$. Evidently, $K_{\text{discrete}}(r)$ and $K(r)$ become indistinguishable when k is 16 or smaller.

In practice, 1024×1024 pixels are a typical size of nanomaterial and other material images. For such an image size, the numerical analysis shows that $k = 16$ is sufficiently small. In the later analysis, the value of k is chosen based on the physical size of a stand-alone nanoparticle (about 5×5 pixels in size), which is even smaller than a 16×16 quadrat in a 1024×1024-pixel image. For other applications,

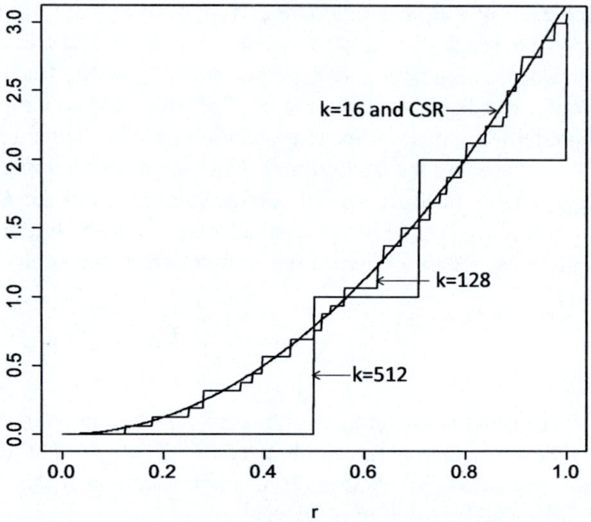

Fig. 5.6 *K* function under CSR and K_{discrete} with different *k*'s. (Reprinted with permission from Dong et al. 2017)

once given an image size, one can use the aforementioned approximation formula to find out how small a *k* needs to be for discretization and then choose its value accordingly.

5.4.2 Adjustment of the Normalizing Parameter

The discretization addresses only part of the problems arising from the particle's aggregation. The other part comes from the fact that imaging instruments only measure local areas of the host material. When particles aggregate, the intensity of particles observed per local area differs. In the material mixing problem, the nanoparticles blended into the host material are of a known quantity (measured in terms of weight or volume). Under the null hypothesis that the nanoparticles are dispersed and distributed homogenously throughout the host material, the expected number of particles appearing under a given view field of TEM, namely, in a TEM image of the same size, should be the same. When the number of particles in an image is substantially different from that of others, this appearance itself is an indication of inhomogeneity.

When using the *K* function, including the discretized *K* function, the normalizing parameter, λ, is the average intensity of points per image. If one observes an image containing particles with bad dispersion (large agglomerates) but reasonable distribution (agglomerates spacing evenly), the fact that this local image has more particles (than some other images) does not alert the *K* function about its implication of inhomogeneity, because the higher local particle intensity will be normalized. Not only unable to alert the implication of inhomogeneity, the use of a local λ in fact tends to cause the discretized K function to deem an image containing large agglomerates as being distributed more uniformly, which is rather counter-intuitive.

This can be better understood by looking at the images in Fig. 5.5a, b. When using the original K function, the two images were deemed the same in terms of spatial point distribution, as the size difference of the point agglomerates is simply ignored and their centroid positions are the same. When using the discretized K function, they produce different assessment outcomes, in which Fig. 5.5b, which contains large agglomerates, is considered more uniformly distributed. This is because with the presence of large agglomerates, the denominator of the discretized K function increases, by the amount of extra number of points that the large agglomerates have over the small agglomerates, resulting in a K value closer to CSR. The numerical analysis confirms this intuition.

So the question becomes what one should use as the normalizing parameter when local images are observed but the global homogeneity is to be assessed. Dong et al. (2017) believe that a global λ needs to be used. Then, the question is how the global λ can be estimated. Dong et al. (2017) think the estimation is going to be application dependent. In the following, we discuss how to do this for the material mixing applications; the idea could be applicable broadly to other applications. In the cases that the global λ cannot be estimated using the content addition parameters as in the mixing process, Dong et al. (2017) recommend using the average of the local intensities associated with all available view fields (i.e., all local images attained in an application).

In the material mixing applications, denote by c the volumetric portion of nanoparticles (or other additive agents) that are mixed into a host material. Assume that a nanoparticle occupies a quadrat of size $k \times k$. When nanoparticles are indeed uniformly mixed, the quadrats in the image have probability of c to be a particle or probability of $1 - c$ to be the background. The closeness measure in the numerator of the K function becomes

$$E(\text{number of extra points within } r \text{ of a randomly chosen point}) = (Q(r) - 1)c. \tag{5.22}$$

Obviously one need to normalize the above expected value by c, leading to a revised K function, i.e., the $\widetilde{K}$ function, such that

$$\widetilde{K} = \frac{E(\text{number of extra points within } r \text{ of a randomly chosen point})}{c}. \tag{5.23}$$

Similar to how the K function is estimated, $\widetilde{K}$ can be estimated as follows:

$$\widehat{\widetilde{K}} = \frac{\frac{1}{n}\sum_{\ell_1=1}^{n}\sum_{\ell_2 \neq \ell_1} w_{\ell_1\ell_2}\mathbb{1}(d_{\ell_1\ell_2} < r)}{c}. \tag{5.24}$$

In summary, the global normalizing constant in the material mixing applications is simply the volumetric portion of the additive agents, which can be readily calculated when one knows how much host material and additive material are used in the mixing as well as their physical densities.

5.4.3 Relation Between Discretized K and $\widetilde{K}$

Dong et al. (2017) use a set of simulated images to understand how $\widetilde{K}$ behaves. The image simulation tool that Dong et al. (2017) employ is the Ising model (Winkler 2003), which is developed for characterizing the dependency among spatial binary data based on the theory of Markov random field. Specifically, Dong et al. (2017) use the `PottsUtils` package in `R`, developed by Feng and Tierney (2014), to generate spatial data samples based on the Ising model and then turn that into a black and white image. One key parameter used in the `PottsUtils` package is called β, which characterizes the aggregating or clustering level of the spatial points. When $\beta = 0$, it means that a given point is independent from its neighboring vertices. The larger the β is, the more clustered the points are. Figure 5.7 presents a number of image examples simulated using the Ising model with different β's.

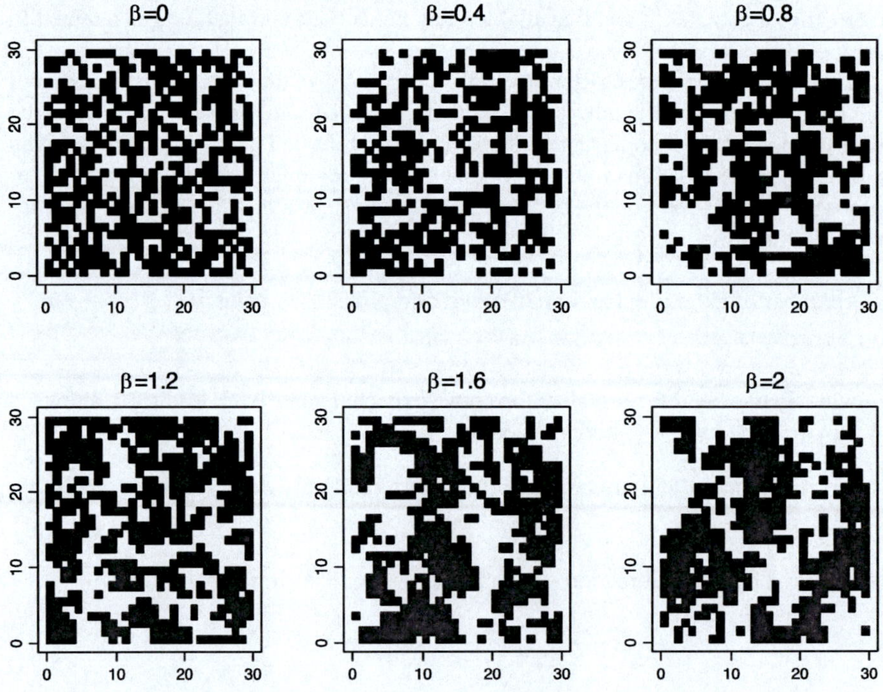

Fig. 5.7 Simulated images using the Ising model with different β values. (Reprinted with permission from Dong et al. 2017)

Dong et al. (2017) conduct the following analysis using the Ising model-based simulated images:

1. Generate a $B \times B$-pixel image using the Ising model, starting with $\beta = 0$;
2. Estimate c, the global point intensity, using the $B \times B$-pixel whole image, which is to be used in $\widetilde{K}$;
3. Random sample a $b \times b$-pixel sub-image from the whole image;
4. Compute the discretized K function for the sub-image;
5. Compute the $\widetilde{K}$ function for the sub-image using the c estimated in Step 2;
6. Plot the discretized K and $\widetilde{K}$ curves on the same plot;
7. Repeat 50 times Steps 3 to 6, so that discretized K and $\widetilde{K}$ form two clusters of curves;
8. Plot the CSR curve on the same plot;
9. Change β to the next value and repeat Steps 1 to 8.

In the analysis, Dong et al. (2017) choose $B = 500$ and $b = 50$. The result is presented in Fig. 5.8. Other value combinations were tried but the insights are the same, so we omit the plots using other value combinations.

Observing the curves in Fig. 5.8, one notices that when $\beta = 0$, both K curves and $\widetilde{K}$ curves are tightly clustered around the CSR curve, as they should be. When β starts to increase, meaning that the points begin to aggregate, the difference between the two sets of curves becomes more and more pronounced. The $\widetilde{K}$ curves appear sensitive responding to even small non-zero values of β. For the same β, the $\widetilde{K}$ deviates more from the CSR curve and has a greater variability. This greater variability in $\widetilde{K}$ curves is expected, as $\widetilde{K}$ responds to both dispersion and distribution effects, whereas the discretized K is shrunk by the presence of large-size agglomerates.

The third observation is that some (nearly half) of the $\widetilde{K}$ curves lies above the CSR curve, while the others below the CSR curve. By contrast, all discretized K curves are above the CSR curve. It turns out that when the local point density of the sub-image is smaller than the global point density c, the resulting $\widetilde{K}$ curve is below the CSR curve, whereas when the local density is larger, the resulting $\widetilde{K}$ is above the CSR curve. This phenomenon of $\widetilde{K}$ is expected, too. In characterizing the mixing homogeneity, one should compare the $\widetilde{K}$ curve to the reference CSR curve—the further it is above the CSR curve or the further it is below the CSR curve, both indicate inhomogeneity.

5.4.4 Nonparametric Test Procedure

When the images of material samples show a difference, practitioners would like to know whether the difference is significant beyond the level of background randomness. The following statistical testing procedure is devised to address this question.

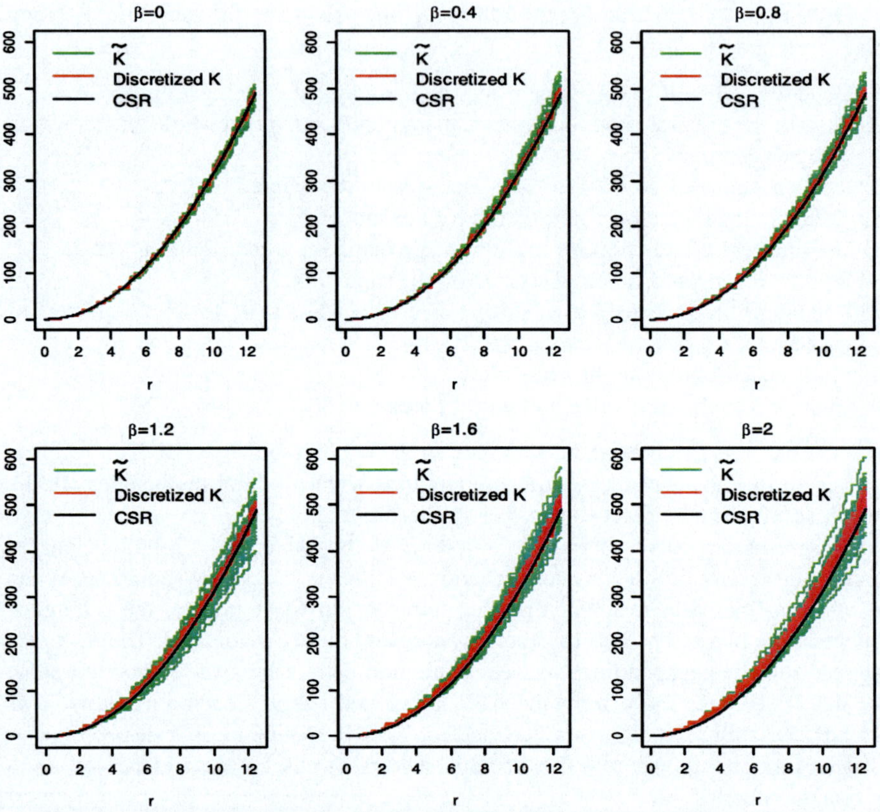

Fig. 5.8 Discretized K versus $\widetilde{K}$ under different degrees of aggregation. (Reprinted with permission from Dong et al. (2017))

Because Dong et al. (2017) conduct pairwise comparisons, it implies that two groups of images are involved. Let i be the group index and j be the image index within a group. Denote the number of images in the i-th group by m_i, so that $i = 1, 2$ and $j = 1, \ldots, m_i$. To each image, a spatial homogeneity characterizing function, including both the original K and the revised $\widetilde{K}$, can be applied, and the outcomes are denoted by K_{ij} and $\widetilde{K}_{ij}$, respectively. Recall that the corresponding symbols with a hat notation, such as $\widehat{\widetilde{K}}_{ij}$, denote the respective estimated values. Let n_{ij} denote the number of particles in the j-th image of the i-th group. Further denote by $n_i = \sum_{j=1}^{m_i} n_{ij}$ the number of particles in the i-th group and by $n_t = \sum_{i=1}^{2} n_i$ the total number of particles in the entire comparison.

Diggle et al. (1991, 2000) propose two test statistics, D_1 and D_2, to be used with Ripley's K function, respectively, for testing the spatial homogeneity difference between multiple groups of images. Dong et al. (2017) use the same test statistics but with the K function replaced by $\widetilde{K}$. The following expressions follow Diggle et al. (1991, 2000)'s original definition, except that K is replaced by $\widetilde{K}$:

$$D_1 = \sum_{i=1}^{2} \int_0^{r_0} \left(\sqrt{\bar{K}_i(r)} - \sqrt{\bar{K}(r)}\right)^2 dr, \tag{5.25}$$

$$D_2 = \sum_{i=1}^{2} n_i \int_0^{r_0} \frac{1}{r^2} (\bar{K}_i(r) - \bar{K}(r))^2 dr, \tag{5.26}$$

where r_0 is the longest distance with which the $\widetilde{K}$ (or K) is evaluated, and

$$\bar{K}_i(r) = \frac{1}{n_i} \sum_{j=1}^{m_i} n_{ij} \widehat{\widetilde{K}}_{ij}(r), \text{ for } i = 1, 2, \text{ and } \bar{K}(r) = \frac{1}{n_t} \sum_{i=1}^{2} n_i \bar{K}_i(r) \tag{5.27}$$

are the average $\widehat{\widetilde{K}}$ within a group and the grand average of the two groups, respectively. The r_0 is often chosen according to Ripley's rule of thumb (Venables and Ripley 2002), which says that r_0 is chosen to be one quarter, in size, of the shorter side of the image window. In computation, the above integrals are approximated by the summation of the integrand over one thousand equally spaced r values.

To compute the p-value of the test statistics, Diggle et al. (1991) initially suggest a bootstrap procedure. In more recent work (Diggle et al. 2000), they argue instead that a permutation procedure works better. Dong et al. (2017) choose to employ a permutation procedure to compute the statistical significance level of both D_1 and D_2. The permutation test entails the following steps:

1. Apply $\widetilde{K}$ function to each image in the two groups and compute $\widehat{\widetilde{K}}_{ij}$, for $i = 1, 2$ and $j = 1, \ldots, m_i$.
2. Compute D_1 and D_2 statistics using the values from Step 1. Refer to either of the statistics as T_{observed}.
3. Permute the images across the two groups. To do so, one can label the images in the two groups sequentially as $\{1, 2, \ldots, m_1, m_1 + 1, \ldots, m_1 + m_2\}$. Randomly shuffle the sequence of the numbers. Take the images whose labels correspond to the first m_1 numbers in a shuffled set and form the new group 1, while the remaining images form the new group 2.
4. Repeat Steps 1 and 2 on the two new groups. Refer to the resulting statistics as T.
5. Repeat Steps 3 and 4 H times and obtain H values of either statistics, namely T_η, for $\eta = 1, \ldots, H$.
6. Compare T_{observed} with $\{T_\eta\}_{\eta=1}^H$. If T_{observed} ranks the e-th largest among $\{T_\eta\}_{\eta=1}^H$, then the resulting p-value is approximated by $\frac{e}{H}$.

In practice, because of the need to reduce measurement cost and time, there may be sometimes only a single image of the material taken under a given condition. This means that two single images are to be compared with one another to differentiate the spatial homogeneity under their respective conditions. One can in fact still use the steps outlined above to conduct the statistical test by following the image partitioning idea in Hahn (2012). The action undertaken is to partition each image into 3×3 sub-images of equal sizes, so that two groups of images are formed with $m_1 = m_2 = 9$. Then, the statistics and steps presented above can be applied.

A final note is that the nonparametric test statistics for both $\widetilde{K}$ and K are in the form of a sum of squares. In light of the observations made in Sect. 5.4.3 that $\widetilde{K}$ has a greater variability around the CSR curve as β starts deviating from zero, this implies that $\widetilde{K}$ is more sensitive to signaling point inhomogeneity when inhomogeneity is present.

5.5 Case Study

Dong et al. (2017) conduct a quantification study of the mixing states using TEM images of nanocomposite materials. To produce the material, the chemicals, butyl acetate, three-functional trimethylolpropane triacrylate and silica nanoparticles, are mechanically mixed into silica nanoparticle suspension. Then the suspension is poured to a bead mill machine to further break up nanoparticles; for more details about a typical bead mill process, please refer to Wang and Forssberg (2006).

The milling time varies from 5 to 90 min. After milling, the nanocomposite is diluted by butyl acetate. The reason for dilution is that the nanoparticles in the original material cannot be properly imaged by TEM. Engineers believe that the nanoparticles' mixing state or the clustering level after the dilution is still a good representation of what is in the original material, although this assumption needs further examination.

A drop of the diluted nanoparticle suspension is casted onto a carbon-coated copper grid and dried at room temperature. The mixing and morphology of the nanoparticles are observed using a `FEI Tecnai F20` TEM. The nanoparticle content as deposited onto the carbon-coated copper-grid sample holder is 0.00124 in terms of volumetric ratio. The diameter of a stand-alone nanoparticle is around 13 nm.

From the images presented later, it is apparently not true that the longer engineers run the bead-milling process, the better dispersed and distributed the nanoparticles will be. One question of interest to engineers is to find out the optimal time length that this bead-milling process needs to run in order to get the most homogeneous dispersion of nanoparticles. Experiments are conducted with the bead-milling process running for multiple time lengths. TEM images of material samples are taken at each time point.

Two sets of experiments are conducted. In the first one, a single TEM image is taken at six time points of 5, 10, 15, 35, 60, and 90 min, respectively. The TEM image is taken at a randomly chosen location of the material sample. In the second experiment, the same process is carried out, but engineers this time want to take images of the material sample at 0 min, i.e., before the bead-milling process runs. Another difference in the second experiment is that multiple images, ranging from 12 to 18 in count, are taken at randomly chosen locations on the sample. To save measurement expenses in the second experiment, two intermediate time points, 15 and 60 min, are removed. There are no substantial reasons why these two time points are chosen for removal; instead it is done based on the engineer's intuition. Dong et al. (2017) analyze the two circumstances separately.

5.5.1 A Single Image Taken at a Given Time Point

Figure 5.9 presents the six single TEM images taken from the material sample at each time point specified above. The images are of 1024 × 1024 pixels. Dong et al. (2017) use `ImageJ` to extract the centroid locations and area information of each particle or agglomerate. The images are discretized by using $k = 5$; the value of k is chosen so that each quadrat is close to the actual size of a stand-alone nanoparticle (about 13 nm in diameter). Because 1024 is not divisible by 5, Dong et al. (2017) discretize a 1020 × 1020 subimage and set those pixels on the boundary to be the background. The normalizing parameter in $\widetilde{K}$, c, is set to be the actual nanoparticle volumetric ratio, that is, $c = 0.00124$.

For each image, Dong et al. (2017) compute the original K function curve, the weighted K_{mm} function curve, the discretized K function curve, and the $\widetilde{K}$ function curve. The `dbmss` package (Marcon et al. 2014) is used to implement the K_{mm} function, whereas the `spatstat` package is used to compute the K function. Dong et al. (2017) estimate $\widetilde{K}$ values based on the K function values. For these real images, Dong et al. (2017) use r in the unit of number of image pixels. Dong et al. (2017) set $r_0 = 1024/4 = 256$. Figure 5.10 presents the seven curves (six curves for the six TEM images plus the curve under CSR) for each function.

The rank order of spatial homogeneity can be concluded from observing how far a specific curve is away from the curve of the ideal model. When using the original Ripley's K and K_{mm}, most of the curves are clustered on the CSR curve, making it difficult to differentiate the mixing states in the corresponding images. In both K and K_{mm} plots, the 90-min curve is a flat line because there is a giant particle agglomerate in the middle with a couple of small particle blocks scattered in the periphery. Because neither K nor K_{mm} makes use of the particle agglomerate size information, the agglomerates are simply reduced into dimensionless points at their centroids whose distances in between are greater than r_0. Consequently, their values are always zero for the range of r shown in the plot. K_{mm} curve does differentiate the 60-min state from the rest of the states, as well as from the 90-min state, whereas the original K fails to differentiate anything but the 90-min state.

When the discretization is applied, the discretized K function improves upon both the original K and K_{mm}, as the differences in the curves corresponding to

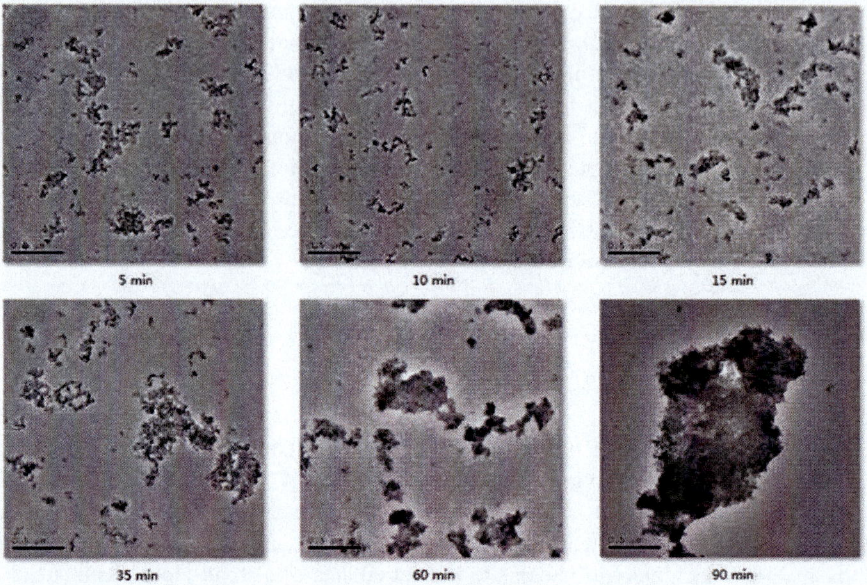

Fig. 5.9 TEM images of the material sample taken at six different time points. (Reprinted with permission from Dong et al. 2017)

different states are more noticeable. Using the distance away from the CSR curve as the criterion, the discretized K suggests that the order of the mixing state, from most homogenous to least, is 5, 10, 15, 60, 35, and then, 90 min.

When both discretization and the new normalizing parameter are applied, the result is the $\widetilde{K}$ function curves. The separation of the $\widetilde{K}$ curves is more pronounced, making it easier to tell the difference between two mixing states. The mixing state order suggested by $\widetilde{K}$ appears different from that suggested by discretized K. Using $\widetilde{K}$, one would conclude that the 10 min running of the bead-milling process produces the most homogenous mixing of the particles, followed by 5, 15, 35, 60 min, and then, 90 min.

Dong et al. (2017) state that when presenting these images to a group of material scientists, the 10 min image is unanimously deemed most preferable, while the preference between 5 and 15 min is evenly split. The preference order over 35, 60, and 90 min is clearly lower than the other three cases; on this aspect, both $\widetilde{K}$ and discretized K reach the same conclusion. But the difference is that $\widetilde{K}$ favors 35 min over 60 min, whereas discretized K does the opposite. Again, the scientists all agree with the rank order that $\widetilde{K}$ produces.

The p-values based on $\widetilde{K}$ for the pairwise comparisons among the six TEM images are presented in Tables 5.1 and 5.2. Table 5.1 includes the results based on D_1, whereas Table 5.2 is based on D_2. Using both test statistics leads to the same conclusion. The difference between 5 min and 10 min is marginally significant,

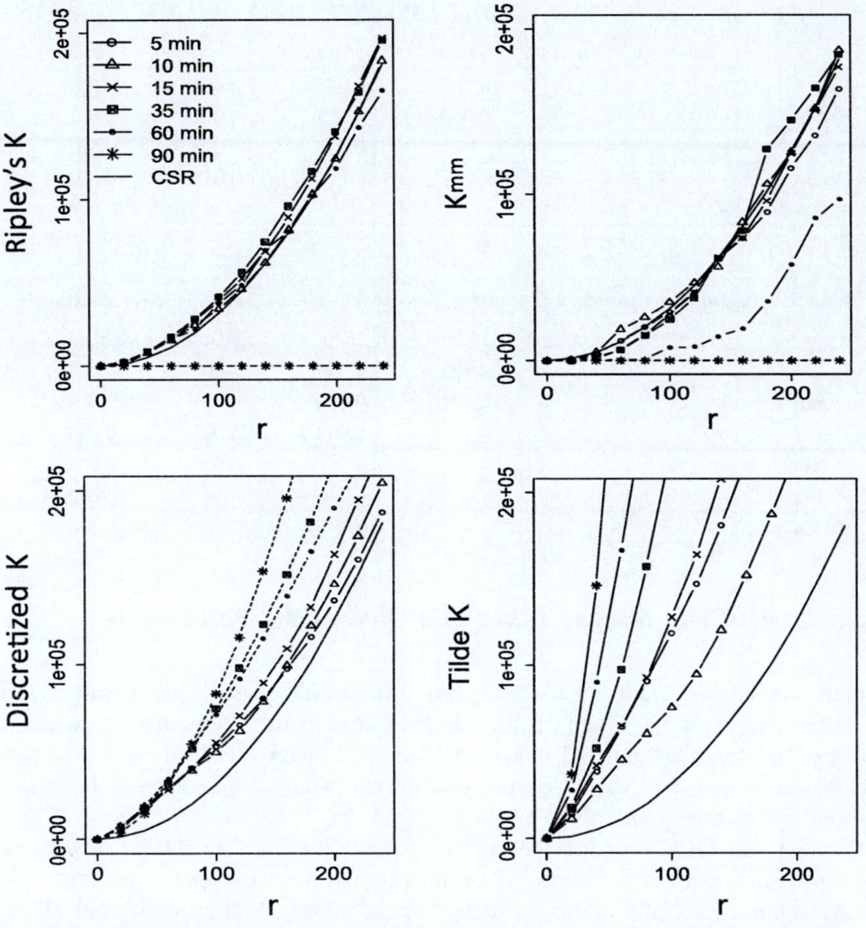

Fig. 5.10 Comparison of the original Ripley's K function, K_{mm} function, discretized K function and $\widetilde{K}$ function. (Reprinted with permission from Dong et al. 2017)

whereas the difference between 5 min and 15 min is insignificant. Not surprisingly, the difference between 10 min and 15 min is more significant than that between 5 min and 10 min. The relatively large p-value between 5 min and 15 min provides a clue of why engineers could not agree among themselves which case to favor. Other pairs of comparison have reasonably small p-values, indicating significant differences. This suggests that the order between 35 min and 60 min does not happen by chance. It seems that $\widetilde{K}$ provides the most sensible outcome that is also easy to interpret.

Table 5.1 Pairwise comparison using the six TEM images: p-values based on test statistic D_1

p	5 min	10 min	15 min	35 min	60 min	90 min
5 min	–	0.111	0.581	0.045	0.001	0.001
10 min	–	–	0.050	0.005	0.001	0.002
15 min	–	–	–	0.033	0.001	0.001
35 min	–	–	–	–	0.015	0.002
60 min	–	–	–	–	–	0.004
90 min	–	–	–	–	–	–

Table 5.2 Pairwise comparison using the six TEM images: p-values based on test statistic D_2

p	5 min	10 min	15 min	35 min	60 min	90 min
5 min	–	0.115	0.492	0.044	0.001	0.001
10 min	–	–	0.031	0.010	0.001	0.001
15 min	–	–	–	0.043	0.001	0.001
35 min	–	–	–	–	0.009	0.001
60 min	–	–	–	–	–	0.002
90 min	–	–	–	–	–	–

5.5.2 Multiple Images Taken at a Given Time Point

In the second study, Dong et al. (2017) take several TEM images at randomly chosen locations on a material sample at each given time point. There are 14 images at 0 min, 13 images at 5 min, 12 images at 10 min, 12 images at 35 min and 18 images at 90 min. Figure 5.11 shows five images each row, which are a subset of the images taken at each time point.

Dong et al. (2017) conduct the pairwise comparison analysis, similar to what was done in the previous subsection. All settings are the same, except that for this study, the test procedure follows that for two groups of images. In the second study, Dong et al. (2017) only use $\widetilde{K}$, as the first study has demonstrated the advantage of $\widetilde{K}$ well over the original K and K_{mm}, and over the discretized K as well.

Figure 5.12 presents the $\widetilde{K}$ curves resulting from individual images. The $\widetilde{K}$ curves from the two groups are differentiated by using two different line types. The presentation of pairwise $\widetilde{K}$ curves involving the images at 0 min is omitted, as it is obvious that before the bead-milling process is applied, the nanoparticles tend to cluster together heavily.

The plots of these pairwise $\widetilde{K}$ curves shed lights in terms of how significantly two groups of images are different from each other. Based on the plots, it is apparent that the images of 5 min is noticeably different from the other three groups. The other groups are generally not that much different, while some of them may be marginally different (e.g., 10 versus 90 min).

Tables 5.3 and 5.4 present the p-values when pairwise comparison is made between two groups of images. In these two tables, the images at 0 min are included in the comparison. Images at 0 min are shown to be indeed different from those

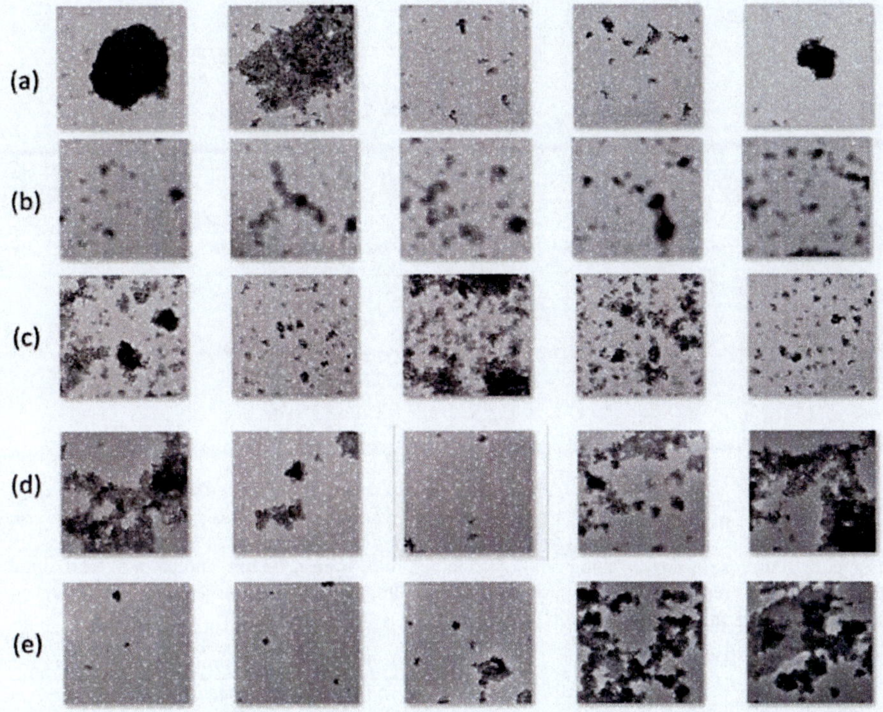

Fig. 5.11 Rows (**a**) through (**e**) are the subsets of images taken at 0, 5, 10, 35, and 90 min, respectively. (Reprinted with permission from Dong et al. 2017)

at 5 and 10 min but not so much different from those at 35 and 90 min. For the images taken at 5, 10, 35, and 90 min, the implications resulting from D_1 and D_2 are consistent with each other and they are also consistent with the plots in Fig. 5.12. Generally speaking, images taken at 5 min show a better mixing state of nanoparticles than any other images (including those at 0 min). Images taken at 10, 35, and 90 min are not that different, while images of 10 min could be marginally different from that of 90 min.

Altogether, this analysis suggests that the bead-milling process does make a difference to the mixing state of nanoparticles. Consistent with the general trend shown in the first experiment, the second experiment also shows that the nanoparticles start to disperse, once the bead-milling operation runs, but will re-cluster if the operation runs beyond certain time.

The difference, though, is that the best mixing state in the second experiment is chosen at 5 min of the bead-milling operation, while that in the first experiment is chosen at 10 min. A closer look at the images taken in the two experiments suggests that the conclusion in the second experiment, based on a group of images, is likely to

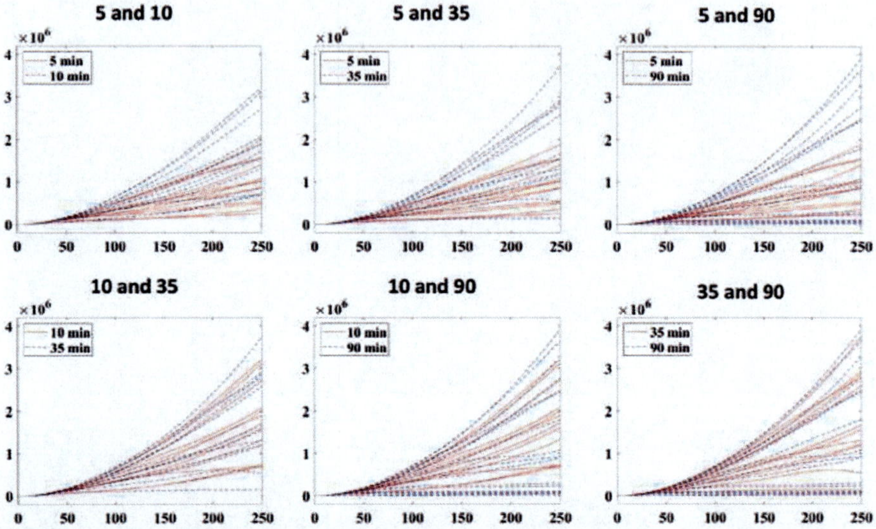

Fig. 5.12 Pairwise $\widetilde{K}$ curves using images taken at 5 min, 10 min, 35 min and 90 min. Horizontal and vertical axes represent r values and the corresponding $\widetilde{K}(r)$ values, respectively. (Reproduced using the images from Dong et al. (2017))

Table 5.3 Pairwise comparison using the five groups of TEM images: p-values based on D_1

p	0 min	5 min	10 min	35 min	90 min
0 min	–	0.005	0.044	0.245	0.557
5 min	–	–	0.006	0.010	0.012
10 min	–	–	–	0.359	0.108
35 min	–	–	–	–	0.469
90 min	–	–	–	–	–

Table 5.4 Pairwise comparison using the five groups of TEM images: p-values based on D_2

p	0 min	5 min	10 min	35 min	90 min
0 min	–	0.001	0.038	0.245	0.563
5 min	–	–	0.016	0.011	0.002
10 min	–	–	–	0.345	0.072
35 min	–	–	–	–	0.442
90 min	–	–	–	–	–

be more robust. The particle mixing states have a great variability over the material sample. Conclusion based on a single image could be misleading.

To see this, consider comparing the right most image in Fig. 5.11, row (b), with the right most image in Fig. 5.11, row (c). Then, consider comparing the right most image in Fig. 5.11, row (b), with the middle image in Fig. 5.11, row (c). These two comparisons would yield opposite conclusions. With more than 12 images per group in the second experiment, such bias, albeit unlikely disappearing altogether, should have been abetted as compared to the circumstance of a single image per time point used in the first experiment.

As mentioned just above, the bead-milling process does break down the agglomerates into smaller pieces of nanoparticle aggregates and disperse them over the material sample. One unexpected phenomenon is that when the bead-milling operation lasts too long, it could create new agglomerates of a large size. Based on the two experiments, it seems unlikely that one needs to run the bead-milling operation for longer than 35 min, as the optimal mixing states take place at an early stage. The optimum is difficult to be pinpointed yet but it most probably happens between 5 and 10 min of the operation. Because of this, the removal of image-taking at 15 min in the second experiment may not matter a lot. Still, in retrospective, it would have been a safer approach, had the engineer removed the image-taking at 35 min but kept that at 15 min.

While it comes as no surprise that using multiple images is more desirable, researchers recognize the extra time and cost associated with the measurement procedure when multiple images are taken and used. Naturally it raises a number of sample-taking related questions, such as how many images are needed to safeguard the quality of the conclusion and what the best sequence may be to take multiple images at a given time point. For instance, is the random sampling the best approach to use? Solution to these questions appears not straightforward and is still under active research.

5.6 Dispersion Analysis of 3D Materials

We note that all the analyses and discussions done so far are based on 2D images. But the real material samples are 3D. Then a question naturally arises—how robust is the conclusion made based on the 2D images when it is extended to the 3D samples?

To answer the above question, we first explain how TEM and SEM images are generated. Please refer to Fig. 5.13 for an illustration of how the two microscopies work. Electrons in TEM irradiate a sample and form an image on the device on the other side of the sample, whereas SEM forms an image using the secondary electrons emitted from the surface of a sample. More discussion on electron microscope imaging can be found in Sect. 2.1.

When TEM uses a solid sample, the sample must be ultra thin, in the order of 100 nm or thinner, prepared by a special ultramicrotomic machine. On the other hand, the sample used for SEM can of any thickness, but its surface must be polished to let the embedded nanoparticles surface. In the rightmost graph of Fig. 5.13, the dimmed portion of a particle above the surface line is removed after the polishing process, but we keep that portion in the graph for illustration.

Due to its imaging mechanism, TEM projects the particles scattered in 3D material samples onto a 2D imaging plane. As pointed out by Li et al. (2014), the 2D projection of multiple layers of particles in a 3D sample may produce artifacts of particle agglomerates even when the particles in the original 3D sample are well separated—see one such example on the 2D plane in the leftmost graph (in its top-

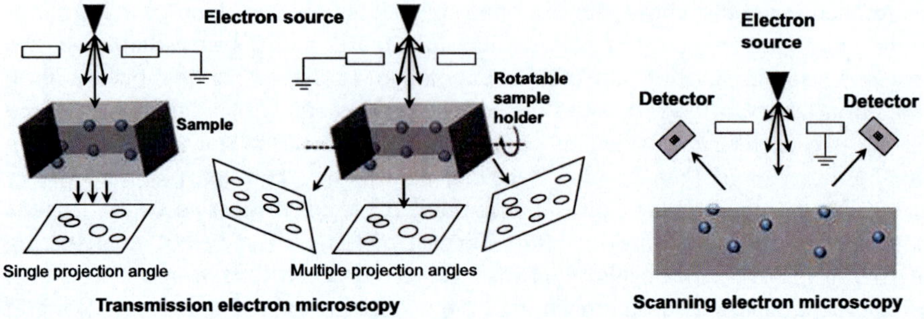

Fig. 5.13 Illustration of the mechanisms of electron microscopies. Left two graphs are for the transmission electron microscopy, whereas the rightmost graph is for the scanning electron microscopy

left corner). This problem does not exist in the SEM cases because the scanning imaging process has to peel the material one layer at a time and there is hence no possibility of multi-layer overlapping.

The question for SEM data analysis is—if the SEM only sees one layer, can its conclusion concerning the quality of particle mixing to be extended to the whole 3D material? Zhou et al. (2014) examine this question. They believe that one will have to take multiple images, preferably from multiple layers in the 3D sample at random. When pooling images from several measurements, Zhou et al. (2014) do not directly make use of the information of whether the images are from multiple spots on the same sample plane or from multiple layers. For this reason, the multi-image analysis conducted in Sect. 5.5.2, albeit for images collected at the same 2D plane, could have been used in the 3D analysis, had the images been collected on multiple layers. The difference is that the analysis in Zhou et al. (2014) is based on the quadrat method and a test statistic similar to Eq. (5.6), whereas the analysis in Sect. 5.5.2 is based on the *K* function.

Zhou et al. (2014) also contemplate the question of how many images are necessary to render the 2D analysis applicable to the 3D materials. Their conclusion is that ten images are sufficient but one could use fewer than ten images generally. One obvious assumption is that the ten images must be collected randomly on multiple layers but doing so is usually a costly proposition.

The 2D projection problem in TEM images may be solved by technology advancement. As illustrated in the middle graph of Fig. 5.13, a TEM can be retrofit with a rotatable sample folder, so that 2D image projections can be attained at multiple projection angles. Even if an artifact of particle agglomerates is formed in some of the projections, it is much less likely that such artifacts would be formed on all 2D projections. With multi-angle image projections, it is possible to reconstruct the 3D objects in the original space. That is in fact the technology basis of electron tomography (Weyland and Midgley 2004), which is available but still used much less prevalently in material science than their medical counterpart. With 3D images available, there are further challenges in handling morphology, location, and dispersion analysis—some of the 3D modeling and analysis issues are summarized in Park and Ding (2019) and please see also discussion about 3D imaging in Sect. 2.1.

Li et al. (2014) study the implication of 2D projection in TEM images. The question that they mean to address is this—assume that the particles in the 3D space do not form agglomerates and the particles are of identical shape and size (equal-sized spheres in Li et al. (2014)), will a dispersion analysis based on a single angle 2D projection be different from the analysis done based on 3D position data, and if so, by how much? The question is practically relevant because although new technology can acquire 3D images, the vast majority of the existing electron micrographs are all single-angle 2D projections, and researchers and practitioners do not want to simply throw them away. More importantly, the 3D technology is still expensive. Even after the invention of the 3D imaging technology, 2D imaging instruments are still routinely used in both academic research and practice, and there does not appear a sign that the 2D technology will be replaced for routine uses anytime soon.

Li et al. (2014) specifically consider the question that when 3D objects are dispersed according to the CSR model, are the objects in the 2D projection also CSR? There they invoke the spatial hardcore process to say that there is no such guarantee.

The hardcore process is to describe particles in the space that are not dimensionless but have a hardcore of a certain size. Because of the existence of this "hardcore," the centroids of the particles cannot be completely random even in 3D. Their position randomness are subject to the constraint that the distance between the centroids of any two particles must be equal to or greater than the sum of their radii.

Denote by l, w, and h the length, width, and height of the 3D material sample under TEM imaging, and by r the radius of the nanoparticles (not to confuse this r with the general r used in the previous sections). Let $\psi(\ell_i)$ be the coordinates of the centroid of the i-th particle, $\ell_i \in \mathbb{L}$, in the 3D space. Li et al. (2014) provide the probability density for the 3D hardcore process as

$$f(\ell) = \begin{cases} \frac{1}{\alpha \times (wlh)}, & \text{if } \|\psi(\ell_i) - \psi(\ell_j)\| > 2r \quad \forall i \neq j, \\ 0 & \text{otherwise,} \end{cases} \tag{5.28}$$

where α is the normalizing constant to make sure that the density function is integrable to one. According to Li et al. (2014), α is also related to the probability that no two particles are closer to each other than $2r$, namely

$$1 - \alpha = \Pr\left(\bigcup_{1 \le i \neq j \le n} E_{ij} \right), \tag{5.29}$$

where $E_{ij} = \{\|\psi(\ell_i) - \psi(\ell_j)\| \le 2r\}$ denotes the event that two randomly distributed particles whose centroid distance is smaller than the sum of the radii of the two particles. Understandably, if no two particles are closer than $2r$, the probability density of the hardcore process is the same as that of the spatial Poisson process (i.e., CSR). The closer α is to one, the more similar the hardcore process is to the spatial Poisson process.

Li et al. (2014) examine the impact of the parameters, l, w, h, and r. They conclude that l and w can usually be large and thus less important in consideration. Using an approximation formula of α, Li et al. (2014) found that h has a profound impact on how well the hardcore process can be approximated by the CSR model. If h is very large relative to r, then α is close to one, suggesting that CSR is a good approximation. But when h is not very large relative to r, α is then substantially smaller than one, indicating that the difference between CSR and the hardcore process cannot be ignored.

This conclusion is hardly surprising, because when particles are scattered in a spacious space (h large), the likelihood that E_{ij} takes place is low, whereas when particles are packed into a cramped space (h small), the likelihood that E_{ij} takes place increases substantially. In the context of nanocomposite material, r is decided by the physical size of nanoparticles and is about 10 nm in Li et al. (2014)'s application, whereas h used in the imaging process is between 60 and 100 nm. There are between 3 and 10% of volume equivalent of nanoparticles to be mixed into the host material. Based on the analysis by Li et al. (2014, Table 1), $h = 100$ nm is not yet large enough relative to the particle size of $r = 10$ nm.

Then the next question that Li et al. (2014) contemplate is which dispersion quantification metric is more robust than others. For that, Li et al. (2014) consider both the quadrat method the distance methods. For the distance-based dispersion functions, Li et al. (2014) consider the K, F, and G functions, whereas for the quadrat method, in addition to the χ^2 index in Eq. (5.6), Li et al. (2014) consider two more indices, which are

- Shannon entropy (SE):

$$\text{SE} = -\sum_{j=1}^{m} p_j \log(p_j),$$

  where $p_j = n_j/n$ is the proportion of particles in the j-th quadrat.
- Skewness (SK):

$$\text{SK} = \frac{m}{(m-1)(m-2)} \sum_{j=1}^{m} \left(\frac{n_j - \bar{n}}{s}\right)^3,$$

  where s is the sample standard deviation, i.e., the square root of the sample variance defined in Eq. (5.5).

Li et al. (2014) conduct their first simulation study to see how much parameter misspecification affects the type I-error. Two particle concentrations are considered—one has 423 particles and the second has 1409 particles, which correspond to the 3 and 10% particle mixing volumes, respectively. The width and length of the sample specimen are fixed at $w = l = 1067$ nm. The height is the parameter that may be misspecified. Using three true h values, i.e., $h_0 = 50, 60$, or 70 nm, Li et al. (2014) simulate a CSR process in the 3D space and then select a decision threshold that give a 0.05 type-I error. Then, they run the same analysis but use an h out of four choices, $\{50, 60, 70, \infty\}$ nm. The case where $h_0 = h$ is skipped in the simulation study for obvious reasons. Li et al. (2014) want to find out how much the resulting type-I error deviates from the 0.05 benchmark. They deem a variation within 0.02 of the benchmark value, i.e., in the range of [0.03, 0.07], as tolerable but anything greater as not.

Tables 5.5 and 5.6 present, respectively, the type-I error rates for the quadrat method and the distance functions. When using the quadrat method, Li et al. (2014) explore three choices of the number of cells used, which are 5×5, 15×15, and 25×25. Accordingly, $m = 25$, $m = 225$, and $m = 625$. When using the distance functions, both the two-sided test and one-sided test are explored. The one-sided test only considers the side in which the spatial points form clusters. This correspond to $K > K_{\text{null}}$ and $G > G_{\text{null}}$ but to $F < F_{\text{null}}$. The two-sided test, on the other hand, considers both sides.

Li et al. (2014) conclude that among the indices used in the quadrat method, SK is most robust. The quadrat method appears to be sensitive to the number of cells used. The general recommendation is not to use too many cells. Among the three distance-based functions, the K function is the only robust option, performing way better than the other two functions.

Li et al. (2014) next study the power of detection, should there be a cluster of particles. For this study, Li et al. (2014) fix $h = 60$ nm and include the three good dispersion metrics identified in the type-I error analysis—the SK metric, the two-sided test based on K, and the one-sided test based on K. Two types of inputs are used to compute the metrics. One is to use the centroids from the particles in the 3D space, whereas the other is to use the centroids from the particles in the 2D projection plane.

To simulate a clustering patten, a clustering ratio, κ, is introduced to quantify the degree of aggregation. The κ is the ratio of the intensity of a suspected clustering area over the intensity of the CSR area, namely, $\kappa = \lambda_{\text{suspicious}}/\lambda_{\text{CSR}}$. Generally, $\kappa = 2$ is considered a mild aggregation, which presents a hard case for clustering detection. Figure 3 in Li et al. (2014) presents two cases of $\kappa = 2$ for which it is difficult to tell whether there is any degree of clustering based on visual inspection. A large κ makes the clustering more obvious and hence easier to be detected.

To help with the detection capability, Li et al. (2014) allow multiple TEM images to be used in decision making. When multiple images are used, a 10% variation in the number of particles per image is allowed.

When using the SK metrics, the number of cells used is set to 50. The partition splits the material sample into two layers, each of which has 5×5 cells.

The analysis results of detection power are presented in Table 5.7. Two observations are immediate—(a) When κ increases, the detection becomes indeed

Table 5.5 Type I error rates when using the quadrat method (*Source*: Li et al. 2014)

		$n = 423$								
		ID[a]			SE			SK		
h_0	h	$m = 25$	$m = 225$	$m = 625$	$m = 25$	$m = 225$	$m = 625$	$m = 25$	$m = 225$	$m = 625$
50	70	0.03	0.02	0.01	0.03	0.02	0.01	0.05	0.03	0.01
	∞	0.01	0.00	0.00	0.01	0.00	0.00	0.04	0.00	0.00
60	50	0.07	0.09	0.12	0.07	0.08	0.11	0.05	0.07	0.09
	70	0.04	0.03	0.02	0.04	0.03	0.02	0.05	0.05	0.03
	∞	0.01	0.00	0.00	0.01	0.00	0.00	0.04	0.01	0.00
70	50	0.08	0.13	0.20	0.08	0.12	0.18	0.05	0.07	0.13
	∞	0.01	0.00	0.00	0.01	0.00	0.00	0.04	0.01	0.00
		$n = 1409$								
		ID			SE			SK		
h_0	h	$m = 25$	$m = 225$	$m = 625$	$m = 25$	$m = 225$	$m = 625$	$m = 25$	$m = 225$	$m = 625$
50	70	0.01	0.00	0.00	0.01	0.00	0.00	0.05	0.03	0.01
	∞	0.00	0.00	0.00	0.00	0.00	0.00	0.04	0.00	0.00
60	50	0.10	0.21	0.33	0.10	0.19	0.30	0.05	0.07	0.10
	70	0.03	0.01	0.00	0.02	0.02	0.01	0.05	0.04	0.02
	∞	0.00	0.00	0.00	0.00	0.00	0.00	0.04	0.00	0.00
70	50	0.16	0.43	0.68	0.16	0.40	0.62	0.05	0.08	0.17
	∞	0.00	0.00	0.00	0.00	0.00	0.00	0.04	0.00	0.00

[a] "ID" refers to the χ^2 statistic in Eq. (5.6), not just the variance-to-mean ratio

Table 5.6 Type I error rates when using the distance-based functions (*Source*: Li et al. 2014)

		$n = 423$					
		K		F		G	
h_0	h	Two-sided test	One-sided test	Two-sided test	One-sided test	Two-sided test	One-sided test
50	70	0.04	0.04	0.09	0.01	0.24	0.01
	∞	0.03	0.01	0.77	0.00	0.00	0.00
60	50	0.06	0.06	0.08	0.11	0.14	0.19
	70	0.04	0.05	0.05	0.03	0.08	0.02
	∞	0.03	0.02	0.61	0.00	0.99	0.00
70	50	0.07	0.07	0.12	0.17	0.27	0.36
	∞	0.03	0.02	0.49	0.00	0.94	0.00
		$n = 1409$					
		K		F		G	
h_0	h	Two-sided test	One-sided test	Two-sided test	One-sided test	Two-sided test	One-sided test
50	70	0.04	0.02	0.96	0.00	0.99	0.00
	∞	0.07	0.00	0.00	0.00	0.00	0.00
60	50	0.05	0.07	0.60	0.70	0.72	0.79
	70	0.04	0.03	0.34	0.00	0.40	0.00
	∞	0.06	0.00	0.00	0.00	0.00	0.00
70	50	0.08	0.10	0.97	0.98	0.99	0.00
	∞	0.05	0.00	0.00	0.00	0.00	0.00

Table 5.7 Detection power of the skewness metric and the two tests based on the K function (*Source*: Li et al. 2014)

		$n = 423$					
		SK		Two-sided K		One-sided K	
κ	# of images	3D particles	2D particles	3D particles	2D particles	3D particles	2D particles
2	1	0.14	0.09	0.20	0.32	0.34	0.40
	5	0.33	0.17	0.46	0.72	0.79	0.85
	10	0.50	0.27	0.69	0.92	0.96	0.98
	15	0.65	0.35	0.82	0.97	0.99	1.00
3	1	0.59	0.33	0.50	0.90	0.63	0.94
	3	0.91	0.69	0.82	1.00	0.92	1.00
	5	0.98	0.88	0.94	1.00	0.98	1.00
4	1	0.86	0.68	0.65	1.00	0.75	1.00
	2	0.98	0.93	0.86	1.00	0.93	1.00
	3	1.00	0.99	0.95	1.00	0.98	1.00
		$n = 1409$					
		SK		Two-sided K		One-sided K	
κ	# of images	3D particles	2D particles	3D particles	2D particles	3D particles	2D particles
2	1	0.51	0.23	0.58	0.74	0.69	0.83
	5	0.95	0.68	0.97	1.00	0.99	1.00
	10	1.00	0.90	1.00	1.00	1.00	1.00
	15	1.00	0.97	1.00	1.00	1.00	1.00
3	1	0.94	0.89	0.86	1.00	0.90	1.00
	3	1.00	1.00	0.99	1.00	1.00	1.00
	5	1.00	1.00	1.00	1.00	1.00	1.00
4	1	1.00	1.00	0.93	1.00	0.95	1.00
	2	1.00	1.00	0.99	1.00	1.00	1.00
	3	1.00	1.00	1.00	1.00	1.00	1.00

easier, and (b) when more images are used, the detection power increases. Both observations are much expected.

Comparing with itself, SK performs better when it uses the centroids from the 3D particles; the advantage over using the centroids of 2D particles is apparent. For the K function, however, both the two-sided and the one-sided tests perform better when using the centroids of the 2D particles as input. Li et al. (2014) state that "*[t]he K-function metrics are uniformly more powerful than SK across all ratios of intensities [κ].*" Between the two test options of the K function, the one-sided test is unsurprisingly more powerful than the two-sided test, as it eliminates a priori one of the possibilities.

The performance of the one-sided test based on the K function is generally satisfactory. Using it to detect a large change in spatial point patterns, like under $\kappa = 3$ or $\kappa = 4$, a single image or a couple of images is sufficient. For detecting a moderate change like under $\kappa = 2$, more images are necessary. The numerical analysis suggests using five to ten images, a conclusion that appears to resonate with what is made by Zhou et al. (2014). When using the other two less powerful methods (SK and two-sided test), having more images definitely helps.

References

Baddeley A (2008) Analysing spatial point patterns in R. Workshop Notes, The Commonwealth Scientific and Industrial Research Organisation (CSIRO) [online]:http://www.csiro.au/resources/pf16h

Baddeley A, Turner R (2005) Spatstat: An R package for analyzing spatial point pattens. Journal of Statistical Software 12(6):1–42

Bailey TC, Gatrell AC (1995) Interactive Spatial Data Analysis. Addison Wesley Longman Limited, Essex, England

Besag J (1977) Comments on Ripley's paper. Journal of the Royal Statistical Soceity, Series B, 39(2):193–195

Diggle PJ (2013) Statistical analysis of spatial and spatio-temporal point patterns. Chapman & Hall/CRC Press, Boca Raton, FL

Diggle PJ, Lange N, Beneš FM (1991) Analysis of variance for replicated spatial point patterns in clinical neuroanatomy. Journal of the American Statistical Association 86(415):618–625

Diggle PJ, Mateu J, Clough HE (2000) A comparison between parametric and non-parametric approaches to the analysis of replicated spatial point patterns. Advances in Applied Probability 32(2):331–343

Dong L, Li X, Qian Y, Yu D, Zhang H, Zhang Z, Ding Y (2017) Quantifying nanoparticle mixing state to account for both particle location and size effects. Technometrics 59:391–403

Feng D, Tierney L (2014) `PottsUtils`: Utility functions of the Potts models. R Package Version 0.3-2 [online]:http://CRAN.R--project.org/package=PottsUtils

Ferreira T, Rasband W (2011) The `ImageJ` User Guide. National Institutes of Health [online]:http://rsb.info.nih.gov/ij/

Hahn U (2012) A studentized permutation test for the comparison of spatial point patterns. Journal of the American Statistical Association 107(498):754–764

Hardy GH (1999) Ramanujan: Twelve Lectures on Subjects Suggested by His Life and Work, 3rd edn. AMS Chelsea Publishing, Providence, RI

Hui L, Smith R, Wang X, Nelson J, Schadler L (2008) Quantification of particulate mixing in nanocomposites. In: Annual Report Conference on Electrical Insulation and Dielectric Phenomena (CEIDP), IEEE, pp 317–320

Li X, Zhang H, Jin J, Huang D, Qi X, Zhang Z, Yu D (2014) Quantifying dispersion of nanoparticles in polymer nanocomposites through transmission electron microscopy micrographs. Journal of Micro and Nano-Manufacturing 2(2):021008

Manas-Zloczower I (1997) Analysis of mixing in polymer processing equipment. Rheology Bulletin 66(1):5–8

Marcon E, Puech F (2003) Evaluating the geographic concentration of industries using distance-based methods. Journal of Economic Geography 3(4):409–428

Marcon E, Lang G, Traissac S, Puech F (2014) dbmss: Distance-based measures of spatial structures. R Package Version 2.1.2 [online]:http://CRAN.R--project.org/package=dbmss

Park C, Ding Y (2019) Automating material image analysis for material discovery. MRS Communications 9(2):545–555

Park C, Huang J, Huitink D, Kundu S, Mallick B, Liang H, Ding Y (2012) A multi-stage, semi-automated procedure for analyzing the morphology of nanoparticles. IIE Transactions Special Issue on Nanomanufacturing 44(7):507–522

Park C, Huang J, Ji J, Ding Y (2013) Segmentation, inference and classification of partially overlapping nanoparticles. IEEE Transactions on Pattern Analysis and Machine Intelligence 35(3):669–681

Penttinen A, Stoyan D, Henttonen HM (1992) Marked point processes in forest statistics. Forest Science 38(4):806–824

Ray SS, Okamoto M (2003) Polymer/layered silicate nanocomposites: A review from preparation to processing. Progress in Polymer Science 28(11):1539–1641

Ripley BD (1976) The second-order analysis of stationary point processes. Journal of Applied Probability 13(2):255–266

Ripley BD (1991) Statistical Inference for Spatial Processes. Cambridge University Press, Cambridge, UK

Smith TE (2016) Notebook on Spatial Data Analysis. [online] http://www.seas.upenn.edu/~ese502/#notebook, Philadelphia, PA.

Venables WN, Ripley BD (2002) Modern Applied Statistics With S, 4th edn. Springer-Verlag, New York

Wang Y, Forssberg E (2006) Production of carbonate and silica nanoparticles in stirred bead milling. International Journal of Mineral Processing 81(1):1–14

Weyland M, Midgley PA (2004) Electron tomography. Materials Today 7(12):32–40

Winkler G (2003) Image Analysis, Random Fields and Dynamic Monte Carlo Methods: A Mathematical Introduction, 2nd edn. Springer-Verlag, Berlin Heidelberg

Zhou Q, Zhou J, De Cicco M, Zhou S, Li X (2014) Detecting 3D spatial clustering of particles in nanocomposites based on cross-sectional images. Technometrics 56(2):212–224

Zou H, Wu S, Shen J (2008) Polymer/silica nanocomposites: Preparation, characterization, properties, and applications. Chemical Reviews 108(9):3893–3957

Chapter 6
Lattice Pattern Analysis

6.1 Basics of Lattice Pattern Analysis

A lattice is a repeating spatial arrangement of points. In geometry, a lattice in $\mathbb{R}^p$ is formally defined as a collection of the points whose coordinates are represented in the following form,

$$L = \boldsymbol{v}_0 + \left\{ \sum_{k=1}^{p} a_i \boldsymbol{v}_i : a_i \in \mathbb{Z} \right\},$$

where $\mathbb{Z}$ is the set of all integers, $\boldsymbol{v}_0 \in \mathbb{R}^p$ is the origin of the lattice, and $B = \{\boldsymbol{v}_i \in \mathbb{R}^p : i = 1, \ldots, p\}$ is the basis. For a fixed basis, a point in the lattice is represented by p integers, $(a_1, a_2, \ldots, a_p)$, and the origin is a zero-vector in the representation. Fig. 6.1 illustrates a lattice in $\mathbb{R}^2$.

The mathematical concept of lattice has been used to describe regular and repeated arrangements of small scale elements within bulk materials. Particularly, in crystallography, a lattice has been used to describe regularly spaced atoms in a crystal material. Locating individual atoms, identifying the lattice of their locations, and probing which atoms deviate spatially from the lattice are of great interest, because the spatial arrangement impacts the properties of crystal materials and atoms whose locations deviate from the lattice coordinates are regarded as defects.

Identifying such irregularly spaced atoms is as important as identifying regularly spaced atoms. The deviation from the lattice often appears only locally around defective regions of the crystal material. One can consider the lattice as the global feature of a crystalline structure, and the deviation from the lattice as the local feature.

Studying the global crystallography has been done with simple instrumentation such as an X-ray diffractometer, where the X-ray diffracted by a crystal material

C. Park, Y. Ding, *Data Science for Nano Image Analysis*, International Series in Operations Research & Management Science 308,
https://doi.org/10.1007/978-3-030-72822-9_6

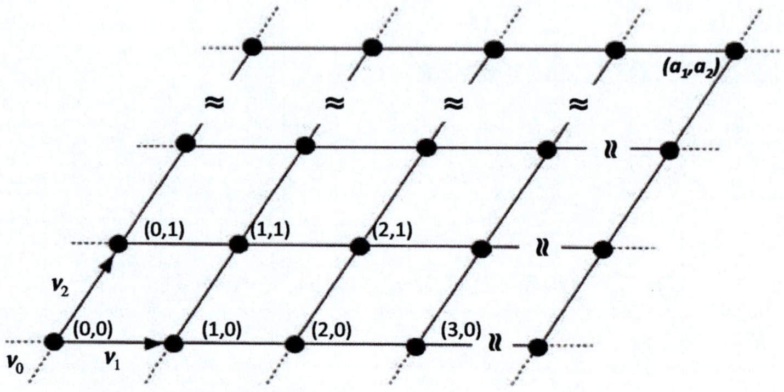

Fig. 6.1 Lattice in $\mathbb{R}^2$, where $\{v_i \in \mathbb{R}^2; i = 1, 2\}$ is the basis and v_0 is the origin

is related to the pair distances between atoms in the crystal. The distribution of the pair distances can be analyzed to signal the presence of lattice deviation (or defects). By the nature of the analysis, however, this simple technique cannot be used to locate precisely the local defects in a crystalline material. To probe both the global and local features, a scanning probe microscope (SPM) or scanning transmission electron microscope (STEM) ought to be used. Improvements in underlying instrumentation and data processing capabilities have allowed the determination of atomic positions with a sub-10 pm precision (Yankovich et al. 2014; Kim et al. 2012), which enables the visualization of chemical and mechanical strains (Kim et al. 2014), and order parameter fields including ferroelectric polarization (Chang et al. 2011; Nelson et al. 2011; Jia et al. 2007, 2011) and octahedral tilts (Jia et al. 2009; Kim et al. 2013; Borisevich et al. 2010a; He et al. 2010; Borisevich et al. 2010b). However, quantifying structural information directly from images is still challenging due to a large number of atoms present and artifacts resulting from imaging imperfection.

The lattice pattern analysis based on material images is to extract the global and local features of a crystalline material from its atomically resolved images. We use Fig. 6.2 to illustrate an STEM image taken at a sub-angstrom resolution and explain how the global and local features are defined in the image.

Individual atomic columns are represented as the bright spots on a dark background. For illustrative purposes, we overlaid a green cross on the location of each atom. Readers may note that most of the atoms lay on a regularly spaced lattice grid. However, this regularity is occasionally broken due to atomic defects; see one of such examples in Fig. 6.2, where a magenta circle pointed by an arrow shows that an atom is supposed to exist but it is in fact missing. The regular lattice grid, and the missing atoms breaking the regularity, are important characteristics defining the properties of the sample material.

Many existing lattice pattern analysis methods process the atomically resolved images sequentially in two steps: First, identify individual atoms, and second, infer the global lattice spacing and account for local defects (Belianinov et al. 2015).

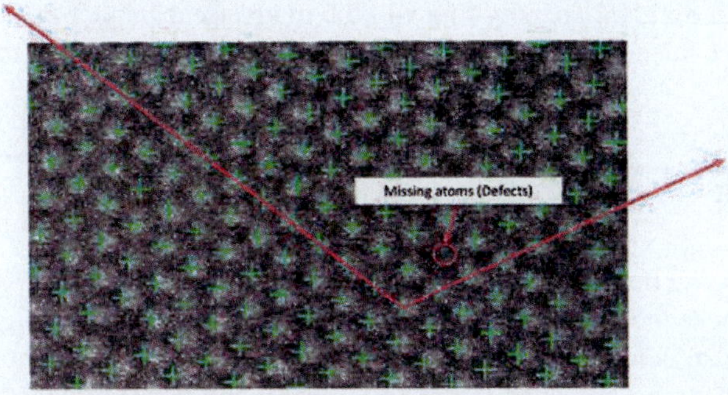

Fig. 6.2 Atomic scale image overlaid with individual atom locations, showing their symmetries and defects. Reprinted with permission from Li et al. (2018)

The first step can be thought as a spot detection problem that locates bright spots on a dark image without a priori information. Popular approaches to spot detection include local filtering such as the top-hat filter (Bright and Steel 1987) and the LoG filter (Sage et al. 2005). In such filtering approaches, a local filtering is first applied to an input image, and the filtered image is then thresholded to locate spots, or the h-dome method is used to locate the local maxima of the filtered image, potentially identifying an atom (Vincent 1993; Smal et al. 2008; Rezatofighi et al. 2012). The filtering step usually relies on two or three crucial tuning parameters, including spot size, distance between spots and threshold intensity, for which some preliminary information is necessary. However, for low contrast images having low signal-to-noise ratios, the spot detection approach is not very accurate. Besides the local filtering methods, Hughes et al. (2010) employed an approximate likelihood estimator for particle location, derived from a Poisson random field model for photon emission imaging. We describe the likelihood-based spot detection approach in Sect. 6.2.

An alternative approach to the two-step, sequential lattice pattern analysis is the integrated approach that solves the two steps simultaneously. Intuitively, solving the two steps together can improve the accuracy of final outcomes, because knowing a global lattice grid helps locate individual atom locations on the grid, while knowing individual atom locations helps identify the global lattice and local defects. Li et al. (2018) formulated the problem of estimating atom locations and identifying their global and local features as a regression problem. The solution to the framed regression problem is also regularized with the use of two sparsity terms—a group sparsity and an individual sparsity. A choice of a specific lattice grid implies possible atomic locations restricted to the grid vertices of the chosen lattice, so selecting a lattice grid translates to selecting a representative group of atoms, which can be guided by the group sparsity penalty term. The group sparsity term penalizes selecting multiple groups since the global lattice grid is unique. On the other hand, not all atoms conform to the global lattice grid due to atomic defects. Li et al. (2018)

then used the individual sparsity to avoid the inadvertent impact of image artifacts and thus reduce false positives. We describe the integrated approach in Sect. 6.3.

6.2 Simple Spot Detection

In this section, we describe a simple spot detection approach for the lattice pattern analysis. Suppose that a material sample is imaged into an $M \times N$ digital image by an electron beam, and the sample consists of T atoms with the tth atom positioned at the pixel location (x_t, y_t) of the image. Ideally, the measurement of the sample would have a sharp intensity peak at each atomic position (x_t, y_t), so that the intensity function, $f(\cdot, \cdot)$, can be expressed as

$$f_\delta(x, y) = \sum_{t=1}^{T} \alpha_t \delta(x - x_t, y - y_t), \tag{6.1}$$

where δ is the Dirac delta function. However, due to inherent electronic lens aberration (Nellist and Pennycook 2000), STEM produces instead a blurred image, i.e., a convolution of the peaks with a Gaussian point spreading function P,

$$f(x, y) = P * f_\delta = \sum_{t=1}^{T} \alpha_t \exp\left(\frac{-(x - x_t)^2 - (y - y_t)^2}{\tau^2}\right) \tag{6.2}$$

where τ^2 is a positive constant and $*$ is the convolution operator. We do not know the true number of atoms in the image, T, and their actual locations (x_t, y_t). Therefore, one can pose an infinite mixture model that considers all possible atom locations, i.e., let $T = \infty$, such that

$$f(x, y) = \sum_{t=1}^{\infty} \alpha_t \exp\left(\frac{-(x - x_t)^2 - (y - y_t)^2}{\tau^2}\right). \tag{6.3}$$

For a digital image, the infinite mixture model is reduced to the following finite mixture model that places a mixture component at every image pixel location (m, n),

$$f(x, y) = \sum_{(m,n)\in \mathbb{Z}_M \times \mathbb{Z}_N} \alpha_{m,n} \exp\left(\frac{-(x - m)^2 - (y - n)^2}{\tau^2}\right), \tag{6.4}$$

where $\mathbb{Z}_M = \{1, 2, \ldots, M\}$, and the value of $\alpha_{m,n}$ is

$$\alpha_{m,n} \begin{cases} > 0 & \text{if an atom exists on pixel } (m, n) \\ = 0 & \text{otherwise.} \end{cases}$$

What is actually measured for a given material sample is a noisy version of f. The measured image $I(x, y)$ at pixel location (x, y) can be expressed as

$$I(x, y) = f(x, y) + \epsilon(x, y), \tag{6.5}$$

where $\epsilon(x, y)$ is an independent white noise. Let $\boldsymbol{Y}$ denote the $M \times N$ matrix of $I(x, y)$'s and $\boldsymbol{A}$ denote the matrix of $\alpha_{m,n}$'s. We also define $\boldsymbol{u}_m$ as the $M \times 1$ vector with its ith element equal to $\exp\left(-(i-m)^2/\tau^2\right)$ and $\boldsymbol{v}_n$ as the $N \times 1$ vector with its jth element equal to $\exp\left(-(j-n)^2/\tau^2\right)$. The measurement model in Eq. (6.5), when aggregated for the whole image, can be expressed in the following vector-matrix form,

$$\boldsymbol{Y} = \boldsymbol{U}_\tau \boldsymbol{A} \boldsymbol{V}_\tau^T + \boldsymbol{E}, \tag{6.6}$$

where $\boldsymbol{U}_\tau = (\boldsymbol{u}_1, \ldots, \boldsymbol{u}_M)$, $\boldsymbol{V}_\tau = (\boldsymbol{v}_1, \ldots, \boldsymbol{v}_N)$ and $\boldsymbol{E}$ is the $M \times N$ noise matrix of $\epsilon(x, y)$'s. Note that the unknown $\boldsymbol{A}$ is supposed to be very sparse because atoms locate only at a few of the pixel locations. The squared loss function for choosing $\boldsymbol{A}$ is

$$L(\boldsymbol{A}; \tau^2, \boldsymbol{Y}) = ||\boldsymbol{Y} - \boldsymbol{U}_\tau \boldsymbol{A} \boldsymbol{V}_\tau^T||_F^2. \tag{6.7}$$

The choice of $\boldsymbol{A}$ can be decided by minimizing the L2 loss, $L(\boldsymbol{A}; \tau^2, \boldsymbol{Y})$, together with an L1 sparsity penalty on $\boldsymbol{A}$. The non-negative elements of the resulting sparse matrix $\boldsymbol{A}$ indicate the individual atom locations, which is the solution to the simple spot detection problem. This L1-regularized minimization problem can be solved by the widely used least absolute shrinkage and selection operator (LASSO) method or the least angle regression (Efron et al. 2004).

6.3 Integrated Lattice Analysis

The simple spot detection does not work well for low contrast images and tends to yield many false detections of atoms, as well as miss detections. A solution to mitigate this issue is to exploit the global lattice of atoms' spatial arrangements to prune out falsely detected atoms, while in the same time, identifying the miss detected atoms that do not appear clearly in low contrast images. Since the global lattice is also unknown, the global lattice and individual atom locations can be jointly estimated via an integrated solution approach. This section describes the integrated approach developed by Li et al. (2018).

Let us first introduce the concept of a lattice group. The spatial locations of atoms in a perfect crystalline material can be completely described by a lattice group. The space of all pixel locations for a 2D digital image of size $M \times N$ is represented by $\mathbb{Z}_M \times \mathbb{Z}_N$. A lattice group $L^{(g)}$ in the space is defined by its coordinate origin

$\boldsymbol{s}_g \in \mathbb{Z}_M \times \mathbb{Z}_N$ and two integer-valued lattice basis, $\boldsymbol{p}_g, \boldsymbol{q}_g \in \mathbb{Z}_M \times \mathbb{Z}_N$,

$$L_g := \{\boldsymbol{s}_g + z_p \boldsymbol{p}_g + z_q \boldsymbol{q}_g \in \mathbb{Z}_M \times \mathbb{Z}_N; z_p, z_q \in \mathbb{Z}_+\},$$

where $\mathbb{Z}_+$ denote the set of all non-negative integers, the subscript $g \in G$ is used to index a lattice group in the collection of possible lattice groups, and G denote the collection. Define $\boldsymbol{A}_{(g)}$ an $M \times N$ matrix with its (m, n)th element as

$$(\boldsymbol{A}_{(g)})_{m,n} = \begin{cases} \alpha_{m,n}, & \text{if } (m, n) \in L_g \\ 0, & \text{otherwise.} \end{cases}$$

The loss function in Eq. (6.7) can be written as

$$L(\boldsymbol{A}; \tau^2, \boldsymbol{Y}) = ||\boldsymbol{Y} - \sum_{g \in G} \boldsymbol{U}_\tau \boldsymbol{A}_{(g)} \boldsymbol{V}_\tau^T ||_F^2. \tag{6.8}$$

Since all atom locations belong to either a single lattice group (for single-crystalline material) or a few lattice groups (in the case of multi-phase materials) among all listed in G, $\boldsymbol{A}$ can be regularized with the use of group sparsity. For example, $\boldsymbol{A}$ can be regularized with the use of a group norm, like what is used in the penalty term of the group lasso method, i.e.,

$$\lambda \sum_{g \in G} ||\boldsymbol{A}_{(g)}||_F. \tag{6.9}$$

The group lasso penalty works like the lasso regularization but at the group level, meaning that either all variables in a group are shrunk to zero or all of them are kept non-zero (Yuan and Lin 2006). The group lasso criterion does not produce within-group sparsity, i.e., a non-zero group norm $||\boldsymbol{A}_{(g)}||_F$ implies all variables in group g are kept as non-zero. This is not appropriate for the lattice analysis problem, because there could be some vacant locations in a chosen lattice group due to atomic defects, so that some elements in the chosen group could be zero. Therefore, simply regularizing $\boldsymbol{A}$ with the group norm would result in over detection. To minimize the faulty detections, a within-group sparsity should be considered. The strategy of imposing both group-level sparsity and within-group sparsity was in fact previously developed for the sparse group lasso method (Simon et al. 2013) and the hierarchical group sparsity or more general graph group sparsity (Huang et al. 2011; Jenatton et al. 2011). Here we follow the specific approach in Huang et al. (2011) to inject the two-level sparsity regularization into our formulation. The sparse group lasso method (Simon et al. 2013) could be an alternative for us to follow, but the sparse group lasso method does not perform well due to several reasons to be discussed in Sect. 6.4.4.

Consider a set of the lattice groups, $\{L_g; g \in G\}$, and define $S_{m,n}$ as a group of singleton (m, n). The singleton groups and the lattice groups form the following inclusion relation—for each $S_{m,n}$, there exists g that satisfies

$$S_{m,n} \subset L_g.$$

In addition, L_g and the entire image $\mathbb{Z}_M \times \mathbb{Z}_N$ has the relation $L_g \subset \mathbb{Z}_M \times \mathbb{Z}_N$. These inclusion relations can be represented by a tree hierarchy that has $\mathbb{Z}_M \times \mathbb{Z}_N$ as a root node, all L_g's as the first level children, and all $S_{m,n} \subset L_g$ as the second level children of L_g. Following Huang et al. (2011, Section 4.3), we represent the sparsity cost for the tree as

$$C(A) = \left\{ \log_2(2|G|) \cdot \sum_{g \in G} ||tr(A_{(g)} A_{(g)}^T)||_0 + \log_2 2K \cdot ||A||_0 \right\}, \tag{6.10}$$

where $K = \max_{g \in G} |L_g|$. Note that the first cost term represents the group-level sparsity, and the second term represents the within-group sparsity. We find A by solving

$$\text{Minimize} \quad L(A; \tau^2, Y) \text{ subject to } C(A) \leq c, \tag{6.11}$$

where $c > 0$ is a tuning parameter. The cost of making group $A_{(g)}$ nonzero is $\log_2(2|G|)$, which is much smaller than the cost for adding an individual, $\log_2 2K$. This suggests that the regulation criterion favors a group selection unless there are strong counter evidences from $L(A; \tau^2, Y)$. The problem is a non-convex optimization problem, and a sub-optimal solution can be achieved using the solution approach described in Sect. 6.4.

6.4 Solution Approach for the Integrated Lattice Analysis

The optimization formulation in Eq. (6.11) is to minimize the square loss, $L(A; \tau^2, Y)$, subject to a structural sparsity term, $C(A)$. For solving this constrained minimization problem, the heuristic approach proposed in Huang et al. (2011), for solving a general structural sparsity regularization problem, is applicable when the structural sparsity term originates from a tree structured or graph structured groupings of data elements. Huang et al. (2011)'s algorithm is applicable to problem (6.11) because $C(A)$ does originate from a tree structured grouping of the atom locations. Li et al. (2018) revised the algorithm for a better computational efficiency, and also presented the consistency and convergence results for the revised algorithm.

Huang et al. (2011)'s algorithm is referred to as the group orthogonal matching pursuit (gOMP) algorithm, which iteratively selects a group or an individual variable in each iteration that improves the squared loss most, while bounding the regularization term below the prescribed threshold. The number of iterations can be equal to the number of non-zero elements in the true $\boldsymbol{A}$. But when all non-zero variables in the true $\boldsymbol{A}$ belong to a single group yet some variables in that group are zero, the group selection then does not fully explain the true $\boldsymbol{A}$ and many individual selections have to be performed, leading potentially to numerous iterations. Li et al. (2018) revised the algorithm by splitting the variable selection iterations into two levels, group-level selection and within-group selection. The idea is that at each iteration one first selects a group of variables and then employs a marginal regression to choose non-zero elements within the chosen group. Algorithm 6.1 describes the details of the algorithm.

Let g_k denote the index of the lattice group selected at iteration k and then let $F^{(k)} = \cup_{l=1}^{k} L_{g_l}$ and

$$\hat{\boldsymbol{A}}^{(k)} = \arg\min L(\boldsymbol{A}; \tau^2, \boldsymbol{Y}) \text{ subject to } \operatorname{supp}(\boldsymbol{A}) \subset F^{(k)},$$

where $\operatorname{supp}(\boldsymbol{A}) = \{(m,n); (\boldsymbol{A})_{m,n} \neq 0\}$. For group selection, Li et al. (2018) followed what was done by Huang et al. (2011), i.e., select $g_k \in G$ that maximizes the following gain ratio,

$$\phi(g_k) = \frac{L(\hat{\boldsymbol{A}}^{(k-1)}; \tau^2, \boldsymbol{Y}) - L(\hat{\boldsymbol{A}}^{(k)}; \tau^2, \boldsymbol{Y})}{C(\hat{\boldsymbol{A}}^{(k)}) - C(\hat{\boldsymbol{A}}^{(k-1)})}. \tag{6.12}$$

The group selection augments $F^{(k-1)}$ by a set of non-zero elements of $\boldsymbol{A}$ to form $F^{(k)} = L_{g_k} \cup F^{(k-1)}$. Then, the marginal regression (Genovese et al. 2012) is applied for obtaining a sparse solution of the following regression model,

$$\boldsymbol{Y} = \boldsymbol{U}_\tau \boldsymbol{A} \boldsymbol{V}_\tau^T + \boldsymbol{E} \text{ subject to } \operatorname{supp}(\boldsymbol{A}) \subset F^{(k)}.$$

For the marginal regression, first compute the marginal regression coefficients,

$$\left(\hat{\boldsymbol{A}}_\gamma^{(k)}\right)_{m,n} = \boldsymbol{u}_m^T \boldsymbol{Y} \boldsymbol{v}_n \text{ if } (m,n) \in F^{(k)} \text{ or 0 otherwise.}$$

Each of the marginal regression coefficients is filtered by a threshold of ρ,

$$\left(\hat{\boldsymbol{A}}_\rho^{(k)}\right)_{m,n} = \left(\hat{\boldsymbol{A}}_\gamma^{(k)}\right)_{m,n} 1\left\{\left(\hat{\boldsymbol{A}}_\gamma^{(k)}\right)_{m,n} \geq \rho\right\}.$$

The iterations of the group selection and the subsequent marginal regression are repeated as long as a sparsity condition $C(\hat{\boldsymbol{A}}_\rho^{(k)}) \leq c$ is satisfied. Li et al. (2018) called the revised algorithm *gOMP-Thresholding*.

The algorithm has multiple tuning parameters, including the list of potential lattice groups $\{L_g; g \in G\}$, the bandwidth τ^2 for a point spreading function, the constant c in the constraint function that decides when the iterations stop, and the

Algorithm 6.1 $gOMP$-Thresholding

Require: parameter τ^2, the list of potential groups $\{L_g; g \in G\}$, stopping criterion c, threshold ρ
Input: input image $\boldsymbol{Y}$
Output: $\boldsymbol{A}$

Initialization $F^{(0)} = \emptyset$ and $\hat{\boldsymbol{A}}_\rho^{(k)} = \boldsymbol{0}$
while $C(\hat{\boldsymbol{A}}_\rho^{(k)}) < c$ **do**
    $k = k + 1$
    Select $g_k \in G$ to maximize $\phi(g_k)$ following the group selection criterion in Eq. (6.12).
    Let $F^{(k)} = L_{g(k)} \cup F^{(k-1)}$.
    Marginal Regression: $\left(\hat{\boldsymbol{A}}_\gamma^{(k)}\right)_{m,n} = \boldsymbol{u}_m^T \boldsymbol{Y} \boldsymbol{v}_n$ if $(m, n) \in F^{(k)}$ or zero otherwise.
    Thresholding: $\left(\hat{\boldsymbol{A}}_\rho^{(k)}\right)_{m,n} = \left(\hat{\boldsymbol{A}}_\gamma^{(k)}\right)_{m,n} 1\left\{\left(\hat{\boldsymbol{A}}_\gamma^{(k)}\right)_{m,n} \geq \rho\right\}$.
end while

Source: Li et al. (2018)

threshold ρ that sets the within-group sparsity. The choices of the tuning parameters and statistical properties of the choices are presented in the next section.

6.4.1 *Listing L_g's and Estimating τ*

One of the input parameters required by the $gOMP$-$Thresholding$ algorithm is the list of the lattice groups L_g that may be found in the input image. Certainly, one can consider all possible lattice groups by exhausting the combinations of $\boldsymbol{s}_g$, $\boldsymbol{p}_g$ and $\boldsymbol{q}_g$. However, the number of combinations is theoretically infinite or would at least be colossal even when only the finite number of uniformly sampled values of $\boldsymbol{s}_g$, $\boldsymbol{p}_g$ and $\boldsymbol{q}_g$ are considered. Such strategy is certainly impractical.

To be practical, one could fix $\boldsymbol{p}_g$ and $\boldsymbol{q}_g$ to certain estimated values while exploring $\boldsymbol{s}_g$—doing so allows us to narrow the number of possible lattice groups down to the range of $\boldsymbol{s}_g$. Let $\hat{\boldsymbol{p}}$ and $\hat{\boldsymbol{q}}$ denote the estimates of $\boldsymbol{p}_g$ and $\boldsymbol{q}_g$, respectively. Due to the lattice periodicity, the range of $\boldsymbol{s}_g$ is restricted to the parallelogram formed by the two basis vectors $\hat{\boldsymbol{p}}$ and $\hat{\boldsymbol{q}}$, that is,

$$\boldsymbol{s}_g \in \{a_p \hat{\boldsymbol{p}} + a_q \hat{\boldsymbol{q}} \in \mathbb{Z}_M \times \mathbb{Z}_N; a_p, a_q \in [0, 1)\},$$

where $\times$ is a cross product operator of two vectors. Note that $||\hat{\boldsymbol{p}} \times \hat{\boldsymbol{q}}||_2$ is the area of the parallelogram formed by the two basis vectors of $\hat{\boldsymbol{p}}$ and $\hat{\boldsymbol{q}}$, equaling to the number of pixel locations in the parallelogram. Since the four vertices of the parallelogram represent the same $\boldsymbol{s}_g$ due to the lattice periodicity, the number of all possible $\boldsymbol{s}_g$ is the area minus redundancy, i.e., $||\hat{\boldsymbol{p}} \times \hat{\boldsymbol{q}}||_2 - 3$. Using $\hat{\boldsymbol{p}}$ and $\hat{\boldsymbol{q}}$ with the range of $\boldsymbol{s}_g$, we can then list all possible lattice groups. Specifically, the gth group is

$$L_g = \{\boldsymbol{s}_g + z_p \hat{\boldsymbol{p}} + z_q \hat{\boldsymbol{q}} \in \mathbb{Z}_M \times \mathbb{Z}_N; z_p, z_q \in \mathbb{Z}\}.$$

Note that $L_g \bigcap L_{g'} = \emptyset$ and $\bigcup_{g \in G} L_g = \mathbb{Z}_M \times \mathbb{Z}_N$. The groups form a non-overlapping partition of $\mathbb{Z}_M \times \mathbb{Z}_N$.

Next we shall discuss the strategies for obtaining good estimates, $\hat{\boldsymbol{p}}$ and $\hat{\boldsymbol{q}}$, which are pre-requisites for getting a practical L_g above. Estimating the two basis vectors $\hat{\boldsymbol{p}}$ and $\hat{\boldsymbol{q}}$ under low signal-to-noise ratios and in presence of missing atoms is difficult. The missing of atoms causes problems because the input image does not then contain a perfect lattice. To tackle these challenges, we propose to use the double fourier transform of an input image I to achieve the estimates $\hat{\boldsymbol{p}}$ and $\hat{\boldsymbol{q}}$.

The double fourier transform is defined by the fourier transform of the square of the fourier transform of I,

$$\mathcal{F}\{|\mathcal{F}\{I\}|^2\},$$

where $\mathcal{F}$ is a 2d fourier transform operator. According to Eq. (6.5), an input image is

$$I(\boldsymbol{x}) = f(\boldsymbol{x}) + \epsilon(\boldsymbol{x}),$$

where $\boldsymbol{x} = (x, y)$ denotes an two-dimensional image coordinate. The main signal $f(\boldsymbol{x})$ is the image intensity contributed by all atoms on the underlying lattice minus the contribution by missing atoms, i.e.,

$$f(\boldsymbol{x}) = P * f_\delta(\boldsymbol{x}) = P * \left(\sum_{\boldsymbol{x}_l \in L_g} \alpha_\delta \delta(\boldsymbol{x} - \boldsymbol{x}_l) - \sum_{\boldsymbol{x}_e \in E_g} \alpha \delta(\boldsymbol{x} - \boldsymbol{x}_e) \right),$$

where $E_g \subset L_g$ is the set of locations for which the atoms are missed. Let $f_\delta^* = \sum_{\boldsymbol{x}_l \in L_g} \alpha_\delta \delta(\boldsymbol{x} - \boldsymbol{x}_l)$ and $e_\delta^* = \sum_{\boldsymbol{x}_e \in E_g} \alpha \delta(\boldsymbol{x} - \boldsymbol{x}_e)$. The fourier transform of the input image is then

$$\mathcal{F}\{I\} = \mathcal{F}\{P\}\mathcal{F}\{f_\delta^*\} - \mathcal{F}\{P\}\mathcal{F}\{e_\delta^*\} + \mathcal{F}\{\epsilon\}.$$

Typically, the cardinality of E_g is ignorably small compared to the cardinality of L_g. Moreover, the locations in E_g are randomly distributed over the entire space of the input image. Since the power spectrum of a signal with randomly located peaks is uniform, and the uniform magnitude is proportional to the cardinality of E_g, the effect of $\mathcal{F}\{P\}\mathcal{F}\{e_\delta^*\}$ on the total fourier coefficient is ignorable. As such, we can simplify the expression of $\mathcal{F}\{I\}$ to be

$$\mathcal{F}\{I\} \approx \mathcal{F}\{P\}\mathcal{F}\{f_\delta^*\} + \mathcal{F}\{\epsilon\}.$$

We would like to assume further that the fourier transform of noise and the fourier transform of signal are nearly orthogonal, which is true for many practical cases since the noise is typically described by high frequency components and the signal is mostly described by low frequency components. Under this orthogonality assumption, $\mathcal{F}\{f_\delta^*\}\mathcal{F}\{\epsilon\} \approx 0$, which means

$$|\mathcal{F}\{I\}|^2 = |\mathcal{F}\{P\}|^2|\mathcal{F}\{f_\delta^*\}|^2 + |\mathcal{F}\{\epsilon\}|^2.$$

The double fourier transform of the input image is the fourier transform of $|\mathcal{F}\{I\}|^2$, which is

$$\mathcal{F}\{|\mathcal{F}\{I\}|^2\} = \mathcal{F}\{|\mathcal{F}\{P\}|^2\} * \mathcal{F}\{|\mathcal{F}\{f_\delta^*\}|^2\} + \mathcal{F}\{|\mathcal{F}\{\epsilon\}|^2\}. \tag{6.13}$$

We would like to further simplify the above expression through several observations and properties noted in the sequel. First, we note that the fourier transform of a Gaussian point spread function P is still a Gaussian point spread function and $|\mathcal{F}\{P\}|^2$ is again a Gaussian point spread function with a wider spreading width. What these two properties suggest is that $\mathcal{F}\{|\mathcal{F}\{P\}|^2\}$ is also a Gaussian point spread function (but with a wider spreading width), and therefore, we denote $\mathcal{F}\{|\mathcal{F}\{P\}|^2\}$ by $\tilde{P}$. Second, we note that the fourier transform is orthonormal transformation, meaning that the real parts and the imaginary parts of the fourier transform of a Gaussian white noise remain Gaussian white noises and suggesting that $|\mathcal{F}\{\epsilon\}|^2$ is a constant multiple of a (non-centered) chi-square random variable with degree of freedom of 2. Thus, $\mathcal{F}\{|\mathcal{F}\{\epsilon\}|^2\}$ is a linear combination of (non-centered) chi-square random variables, which we denote by $\tilde{\epsilon}$. Note that $\tilde{\epsilon}$ is still independent white noises since the fourier transform is orthonormal. Making use of all properties stated above, we can simplify the previous expression to

$$\mathcal{F}\{|\mathcal{F}\{I\}|^2\} = \tilde{P} * \mathcal{F}\{|\mathcal{F}\{f_\delta^*\}|^2\} + \tilde{\epsilon}.$$

The fourier transform of f_δ^* is

$$\begin{aligned}\mathcal{F}\{f_\delta^*\}(\boldsymbol{u}) &= \sum_{\boldsymbol{x}_l \in L_g} \alpha_\delta \int \delta(\boldsymbol{x} - \boldsymbol{x}_l) \exp\{-j\boldsymbol{u}^T\boldsymbol{x}\} \\ &= \sum_{\boldsymbol{x}_l \in L_g} \alpha_\delta \exp\{-j\boldsymbol{u}^T\boldsymbol{x}_l\},\end{aligned}$$

and its square is

$$|\mathcal{F}\{f_\delta^*\}(\boldsymbol{u})|^2 = \alpha_\delta^2 \left(\sum_{\boldsymbol{x}_l \in L_g} \cos(\boldsymbol{u}^T\boldsymbol{x}_l)\right)^2 + \alpha_\delta^2 \left(\sum_{\boldsymbol{x}_l \in L_g} \sin(\boldsymbol{u}^T\boldsymbol{x}_l)\right)^2.$$

Note $\boldsymbol{x}_l \in L_g$ is represented by

$$\boldsymbol{x}_l = \boldsymbol{s}_g + z_{p,l}\boldsymbol{p}_g + z_{q,l}\boldsymbol{q}_g \text{ for } z_{p,l}, z_{q,l} \in \mathbb{Z}. \tag{6.14}$$

The square of the fourier transform is simplified to

$$
\begin{aligned}
|\mathcal{F}\{f_\delta^*\}(\boldsymbol{u})|^2 &= \alpha_\delta^2 \left(\sum_{z_{p,l},z_{q,l}\in\mathbb{Z}} \cos(\boldsymbol{u}^T(\boldsymbol{s}_g + z_{p,l}\boldsymbol{p}_g + z_{q,l}\boldsymbol{p}_g)) \right)^2 \\
&+ \alpha_\delta^2 \left(\sum_{z_{p,k},z_{q,k}\in\mathbb{Z}} \sin(\boldsymbol{u}^T(\boldsymbol{s}_g + z_{p,k}\boldsymbol{p}_g + z_{q,k}\boldsymbol{p}_g)) \right)^2 \\
&= \alpha_\delta^2 \sum_{z_{p,l},z_{q,l},z_{p,k},z_{q,k}} 1 + 2\cos(\boldsymbol{u}^T((z_{p,l} - z_{p,k})\boldsymbol{p}_g + (z_{q,l} - z_{q,k})\boldsymbol{q}_g)) \\
&= \alpha_\delta^2 \sum_{z'_p,z'_q\in\mathbb{Z}} \left(\frac{1}{z'_p z'_q} \right) (1 + 2\cos(\boldsymbol{u}^T(z'_p\boldsymbol{p}_g + z'_q\boldsymbol{q}_g))).
\end{aligned}
$$

Let $\tilde{\boldsymbol{x}}_l = z'_p\boldsymbol{p}_g + z'_q\boldsymbol{q}_g$ and $\tilde{L}_g = \{z'_p\boldsymbol{p}_g + z'_q\boldsymbol{q}_g; z'_p, z'_q \in \mathbb{Z}\}$. The previous expression for $|\mathcal{F}\{f_\delta^*\}(\boldsymbol{u})|^2$ can be written as

$$
|\mathcal{F}\{f_\delta^*\}(\boldsymbol{u})|^2 \propto \sum_{\tilde{\boldsymbol{x}}_l\in\tilde{L}_g} \frac{1}{||\tilde{\boldsymbol{x}}_l||_2}(1 + 2\cos(\boldsymbol{u}^T\tilde{\boldsymbol{x}}_l)).
$$

Using this result, one can derive the double fourier transform of f_δ^* as

$$
\mathcal{F}\{|\mathcal{F}\{f_\delta^*\}|^2\}(\boldsymbol{\omega}) \propto \sum_{\tilde{\boldsymbol{x}}_l\in\tilde{L}_g} \frac{1}{||\boldsymbol{\omega}||_2}\delta(\boldsymbol{\omega} - \tilde{\boldsymbol{x}}_l).
$$

Therefore, the double fourier transform image of $\boldsymbol{Y}$ can be approximated by

$$
\mathcal{F}\{|\mathcal{F}\{I\}|^2\}(\boldsymbol{\omega}) = h\tilde{P} * \sum_{\tilde{\boldsymbol{x}}_l\in\tilde{L}_g} \frac{1}{||\boldsymbol{\omega}||_2}\delta(\boldsymbol{\omega} - \tilde{\boldsymbol{x}}_l) + \tilde{\epsilon}.
$$

where h is an constant. Note that the double fourier transform has peaks spaced every $\tilde{\boldsymbol{x}}_l \in \tilde{L}_g$, where $\tilde{L}_g$ has exactly the same basis vectors as the original lattice group L_g of the input image and is invariant to any spatial shift $\boldsymbol{s}_g$ of the lattice locations. In addition, the peaks in the double fourier transform are much more amplified in lower frequency bands that correspond to the smaller $||\boldsymbol{\omega}||_2$, while the noise $\tilde{\epsilon}$ is independently and identically distributed over $\boldsymbol{\omega}$. Considering all these, it is apparent that the lower frequency region of the double fourier transform reveals the original lattice basis vectors with a very high SNR ratio. Figure 6.3 show an example image and its double fourier transform, which is consistent with the pattern.

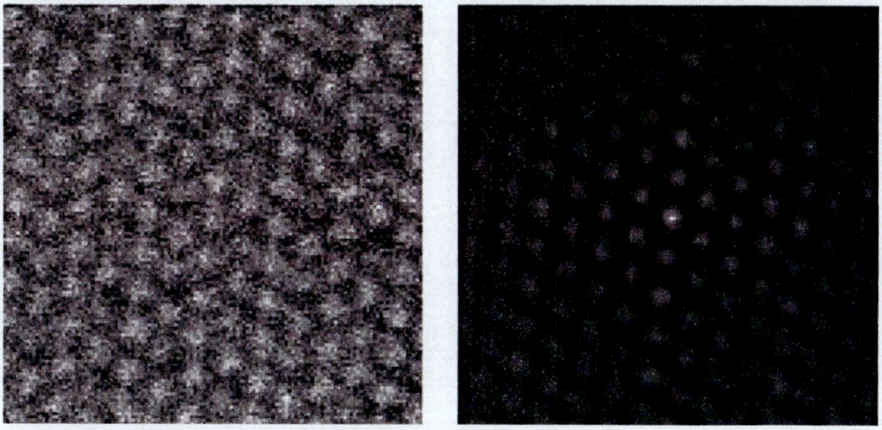

Fig. 6.3 Comparison of (**a**) an input image and (**b**) its double fourier transform. Reprinted with permission from Li et al. (2018)

We could employ an existing spot detection algorithm, in particular, the determinant of the Hessian (Bay et al. 2008), on the double fourier transform image to estimate $\boldsymbol{p}_g$ and $\boldsymbol{q}_g$ as well as the spreading width of the point spreading function $\tilde{P}$ on the double fourier transform domain, which is denoted by $\tilde{\tau}$. Based on the properties of the fourier transform of Gaussian point spread function, one knows that the spreading width $\tilde{\tau}$ is $\sqrt{2}$ times wider than the spreading width τ of the point spreading function P in the original input image. Naturally, once $\tilde{\tau}$ is estimated, one can get $\tau \approx \frac{1}{\sqrt{2}}\tilde{\tau}$.

6.4.2 Choice of Stopping Condition Constant c and Related Error Bounds

We here present the error bound of the solution of the algorithm relative to the ground truth. We start off by introducing some notations. Let $\bar{\boldsymbol{A}}$ denote the true signal to be estimated. For all $F \subset \mathbb{Z}_M \times \mathbb{Z}_N$, define

$$\eta_+(F) = \sup\left\{\frac{1}{MN}||\boldsymbol{U}_\tau \boldsymbol{A} \boldsymbol{V}_\tau^T||_F^2/||\boldsymbol{A}||_F^2 \,:\, \mathrm{supp}(\boldsymbol{A}) \subset F\right\}, \text{ and}$$

$$\eta_-(F) = \inf\left\{\frac{1}{MN}||\boldsymbol{U}_\tau \boldsymbol{A} \boldsymbol{V}_\tau^T||_F^2/||\boldsymbol{A}||_F^2 \,:\, \mathrm{supp}(\boldsymbol{A}) \subset F\right\}.$$

Moreover, for $c > 0$, define

$$\eta_+(c) = \sup\{\eta_+(\mathrm{supp}(\boldsymbol{A})) : C(\boldsymbol{A}) < c\},$$
$$\eta_-(c) = \inf\{\eta_-(\mathrm{supp}(\boldsymbol{A})) : C(\boldsymbol{A}) < c\}, \text{ and}$$
$$\eta_0 = \sup\{\eta_+(L_g) : g \in G\}.$$

Theorem 6.1 *Consider the true signal* $\bar{\boldsymbol{A}}$ *and* ϵ *such that*

$$\epsilon \in (0, ||\boldsymbol{Y}||_F^2 - ||\boldsymbol{Y} - \boldsymbol{U}_\tau \bar{\boldsymbol{A}} \boldsymbol{V}_\tau^T||_F^2].$$

If the choice of c satisfies

$$c \geq \frac{\eta_0 C(\bar{\boldsymbol{A}})}{\nu \eta_-(c + C(\bar{\boldsymbol{A}}))} \log \frac{||\boldsymbol{Y}||_F^2 - ||\boldsymbol{Y} - \boldsymbol{U}_\tau \bar{\boldsymbol{A}} \boldsymbol{V}_\tau^T||_F^2}{\epsilon} \text{ for } \nu \in (0, 1],$$

with probability $1 - p$,

$$||\hat{\boldsymbol{A}}_\rho^{(k)} - \bar{\boldsymbol{A}}||_F^2 \leq \frac{10||\boldsymbol{U}_\tau \bar{\boldsymbol{A}} \boldsymbol{V}_\tau^T - \mathbb{E}[\boldsymbol{Y}]||_F + 37\sigma^2(c + \eta_0) + 29\sigma^2 \log(6/p) + 2.5\epsilon}{MN\eta_-(c + C_0 + C(\bar{\boldsymbol{A}}))}$$

The proof of Theorem 6.1 is straightforward by using Theorems 6 and 9 of Huang et al. (2011). The theorem implies that the *gOMP-Thresholding* algorithm's solution output, $\hat{\boldsymbol{A}}_\rho^{(k)}$, is within the stated error bound relative to the ground truth $\bar{\boldsymbol{A}}$ when c is chosen properly, i.e., c satisfies the condition stated in the theorem. However, the theorem does not provide any practical guidance in terms of how to choose c because the condition for c therein is not computable.

On the other hand, the choice of c is related to the number of the lattice groups selected to describe an input image, since every iteration of the algorithm selects exactly one lattice group and c determines the number of iterations to run. When the number of lattice groups in the input image is known, that number can be used to determine c. For example, when a single crystalline material is imaged, there is only one lattice type with basis vectors $\boldsymbol{p}_g$ and $\boldsymbol{q}_g$. The number of atoms on the lattice group within an $M \times N$ digital image is

$$\frac{MN}{||\boldsymbol{p}_g \times \boldsymbol{p}_g||}.$$

The stopping condition c should then be set for one group selection, that is

$$c = \log_2(2|G|) + \log_2(2K)\frac{MN}{||\boldsymbol{p}_g \times \boldsymbol{p}_g||},$$

This choice of c is used in our numerical experiments.

6.4.3 Choice of Threshold ρ

In this section, we describe how to choose the threshold of ρ. Recall that ρ is applied to the marginal regression outcome $\hat{\boldsymbol{A}}_{\gamma}^{(k)}$ for enforcing a within-group sparsity. Let ρ_j denote the value of the jth largest element of $\hat{\boldsymbol{A}}_{\gamma}^{(k)}$, and then define

$$S_j = \{(m, n) \in \mathbb{Z}_M \times \mathbb{Z}_N : \left(\hat{\boldsymbol{A}}_{\gamma}^{(k)}\right)_{m,n} \geq \rho_j\}. \tag{6.15}$$

In addition, let $\boldsymbol{y} = vec(\boldsymbol{Y})$, let $\boldsymbol{X}_j$ denote $MN \times |S_j|$ matrix with columns $\{\boldsymbol{v}_n \otimes \boldsymbol{u}_m : (m, n) \in S_j\}$, where $\otimes$ denotes a Kroneck product, and let $\boldsymbol{\Sigma}_j = \boldsymbol{X}_j^T \boldsymbol{X}_j$ and $\boldsymbol{H}_j = \boldsymbol{X}_j(\boldsymbol{X}_j^T \boldsymbol{X}_j)^{-1} \boldsymbol{X}_j^T$. Similarly, when $\bar{S} = \text{supp}(\bar{\boldsymbol{A}})$ denote the support of the ground truth solution, $\bar{\boldsymbol{X}}$ denote $MN \times |\bar{S}|$ matrix with columns $\{\boldsymbol{v}_n \otimes \boldsymbol{u}_m : (m, n) \in \bar{S}\}$ and $\bar{\boldsymbol{\Sigma}} = \bar{\boldsymbol{X}}^T \bar{\boldsymbol{X}}$.

Theorem 6.2 *Let* $\bar{a} = \inf\{(\bar{\boldsymbol{A}})_{m,n} : (m, n) \in \bar{S}\}$ *and* $\bar{b} = \inf\{||\bar{\boldsymbol{\Sigma}} \boldsymbol{\mu}||_2 : ||\boldsymbol{\mu}||_2 = 1\}$. *For* $q \geq 1$, *set*

$$j^* = \min\{j : Del(j) < \sigma \delta_k\}. \tag{6.16}$$

where $Del(j) = ||(\boldsymbol{H}_{j+1} - \boldsymbol{H}_j)\boldsymbol{y}||_F^2$, $\delta_k = q\sqrt{2 \log ||\hat{\boldsymbol{A}}_{\gamma}^{(k)}||_0}$, *and* σ *is the standard deviation of a Gaussian observation noise in* $\boldsymbol{Y}$. *If the following condition holds*

$$\bar{a} \geq 2q\sigma \bar{b}^{-1/2} \sqrt{\frac{2 \log ||\hat{\boldsymbol{A}}_{\gamma}^{(k)}||_0}{MN}} \text{ and } \bar{S} \subset supp(\hat{\boldsymbol{A}}_{\gamma}^{(k)}), \tag{6.17}$$

then $S_{j^*} = \bar{S}$ *with probability no less than* $1 - 4 \log ||\hat{\boldsymbol{A}}_{\gamma}^{(k)}||_0^{-q^2}$.

The proof of the theorem is straightforward by using Theorem 5 of Slawski and Hein (2013). Theorem 6.2 provides the statistical guarantee of support recovery for the *gOMP-Thresholding* algorithm with the choice of threshold ρ_{j^*}. But obtaining j^* would require the knowledge of noise level σ, which is unfortunately unknown. The practical meaning of j is the number of atoms in image $\boldsymbol{Y}$, and j^* is the optimal choice of this number. In practice, we can substitute the estimate of the noise, $\hat{\sigma}$, to get a naive pug-in estimate of the number of atoms, i.e.,

$$\hat{j}^* = \min\{j : Del(j) < \hat{\sigma} \delta_k\}, \tag{6.18}$$

where

$$\hat{\sigma}^2 = \frac{||\boldsymbol{Y} - \boldsymbol{U}_\tau \hat{\boldsymbol{A}}_{\gamma}^{(k)} \boldsymbol{V}_\tau^T||_F^2}{MN - 1}.$$

The corresponding threshold estimate is $\rho_{\hat{j}*}$. The naive estimation produces rather satisfactory results for all of the numerical examples that will be presented later in this chapter.

6.4.4 Comparison to the Sparse Group Lasso

As an alternative to the integrated lattice analysis approach described above, one can consider the following sparse group lasso (SGL) formulation,

$$\text{Minimize} \quad ||\boldsymbol{Y} - \sum_{g \in G} \boldsymbol{U}_\tau \boldsymbol{A}_{(g)} \boldsymbol{V}_\tau^T||_F^2 + \lambda_1 \sum_{g \in G} ||\boldsymbol{A}_{(g)}||_F + \lambda_2 ||\boldsymbol{A}||_1$$

where $||\boldsymbol{A}||_1 = \sum_{m=1}^{M} \sum_{n=1}^{N} |(\boldsymbol{A})_{m,n}|$ is the elementwise L1 norm of $\boldsymbol{A}$. Chatterjee et al. (2012) showed that the SGL regularizer is a special case of the regularization in the integrated lattice approach, which uses the hierarchical tree-induced sparsity norm (Liu and Ye 2010; Jenatton et al. 2011). They provided explicit bounds for the consistency of SGL and Liu and Ye (2010) proposed a sub-gradient approach to solve the SGL problem.

We tested the SGL algorithm and the integrated lattice analysis approach using a test image as shown in Fig. 6.4, where all atoms belong to a single lattice but there are missing locations. We used the MATLAB package SLEP for implementing SGL, which uses the sub-gradient solution algorithm (Liu and Ye 2010). In this sub-gradient solution algorithm, two tuning penalty parameters, λ_1 and λ_2, play crucial roles impacting the quality of the results. Performing a popular cross-validation

Fig. 6.4 The test image for comparing the SGL algorithm and the integrated lattice analysis approach. Reprinted with permission from Li et al. (2018)

selection that exhaustively searches in the two dimensional space of (λ_1, λ_2) is computationally demanding. Instead, we used the *alternative search* (She 2009, 2010), which runs in two steps, that is, first choosing λ_2 while fixing λ_1 to a small constant and then choosing λ_1 while fixing λ_2 at the value chosen in the first step. In the first step, when λ_1 is set to a small magnitude, SGL performs like in a group lasso method, i.e., it performs group selection but not the within-group selection, which is comparable to the group selection step in the *gOMP-Thresholding* (corresponding to Line 4 of Algorithm 1). Figure 6.5 presents the numerical outcome of SGL and that of *gOMP-Thresholding* with no thresholding step and both outcomes are comparable.

Once λ_2 is chosen, λ_1 is then fine tuned using a selective cross validation (She et al. 2013). We compared in Figs. 6.6 and 6.7 the SGL outcome under fine-tuned parameters and the *gOMP-Thresholding* outcome (i.e., *gOMP* with the thresholding step). SGL made two false detections, whereas *gOMP-Thresholding* made only one false detection. We observed from many other numerical cases that choosing a good λ_2 for SGL was not a straightforward task, while the *gOMP-Thresholding* algorithm has an easy-to-use threshold selector as presented in Sect. 6.4.3.

6.5 Numerical Examples with Synthetic Datasets

This section presents numerical experiments based on synthetic images to show how the integrated lattice analysis method performs. A synthetic image of 75×75 pixels is generated to house 121 (11×11) atoms if the atoms locate on all lattice grid locations with no vacancy.

We generated 250 random variants of the synthetic image. Three factors are considered while generating the 100 images. The first factor is the number of atom vacancies placed, which varies with the choices of $\{5, 10, 15, 20, 25\}$.

The second factor is the spatial pattern of atom vacancies. We considered five different patterns. The uniform mode is where atom vacancies were uniformly sampled among the 121 atom sites. The uniform mode is considered as mode 0. The other four modes (i.e., modes 1, 2, 3 and 4) are where atoms were randomly sampled among a subset of the 121 atom sites. Figure 6.8 shows the subsets for the four modes. In the figure, most of the 121 lattice grid locations are occupied by white blobs, but some grid locations do not show white blobs. The locations with no white blobs correspond to the subset of potential vacancies, and the atom vacancies are selected among the subset.

The third design factor while creating the simulation is the observation noise level. We applied the Gaussian white noises with its variance chosen from the set of $\{0.05, 0.15, 0.25, 0.35, 0.45, 0.55, 0.65, 0.75, 0.85, 0.95\}$. In this simulation study, the intensity scales of all synthetic images were normalized to [0, 1], for which the true signal's variance σ^2_{sig} is around 0.075. The true signal variance was computed

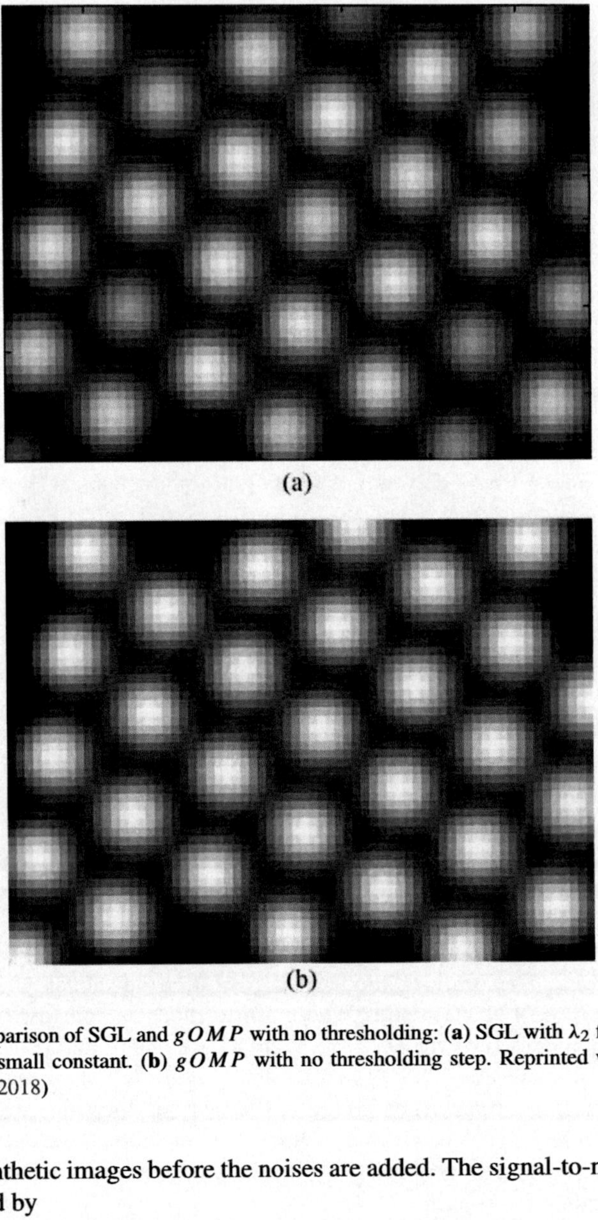

Fig. 6.5 Comparison of SGL and $gOMP$ with no thresholding: (a) SGL with λ_2 fine-tuned while fixing λ_1 to a small constant. (b) $gOMP$ with no thresholding step. Reprinted with permission from Li et al. (2018)

using the synthetic images before the noises are added. The signal-to-noise ratio can be calculated by

$$10\log_{10}(\sigma_{sig}^2) - 10\log_{10}(\sigma_{noise}^2),$$

where σ_{sig}^2 and σ_{noise}^2 are the variances of the true signal and the noise, respectively. For the set of noise variances chosen for the simulation, the SNR values are 1.76, −3.01, −5.23, −6.69, −7.78, −8.65, −9.38, −10.00, −10.54, and −11.03

(a)

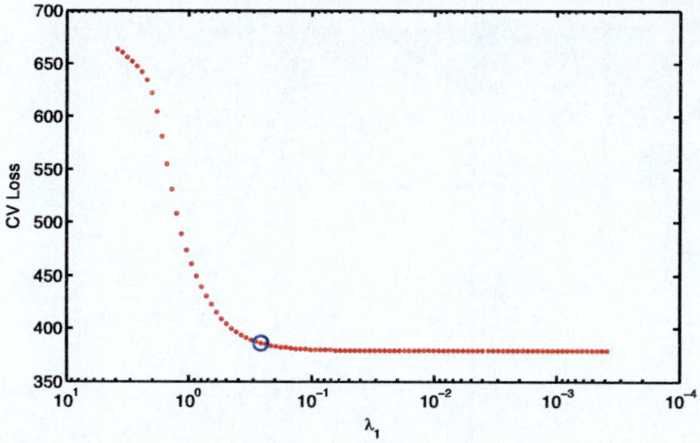

(b)

Fig. 6.6 Results of the sparse group lasso (SGL) with both λ_1 and λ_2 fine tuned: (**a**) Individual atom locations identified are marked with circles and the false positives are marked with crosses. (**b**) SCV loss versus λ_1 plot. The circle indicates the minimum SCV loss. Reprinted with permission from Li et al. (2018)

decibels, respectively. Apparently many of the synthetic images generated have extremely low contrasts. Some of the noisy images are illustrated in Fig. 6.9.

Combining the different choices of the three factors leads to a total of 250 different simulation designs. For each design, we ran 50 experiment replications. The false positives and false negatives of atom detections were counted and averaged over the replicated experiments. The variance of the average counts due to each factor was 137.5 for the first factor (number of vacancies), 7.4 for the second factor (spatial pattern of vacancies), and 930.9 for the third factor (noise variance level). The average counts were the least influenced by the second factor (spatial patterns of atom vacancies), whose variance was comparable to the random variation of the average counts over the 50 replicated experiments.

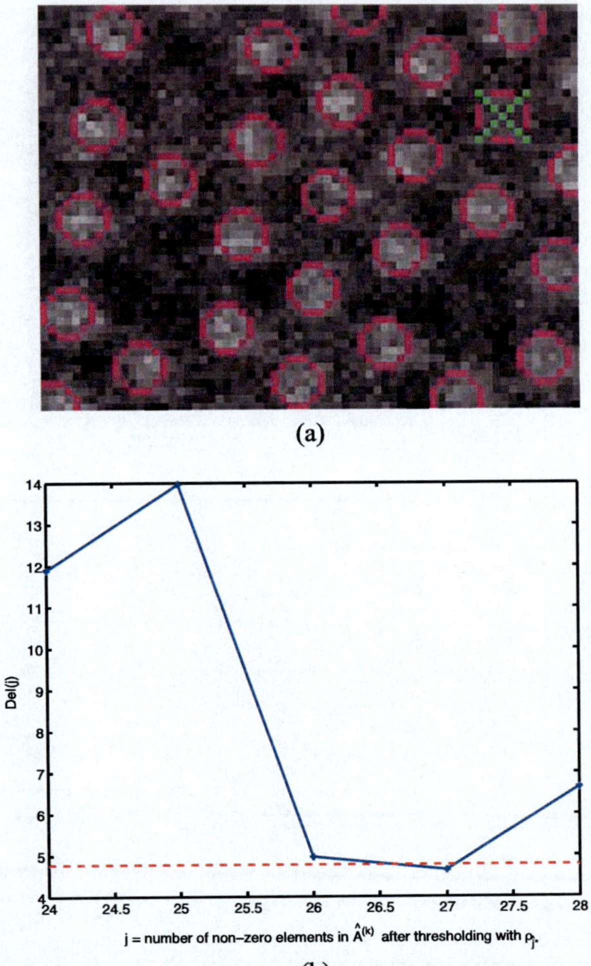

Fig. 6.7 Results of the *gOMP-Thresholding* algorithm with ρ chosen to be $\rho_{\hat{j}*}$: **(a)** Individual atom locations identified are marked with circles and false positives are marked with crosses. **(b)** $Del(j)$ versus $\hat{\sigma}\delta_k$ (horizontal bar) is plotted for threshold selection. Following Eq. (6.16), The $\hat{j}^*$ is selected to be the first j that achieves $Del(j)$ below the horizontal bar. Reprinted with permission from Li et al. (2018)

Figure 6.10 presents the average counts of the false positives and false negatives for different combinations of the first and third factors. The numbers of the false positives and false negatives can be kept very low and steady when the noise variance is below 0.6. However, the numbers increased significantly when the noise variance goes beyond 0.6. The major reason of the increase is attributable to the failure to estimate the two lattice basis vectors, $\boldsymbol{p}_g$ and $\boldsymbol{q}_g$, under the high image noises. To see this argument more clearly, we calculated the L2 norm of the difference between the estimates and the true values of each of the two lattice basis vectors, and used the sum of the two L2 norms to quantify the bias of the resulting estimates. Figure 6.11 presents the estimation biases for the same combinations as in Fig. 6.10 of noise variances and numbers of vacancies. Figure 6.11 f demonstrates the same pattern as observed in Fig. 6.10, i.e., the biases were zero when the noise

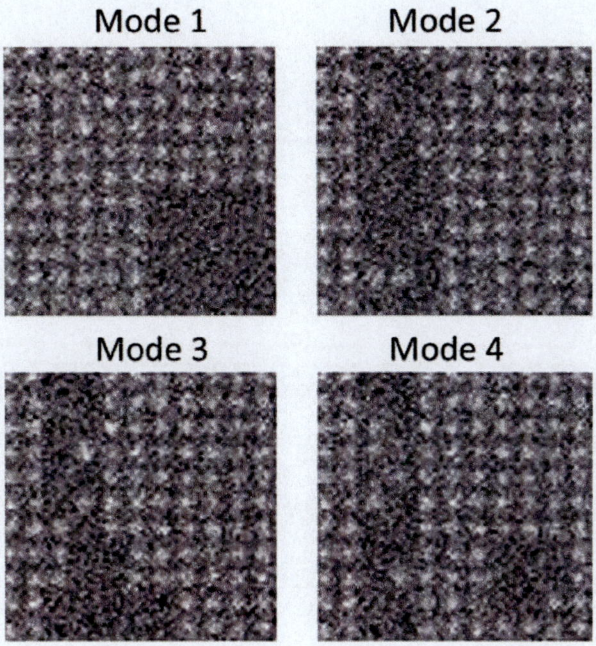

Fig. 6.8 Different spatial patterns of atom vacancy. Reprinted with permission from Li et al. (2018)

variances are below 0.6 but increased considerably for higher noise variances. The lowest noise variances that yield a positive bias depends on the number of atom vacancies. Understandably, in the presence of more vacancies, the bias tends to appear even for lower noise variances. When the estimation biases of the basis vectors are large enough, a group of potential atom locations restricted by the basis vectors become misaligned with the true atom locations, which in turn causes the rise in false positives and false negatives.

6.6 Lattice Analysis for Catalysts

We in this section demonstrate how the integrated lattice pattern analysis approach is applied to the atomic-level structural determination of Mo-V-M-O oxide materials. For this demonstration, we use three STEM images of the Mo-V-M-O catalysts that were synthesized at the Oak Ridge National Laboratory, following the synthesis method reported in He et al. (2015). The images were produced also at the Oak Ridge National Laboratory at the sub-angstrom spatial resolution using a high angle annular dark field scanning transmission electron microscope (HAADF-STEM). The images were labeled as `B5`, `H5`, and `H10`, respectively. The signal-to-noise

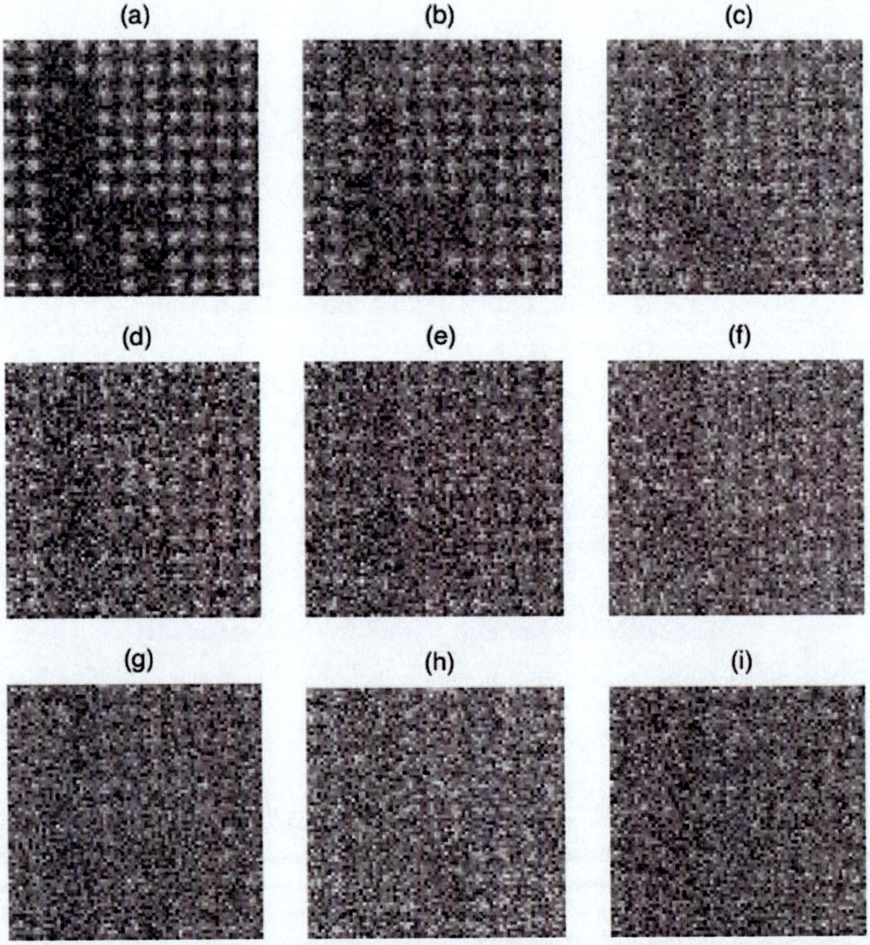

Fig. 6.9 Demonstration of noisy synthetic images used in the simulation study. **(a)** Noise variance = 0.05. **(b)** Noise variance = 0.15. **(c)** Noise variance = 0.35. **(d)** Noise variance = 0.45. **(e)** Noise variance =0.55. **(f)** Noise variance = 0.65. **(g)** Noise variance = 0.75. **(h)** Noise variance = 0.85. **(i)** Noise variance =0.95. Reprinted with permission from Li et al. (2018)

ratios of the three images were estimated to be 7.38, 8.74, and 0.91 decibels, respectively. For estimating the SNR ratios, we handpicked several image foreground areas where atoms are present and several background areas with no presence of atoms. We computed the variance of the image intensities in the foreground areas and the variance of the intensities in the background areas. Assuming independence between the foreground signals and the background noises, we estimate the true signal variance by subtracting the background variance from the foreground variance, while treating the background variance as the noise variance. Apparently, the last image has a very low SNR value.

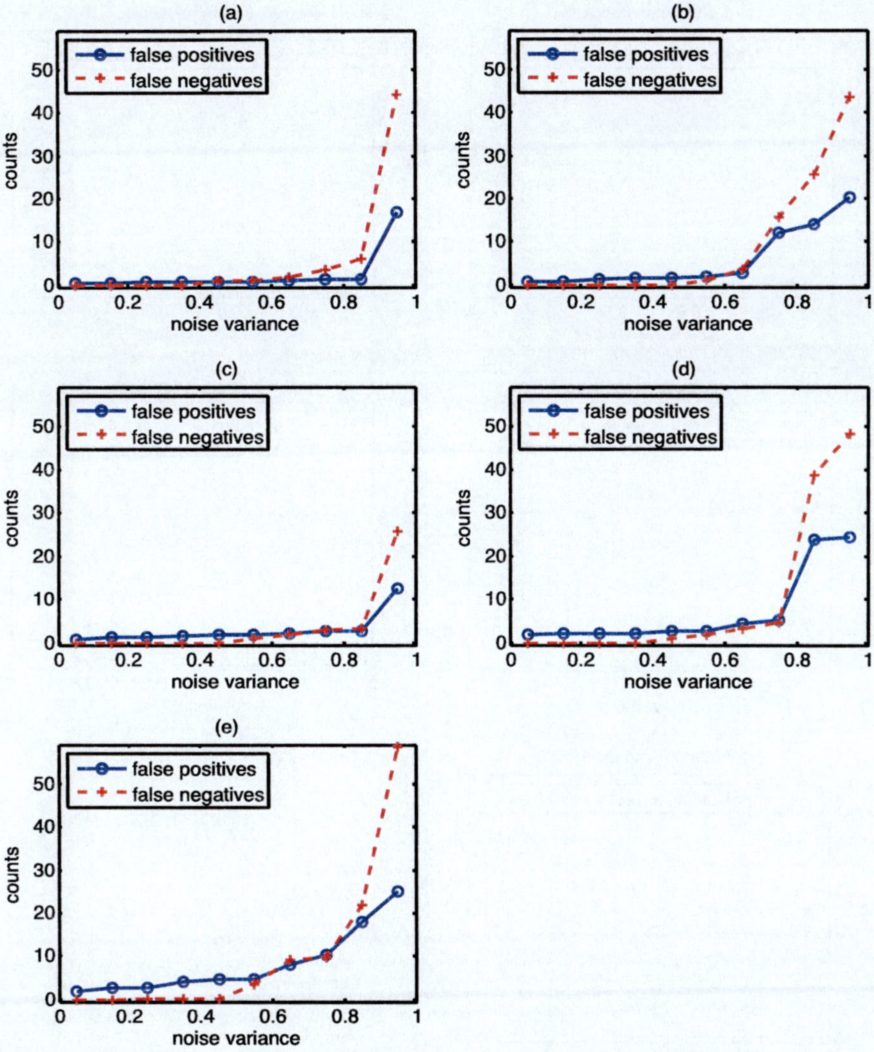

Fig. 6.10 False positive and false negatives of the estimated atom locations in different simulation designs. Reprinted with permission from Li et al. (2018)

The B5 image in Fig. 6.12 was used to demonstrate the capability of the integrated approach in terms of identifying the global lattice grid of atoms under local image distortions. Although the precision of the scanning electron microscope is rather high, the precise control of the electron probe at the sub-angstrom level is still challenging. As a result, the actual probe location may deviate from the probe location assigned for imaging, causing mild image distortions (Sang et al. 2016a).

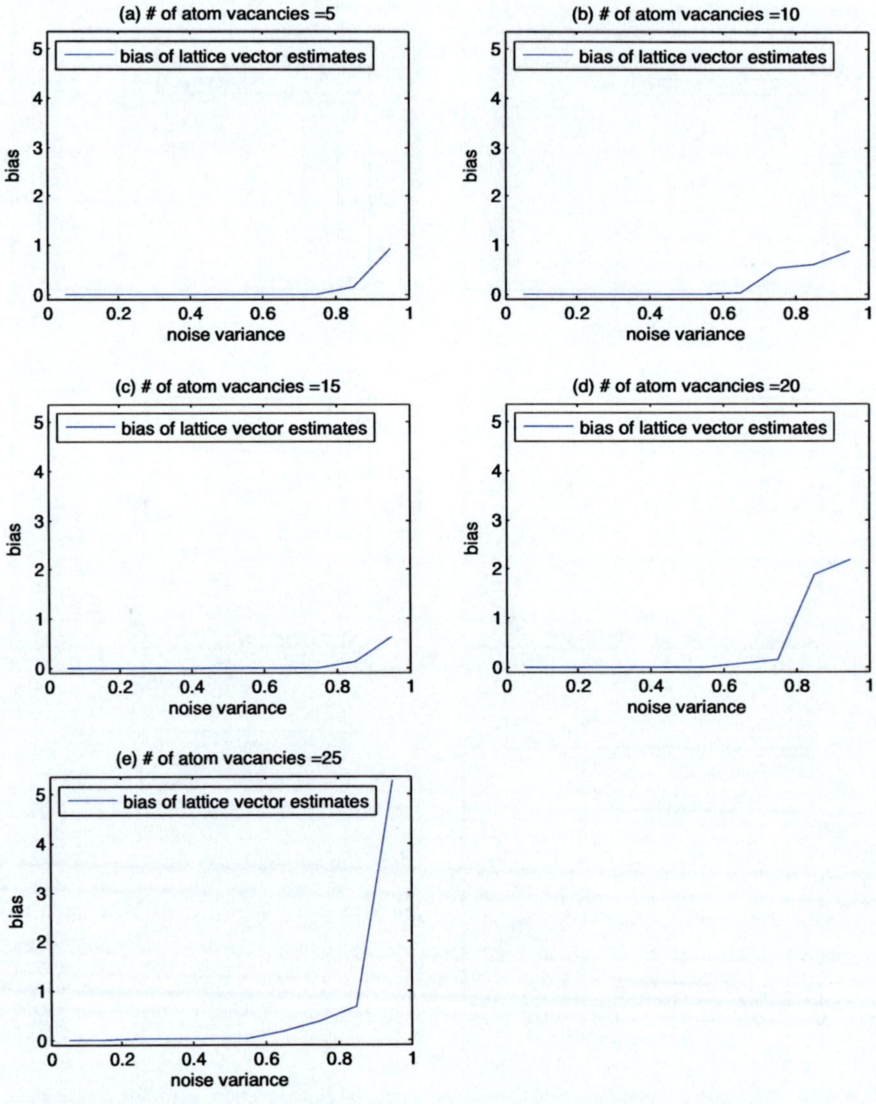

Fig. 6.11 Estimation bias of the lattice basis vectors $\boldsymbol{p}_g$ and $\boldsymbol{q}_g$ in different simulation designs. The estimation bias is quantified by the sum of the estimation biases of the two individual basis vectors, whereas an individual bias is the L2 norm of the difference between the estimated basis vector and the true basis vector. (**a**) # of atom vacancies = 5. (**b**) # of atom vacancies = 10. (**c**) # of atom vacancies = 15. (**d**) # of atom vacancies =20. (**e**) # of atom vacancies = 25. Reprinted with permission from Li et al. (2018)

Fig. 6.12 The B5 image. Reprinted with permission from Li et al. (2018)

When such image distortions occurred during the imaging of a crystal material with a single lattice grid, they could create the illusion that the sample material consists of multiple lattice grids, complicating the task of lattice identification.

With the group-level sparsity regularization used in the integrated lattice analysis approach, the lattice identification is rather robust against local image distortions. Figure 6.13 shows the outcome of the integrated approach applied to B5, in which the estimated lattice positions were overlaid on top of the original image. One can observe that most of the lattice positions identified (solid dots) match well with the actual atom locations (bright, white spots). A few exceptions are on the top left portion of the figure, where atoms are slightly off from the estimated lattice positions. For a better presentation, we magnified some parts of the figure and show the magnification in three insert boxes. In the middle box, atoms deviate more noticeably from the estimated lattice grid, but the deviations disappear or are much reduced in the upper and lower boxes. This deviation illustrates a local image distortion resulting from the scan "ramp-up" (Sang et al. 2016a). Detecting this type of local distortion through simple visual inspection of the original image is very labor-intensive and also subject to human errors. The integrated approach is capable of identifying a global lattice grid correctly under such local image distortions. This capability is practically meaningful since it can be used as a robust atomic lattice identification method. Moreover, the calculation of the deviations from a global lattice grid can provide important feedback for controlling more precisely the electron probe.

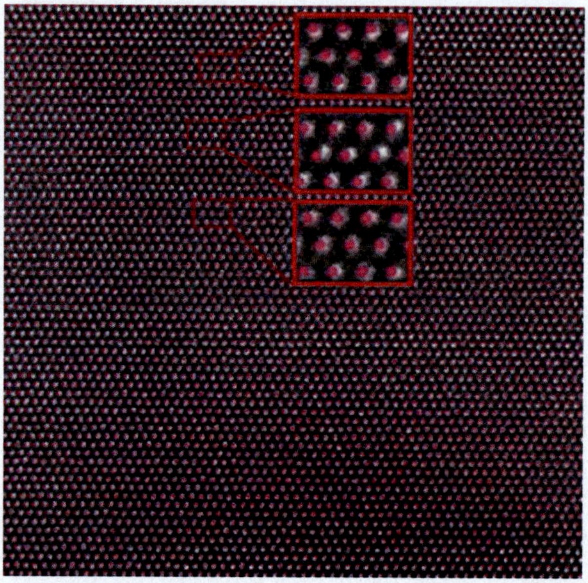

Fig. 6.13 The estimated lattice overlaid on tope of the B5 image. Reprinted with permission from Li et al. (2018)

Atom vacancies, regarded as one type defect in a crystal lattice, are another structure feature closely related to material properties. Adler (2004) reported that distribution of atom vacancies are intrinsically coupled with magnetic, electronic, and transport properties of solid-oxides . We applied the gOMP-thresholding algorithm to the H5 and H10 images of the Mo-V-M-O oxide materials for detecting atomic vacancies on the lattice. Figure 6.14 shows the outcome of atom and vacancy detections on the H5 image. The gOMP-thresholding algorithm identified one lattice group and detected the atomic vacancies correctly. The number of false negatives was zero, while there is one false positive. Figure 6.15 shows the outcome of atom detections on the H10 image. H10 has more complex patterns, of which a half of the image has atoms but the other half is only the background without atoms. Our method yielded an impressive detection outcome with only six false positives out of the total of 170 detections and did not produce any false negatives. By contrast, the SGL method produced more false positives and false negatives for both test images—more specifically, 93 false negatives and 0 false positives for H5 and 8 false negatives and 2 false positives for H10. We believe that the SGL result can be improved with a better choice of its tuning parameters. However, the parameter tuning does not appear straightforward. Even the cross-validation choice did not work very well, despite its high computational demand. In summary, the gOMP-Threshold algorithm was successful in estimating atom vacancies in the low-contrast STEM images, complementing the capability of the first principle-based approaches such as the density function theory (DFT). A follow-up work could be combining

(a)

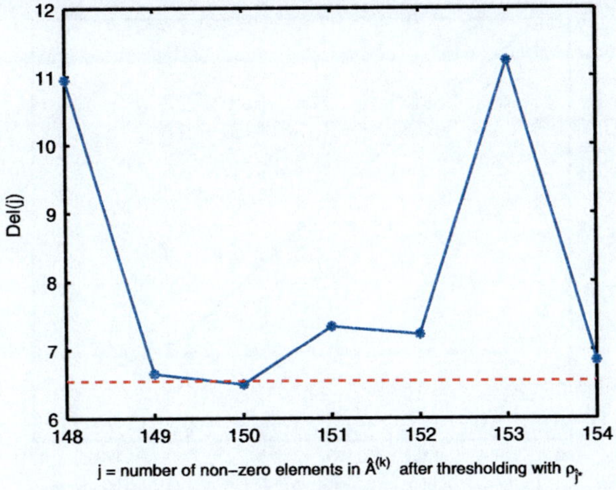

(b)

Fig. 6.14 Results of the integrated approach applied to `H5` with threshold ρ chosen to be $\rho_{\hat{j}*}$: **(a)** Individual atom locations identified are marked with circles, while the false positives are marked with crosses (to avoid any effects of atoms cropped around the image boundary, the image region outside the black dashed bounding box was not analyzed.) **(b)** $Del(j)$ versus $\hat{\sigma}\delta_k$ (horizontal bar) is plotted for threshold selection. Following Eq. (6.16), $\hat{j}^*$ is selected to be the first j that achieves $Del(j)$ below the horizontal bar. Reprinted with permission from Li et al. (2018)

gOMP-thresholding and DFT to investigate atomic-defect configurations, which is a very important topic for future nano-electronic devices and catalytic applications (Sang et al. 2016b).

Used for determining the structure-property relations at the atomic scale, the importance of automated atomic structure identification is on the rise with the advent of genomic libraries, e.g., the NIST Materials Genome Initiative (Dima et al. 2016). Results for the `H5` and `H10` images showcase the promising capability of the integrated approach for fulfilling this broad objective.

(a)

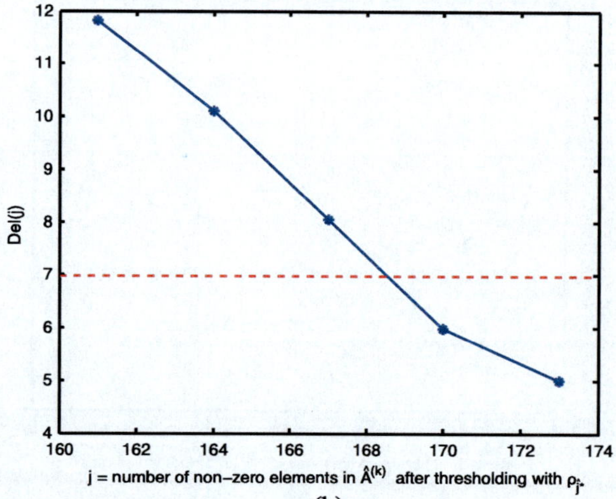

(b)

Fig. 6.15 Results of the integrated approach applied to H10 with threshold ρ chosen to be $\rho_{\hat{j}*}$: **(a)** Individual atom locations identified are marked with circles, while the false positives are marked with crosses (to avoid any effects of atoms cropped around the image boundary, the image region outside the black dashed bounding box was not analyzed.) **(b)** $Del(j)$ versus $\hat{\sigma}\,\delta_k$ (horizontal bar) is plotted for threshold selection. Following Eq. (6.16), $\hat{j}^*$ is selected to be the first j that achieves $Del(j)$ below the horizontal bar. Reprinted with permission from Li et al. (2018)

In terms of computation, the integrated approach took 18 min for analyzing the H5 and H10 images. The majority of the computing time was used to evaluate the threshold value described in Eq. (6.18). For the B5 image, we have not included the thresholding step, because we only need to estimate the global lattice grid with no need for estimating atom vacancies. The computation time without thresholding was 11 s.

6.7 Closing Remark

The integrated lattice pattern analysis approach allows automated analysis of an atomically resolved images for locating individual atoms and identifying their spatial symmetries and defects. The approach is a good candidate for analyzing atomic-scale images and extracting the lattice structural information of materials. A feature database on structural information can be constructed based on the analysis of existing atomic-scale images using the integrated approach. Such database, if constructed, could lay the foundation for enabling structural studies of materials from microscope image data. This is an exciting direction for developing quantitative methodologies to analyze atomically resolved materials data. We expect these types of methods to become more popular in processing catalysts, 2D materials, complex oxides, and other varieties of engineering and scientifically relevant materials.

References

Adler SB (2004) Factors governing oxygen reduction in solid oxide fuel cell cathodes. Chemical Reviews 104(10):4791–4844

Bay H, Ess A, Tuytelaars T, Van Gool L (2008) Speeded-up robust features (SURF). Computer Vision and Image Understanding 110(3):346–359

Belianinov A, He Q, Kravchenko M, Jesse S, Borisevich AY, Kalinin SV (2015) Identification of phases, symmetries and defects through local crystallography. Nature Communications 6:7801.1–7801.8

Borisevich AY, Chang HJ, Huijben M, Oxley MP, Okamoto S, Niranjan MK, Burton J, Tsymbal E, Chu YH, Yu P, Ramesh R, Kalinin SV, Pennycook SJ (2010a) Suppression of octahedral tilts and associated changes in electronic properties at epitaxial oxide heterostructure interfaces. Physical Review Letters 105(8):087204

Borisevich AY, Ovchinnikov OS, Chang HJ, Oxley MP, Yu P, Seidel J, Eliseev EA, Morozovska AN, Ramesh R, Pennycook SJ, Kalinin SV (2010b) Mapping octahedral tilts and polarization across a domain wall in BiFeO3 from Z-contrast scanning transmission electron microscopy image atomic column shape analysis. ACS Nano 4(10):6071–6079

Bright DS, Steel EB (1987) Two-dimensional top hat filter for extracting spots and spheres from digital images. Journal of Microscopy 146(2):191–200

Chang HJ, Kalinin SV, Morozovska AN, Huijben M, Chu YH, Yu P, Ramesh R, Eliseev EA, Svechnikov GS, Pennycook SJ, Borisevich AY (2011) Atomically resolved mapping of polarization and electric fields across Ferroelectric/Oxide interfaces by Z-contrast imaging. Advanced Materials 23(21):2474–2479

Chatterjee S, Steinhaeuser K, Banerjee A, Chatterjee S, Ganguly A (2012) Sparse group lasso: Consistency and climate applications. In: Proceedings of the 2012 SIAM International Conference on Data Mining, Siam, pp 47–58

Dima A, Bhaskarla S, Becker C, Brady M, Campbell C, Dessauw P, Hanisch R, Kattner U, Kroenlein K, Newrock M, Peskin A, Plante R, Li SY, Rigodiat PF, Amaral GS, Trautt Z, Schmitt X, Warren J, Youssef S (2016) Informatics infrastructure for the materials genome initiative. Jom 68(8):2053–2064

Efron B, Hastie T, Johnstone I, Tibshirani R (2004) Least angle regression. The Annals of Statistics 32(2):407–499

Genovese CR, Jin J, Wasserman L, Yao Z (2012) A comparison of the lasso and marginal regression. Journal of Machine Learning Research 13:2107–2143

He J, Borisevich A, Kalinin SV, Pennycook SJ, Pantelides ST (2010) Control of octahedral tilts and magnetic properties of perovskite oxide heterostructures by substrate symmetry. Physical Review Letters 105(22):227203

He Q, Woo J, Belianinov A, Guliants VV, Borisevich AY (2015) Better catalysts through microscopy: Mesoscale M1/M2 intergrowth in Molybdenum-Vanadium based complex oxide catalysts for Propane ammoxidation. ACS Nano 9(4):3470–3478

Huang J, Zhang T, Metaxas D (2011) Learning with structured sparsity. Journal of Machine Learning Research 12(Nov):3371–3412

Hughes J, Fricks J, Hancock W (2010) Likelihood inference for particle location in fluorescence microscopy. The Annals of Applied Statistics 4(2):830–848

Jenatton R, Audibert JY, Bach F (2011) Structured variable selection with sparsity-inducing norms. Journal of Machine Learning Research 12:2777–2824

Jia CL, Nagarajan V, He JQ, Houben L, Zhao T, Ramesh R, Urban K, Waser R (2007) Unit-cell scale mapping of ferroelectricity and tetragonality in epitaxial ultrathin ferroelectric films. Nature Materials 6(1):64–69

Jia CL, Mi S, Faley M, Poppe U, Schubert J, Urban K (2009) Oxygen octahedron reconstruction in the SrTiO 3/LaAlO 3 heterointerfaces investigated using aberration-corrected ultrahigh-resolution transmission electron microscopy. Physical Review B 79(8):081405

Jia CL, Urban KW, Alexe M, Hesse D, Vrejoiu I (2011) Direct observation of continuous electric dipole rotation in flux-closure domains in ferroelectric Pb (Zr, Ti) O3. Science 331(6023):1420–1423

Kim YM, He J, Biegalski MD, Ambaye H, Lauter V, Christen HM, Pantelides ST, Pennycook SJ, Kalinin SV, Borisevich AY (2012) Probing oxygen vacancy concentration and homogeneity in solid-oxide fuel-cell cathode materials on the subunit-cell level. Nature Materials 11(10):888–894

Kim YM, Kumar A, Hatt A, Morozovska AN, Tselev A, Biegalski MD, Ivanov I, Eliseev EA, Pennycook SJ, Rondinelli JM, Kalinin SV, Borisevich AY (2013) Interplay of octahedral tilts and polar order in BiFeO3 films. Advanced Materials 25(17):2497–2504

Kim YM, Morozovska A, Eliseev E, Oxley MP, Mishra R, Selbach SM, Grande T, Pantelides S, Kalinin SV, Borisevich AY (2014) Direct observation of ferroelectric field effect and vacancy-controlled screening at the BiFeO3/LaxSr1-xMnO3 interface. Nature Materials 13(11):1019–1025

Li X, Belianinov A, Dyck O, Jesse S, Park C (2018) Two-level structural sparsity regularization for identifying lattices and defects in noisy images. The Annals of Applied Statistics 12(1):348–377

Liu J, Ye J (2010) Moreau-yosida regularization for grouped tree structure learning. In: Advances in Neural Information Processing Systems, pp 1459–1467

Nellist P, Pennycook S (2000) The principles and interpretations of annular dark-field Z-contrast imaging. Advances in Imaging and Electron Physics 113:148–204

Nelson CT, Winchester B, Zhang Y, Kim SJ, Melville A, Adamo C, Folkman CM, Baek SH, Eom CB, Schlom DG, Chen LQ, Pan X (2011) Spontaneous vortex nanodomain arrays at ferroelectric heterointerfaces. Nano Letters 11(2):828–834

Rezatofighi SH, Hartley R, Hughes WE (2012) A new approach for spot detection in total internal reflection fluorescence microscopy. In: 2012 9th IEEE International Symposium on Biomedical Imaging (ISBI), Ieee, pp 860–863

Sage D, Neumann FR, Hediger F, Gasser SM, Unser M (2005) Automatic tracking of individual fluorescence particles: application to the study of chromosome dynamics. IEEE Transactions on Image Processing 14(9):1372–1383

Sang X, Lupini AR, Unocic RR, Chi M, Borisevich AY, Kalinin SV, Endeve E, Archibald RK, Jesse S (2016a) Dynamic scan control in STEM: Spiral scans. Advanced Structural and Chemical Imaging 2(1):6

Sang X, Xie Y, Lin MW, Alhabeb M, Van Aken KL, Gogotsi Y, Kent PR, Xiao K, Unocic RR (2016b) Atomic defects in monolayer Titanium Carbide (Ti3C2T x) MXene. ACS Nano 10:9193–9200

She Y (2009) Thresholding-based iterative selection procedures for model selection and shrinkage. Electronic Journal of Statistics 3:384–415

She Y (2010) Sparse regression with exact clustering. Electronic Journal of Statistics 4:1055–1096

She Y, Wang J, Li H, Wu D (2013) Group iterative spectrum thresholding for super-resolution sparse spectral selection. IEEE Transactions on Signal Processing 61(24):6371–6386

Simon N, Friedman J, Hastie T, Tibshirani R (2013) A sparse-group lasso. Journal of Computational and Graphical Statistics 22(2):231–245

Slawski M, Hein M (2013) Non-negative least squares for high-dimensional linear models: Consistency and sparse recovery without regularization. Electronic Journal of Statistics 7:3004–3056

Smal I, Niessen W, Meijering E (2008) A new detection scheme for multiple object tracking in fluorescence microscopy by joint probabilistic data association filtering. In: 2008 5th IEEE International Symposium on Biomedical Imaging: From Nano to Macro, Ieee, pp 264–267

Vincent L (1993) Morphological grayscale reconstruction in image analysis: Applications and efficient algorithms. IEEE Transactions on Image Processing 2(2):176–201

Yankovich AB, Berkels B, Dahmen W, Binev P, Sanchez SI, Bradley SA, Li A, Szlufarska I, Voyles PM (2014) Picometre-precision analysis of scanning transmission electron microscopy images of platinum nanocatalysts. Nature Communications 5:4155.1–4155.7

Yuan M, Lin Y (2006) Model selection and estimation in regression with grouped variables. Journal of the Royal Statistical Society: Series B (Statistical Methodology) 68(1):49–67

Chapter 7
State Space Modeling for Size Changes

7.1 Motivating Background

A motivating application for the modeling technique, to be discussed in this chapter, comes from the monitoring and control need in a nanoparticle self-assembly process, which produces nanocrystals from small building blocks such as atoms and molecules. In such self-assembly process, atoms and molecules are spontaneously arranged into order structures at the nanoscale. It is considered a promising method of producing nanocrystals in large quantities (Li et al. 1999; Boal et al. 2000). To produce nanocrystals with desired sizes and shapes, its growth process should be monitored and controlled (Grzelczak et al. 2010), but accomplishing this goal is challenging, due to the existence of multiple growth mechanisms (Zheng et al. 2009), complex interactions among hundreds of nanoscale particles (Park et al. 2015), and after all, the stochastic nature of the growth processes. Critical to the mission of achieving in-process control is a recent technology innovation in nanoscale metrology, the in situ TEM (Zheng et al. 2009). An in situ TEM uses a special sample holder in which a nanocrystal growth process takes place, allowing motion pictures to be taken while the nanocrystals in the sample holder are initiating, crystalizing, and morphing into different sizes and shapes.

7.1.1 The Problem of Distribution Tracking

To track the evolution of nanocrystal growth, or more generally, the evolution of any nano objects, there are two common approaches: distribution tracking versus object tracking. The approach of object tracking is to track individual material objects or structures over different image frames, whereas the approach of distribution tracking is to track the collective behavior, or the growth trajectory of the group, represented

C. Park, Y. Ding, *Data Science for Nano Image Analysis*, International Series in Operations Research & Management Science 308,
https://doi.org/10.1007/978-3-030-72822-9_7

through a dynamic, time-varying probability density function. The topic of object tracking is discussed in Chap. 10, so this chapter focuses on the topic of distribution tracking.

The need for distribution tracking are for two reasons: (a) In some applications, the nano objects may not be easily traceable. For example, in liquid phase experiments with a flow-through setup, samples of materials are continuously pumped into a TEM using microfluidics to flow through the imaging area. Consequently, material samples observed at different times are different materials. Tracking individual nano objects is thus infeasible. Even for experiments that do not use the flow-through setup, it is not always easy to tell whether the objects observed in the previous image frame are the same as in the next image frame. (b) In some other applications, domain experts (material scientists) may care more about the collective change of material samples rather than the changes in individual objects. Under certain circumstances, the distribution change can be connected, more definitively, with the governing dynamics of crystal growth, whereas the change exhibiting in a single object may not be representative.

Consider specifically the problem of tracking the distribution of nanocystal sizes in a growth process. A nanocrystal growth is not stationary but goes through multiple growth stages, from nucleation to growth to equilibrium, driven by various chemical and thermodynamic forces that are all stochastic in nature and span the range from chemical reactions to mesoscale crystal growth phenomena. To ensure robustness and wide applicability, nonparametric modeling of the density functions is inevitable. Empirical analyses by domain experts demonstrate that the density function of nano objects can change from a multi-modal, asymmetric function in the early stages of growth to a uni-modal, symmetric one in the late stages (Zheng et al. 2009; Woehl et al. 2013). It is difficult to specify a parametric function that can adequately describe different growth mechanisms in a multi-stage growth process. Nonparametric approaches do not presume density function types but rather let the data guide the estimation of the probability density function, and thus becomes the preferred approach.

7.1.2 Nanocrystal Growth Video Data

As an emerging technology, the in situ TEMs are not yet widely available and TEM videos available in the public domain are just a few. In this chapter, we use three clips of in situ TEM video: two clips published by Zheng et al. (2009) and one clip published by Woehl et al. (2013). The three videos clips capture 76.6, 42.5, and 112 s of a respective nanocrystal growth process, and there are 1149, 637, and 112 image frames in the respective clips. We label them as Video 1, Video 2 and Video 3, respectively. Figure 7.1 presents four frames of Video 1, capturing the growth of platinum nanocrystals.

When an image frame of the nanocrystal growth process is recorded by an in situ TEM, one first processes the image and extracts the nanocrystal information,

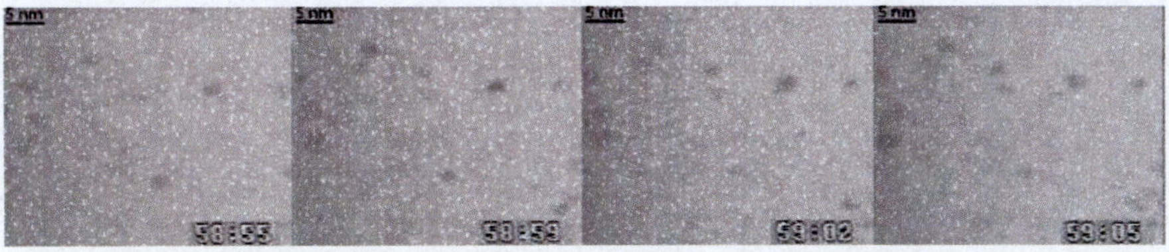

Fig. 7.1 Four frames from the in situ TEM video studied in Zheng et al. (2009). The dark spots are nanocrystals. (Reprinted with permission from Qian et al. 2019)

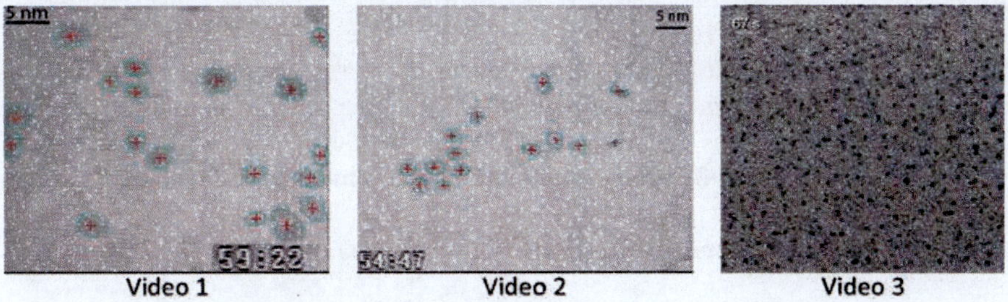

Fig. 7.2 The nanocrystal detection results of a single frame, one each from the three TEM video clips. The green line shows a nanocrystal's edge and the red '+' shows a nanocrystal's center. (Reprinted with permission from Qian et al. 2019)

which is the number and the corresponding size of the nanocrystals in the frame. The specific tool for processing individual images is from Qian et al. (2016), a method particularly potent of handling noisy TEM images with low contrast. The preprocessing result of a single frame from each video clip is shown in Fig. 7.2. Of the three video clips, Video 1 and 2 are of 290 × 242 pixels in size and Video 3 is of 496 × 472 pixels. Considering their relatively small image size, the image preprocessing can be done fairly quickly. For Video 1 and Video 2, the image processing takes only 0.04 s per frame and for Video 3 it takes 0.2 s per frame.

The morphological features extracted from a TEM video can be both sizes and shapes of nanocrystals. In this chapter, we focus on particle size, because all the TEM videos at hand contain nanocrystals of rather uniformly round shapes in the duration of growth that was captured by the videos.

After all nanocrystals in the frame of time t are detected, one calculates $A^{\ell}(t)$, the area of the ℓ-th nanocrystal at time t, for $\ell = 1, \ldots, N_t$, where N_t is the total number of nanocrystals in the frame of time t. Following Woehl et al. (2013), one can use $A^{\ell}(t)$ to compute the radius of the ℓ-th nanocrystal, namely $r^{\ell}(t) = \sqrt{A^{\ell}(t)/\pi}$, to represent the size of each nanocrystal. The radii for the N_t nanocrystals can averaged to $\bar{r}(t)$.

Finally, $r^{\ell}(t)$ is normalized by $\bar{r}(t)$ to obtain the normalized radius $x^{\ell}(t)$, such that $x^{\ell}(t) = r^{\ell}(t)/\bar{r}(t)$. The distribution of the normalized particle radii are used to represent the normalized particle size distribution (NPSD).

NPSD is used as the observational input to the subsequent modeling. The reason that NPSD is used is because studies show that NPSD provides a better indicator than the average absolute size to predict and detect change points in nanocrystal growth (Zheng et al. 2009; Qian et al. 2017). Such modeling choice is also consistent with the domain science's convention and treatment (Lifshitz and Slyozov 1961; Aldous 1999; Woehl et al. 2013).

To facilitate the subsequent computation in estimation and updating, we bin the observations to create a histogram and then use the histogram as the input to the dynamic state space model. We limit the range of $x^{\ell}(t)$ to $[0, 2.0]$, as the nanocrystals twice as large as the average size are very few at any given time. We divide the range into m intervals of equal size δ. Here we use a constant $m = 21$ throughout the monitoring process and denote by x_i the normalized particle size corresponding to the center of the ith interval, $i = 1, \ldots, m$. The reasons behind binning the observations are further elaborated in Sect. 7.4.1, and a sensitivity analysis is conducted in Sect. 7.6.2 concerning the number of intervals used in the input histogram.

The resulting histogram for the frame of time t is denoted by the vector of

$$\mathbf{Y}_t = [Y_{1t}, Y_{2t}, \ldots, Y_{mt}]^T,$$

where Y_{it} is the number of the observed $x^{\ell}(t)$'s falling into the ith interval of the histogram.

7.2 Single Frame Methods

One straightforward treatment of the dynamic TEM data is simply to process a single image frame at a time, without worrying about the dynamic evolution of nano objects over time. The corresponding data science task is to take the nanoparticle size information from the preprocessing step (outlined in Sect. 7.1.2), estimate a nonparametric pdf of particle size for that specific image frame, and then, repeat the same action for each and every image frame.

7.2.1 *Smoothed Histograms*

The histogram data $\mathbf{Y}_t$ consists of m counts, $Y_{1t}, Y_{2t}, \ldots, Y_{mt}$, which follow a multinomial distribution,

$$p(Y_{1t}, Y_{2t}, \ldots, Y_{mt} | f_{1t}, f_{2t}, \ldots, f_{mt}) = \prod_{i=1}^{m} (f_{it})^{Y_{it}}, \quad \sum_{i=1}^{m} f_{it} = 1,$$

where f_{it} is the probability of a particle size being in the ith interval. Let $\mathbf{f}_t = \{f_{1t}, f_{2t}, \ldots, f_{mt}\}$. The log likelihood is

$$L(\mathbf{f}_t|\mathbf{Y}_t) \propto \sum_{i=1}^{m} Y_{it} \log(f_{it}), \quad \sum_{i=1}^{m} f_{it} = 1. \tag{7.1}$$

Maximizing Eq. (7.1) yields the maximum likelihood estimation of $\mathbf{f}_t$, which is $\hat{f}_{it} = Y_{it}/N_t$. The problem of this MLE is that the resulting estimation does not look like a smoothed density function. Rather the resulting $\hat{\mathbf{f}}_t$ reacts too eagerly to the randomness in observations and tends to overfit the noisy data. For instance, when some Y_{it}'s are zeros, the estimation in Eq. (7.1) can produce multi-modal pdfs even if the underlying true distribution is uni-modal.

Simonoff (1983) introduces a penalized approach to produce a smoothed estimation of the histogram. Simonoff (1983)'s intention is to estimate the probabilities in a sparse contingency table; by "sparse," it means that many of the cells in the table are zeros. It is apparent that the histogram of nanoparticle size is simply a one-dimensional frequency table.

The smoothed histogram is estimated through the maximization of the following penalized log-likelihood function,

$$L(\mathbf{f}_t|\mathbf{Y}_t) = \sum_{i=1}^{m} Y_{it} \log f_{it} - \gamma \sum_{i=1}^{m} \left[\log\left(\frac{f_{it}}{f_{(i+1)t}}\right)\right]^2, \quad \sum_{i=1}^{m} f_{it} = 1, \tag{7.2}$$

where $\gamma \geq 0$ is the penalty parameter. The penalty is to force the adjacent probabilities to be close to each other, i.e., their ratio towards one. Simonoff (1983) interprets Eq. (7.2) as the logarithm of the posterior distribution, had the prior distribution been properly chosen to be the penalty function. In this sense, maximizing the log-likelihood function in Eq. (7.2) yields a maximum posterior estimator.

Simonoff (1983) determines the value of γ by minimizing the following mean squared error (MSE):

$$\text{MSE} = \sum_{i=1}^{m} \left\{ f_{it} \sum_{j=1}^{m} h_{ij} \left[Y_{jt} - N_t f_{jt} + 2\gamma \log\left(\frac{f_{(j-1)t} f_{(j+1)t}}{f_{jt}^2}\right)\right]\right\}^2, \tag{7.3}$$

where h_{ij} is the (i, j)th element of $\mathbf{H}$, defined such that

$$\mathbf{H} = \begin{bmatrix} 2\gamma + N_t f_{1t} & -2\gamma & 0 & \cdots & 0 \\ -2\gamma & 4\gamma + N_t f_{2t} & -2\gamma & \cdots & 0 \\ 0 & \ddots & \ddots & \ddots & 0 \\ 0 & \cdots & -2\gamma & 4\gamma + N_t f_{(m-1)t} & -2\gamma \\ 0 & \cdots & 0 & -2\gamma & 2\gamma + N_t f_{mt} \end{bmatrix}^{-1}. \tag{7.4}$$

As the value of γ depends on both the data, $\mathbf{Y}_t$, and the underlying probability, $\mathbf{f}_t$, Simonoff (1983) suggests determining γ and $\mathbf{f}_t$ by optimizing Eqs. (7.2) and (7.3) jointly. First, given γ, maximize Eq. (7.2) with the Newton-Raphson method, and then, update γ by minimizing the MSE in Eq. (7.3) using the $\mathbf{f}_t$ just estimated. At the convergence of this iteration, one can take the values of $\mathbf{f}_t$ and γ.

7.2.2 Kernel Density Estimation

In mathematics, the kernel, also called integral kernel or kernel function, is a function of two variables that defines an integral transform, such as the function $k(\cdot, \cdot)$ in

$$(Tg)(x) = \int_X k(x, x')g(x')dx', \tag{7.5}$$

where $(Tg)(\cdot)$ is an operator working on g. The above integral transform takes a function, $g(x')$, as the input and converts it into another function, $Tg(x)$. In such an integral transform, the function, $k(x, x')$, acts like the nucleus of the transform and is thus called the *kernel* of the integral transform. Another interpretation of $k(x, x')$ is that $k(x, x')$ is a weighting function used in an integration. This weighting function interpretation is used in many statistical applications such as kernel density estimation.

Given a set of random samples, $\{x_1, \ldots, x_{N_t}\}$, drawn from a probability density $f_t(x)$ at time t, and one wishes to estimate $f_t(\cdot)$. A nonparametric density estimation at x_0 can be obtained by

$$\hat{f}_t(x_0) = \frac{\#\{x_i, \text{ s.t. } x_i \in \mathrm{N}_t(x_0)\}}{N_t \cdot \lambda} \tag{7.6}$$

where $\mathrm{N}_t(x_0)$ defines the neighborhood around x_0 of width λ and $\#\{\cdot\}$ denotes the cardinality of a set. When x_0 is moved through the domain of x, one can get an estimation of the density function.

If x_0's are limited to a number of the chosen sites so that $\mathrm{N}_t(x_0)$'s are disjoint, then, the resulting $\hat{f}_t(\cdot)$ is a histogram, rather than a smooth function. In order to get a smooth density estimation, people devise the kernel density estimation, i.e.,

$$\hat{f}_t(x) = \frac{1}{N_t\lambda}\sum_{i=1}^{N_t} k\left(\frac{x_i - x}{\lambda}\right). \tag{7.7}$$

The popular kernel functions used in density estimation, $k(\cdot)$, include Gaussian, Epanechnikov, uniform, among others. In Eq. (7.7), note that the subscript "0" in x_0 is dropped.

In the kernel density estimation, λ determines the smoothness of the resulting density function. A small λ means that the estimation uses the data falling into a narrow window while estimating the density value at x, and thus the corresponding density function is wiggly, whereas a large λ means that the estimation uses the data falling into wide window to estimate the density value at x, and as such, the corresponding density function is smoother. Sheather and Jones (1991) explain the method for estimating the bandwidth parameter. The idea is to minimize the asymptotic mean integrated squared error (Silverman 1986, AMISE):

$$\text{AMISE}(\lambda) = (N_t\lambda)^{-1}R(k) + \frac{1}{4}\lambda^4\sigma_k^4 R(f_t''), \tag{7.8}$$

where $R(g) = \int g^2(x)dx$ for a function g (g can be k or f_t'') and $\sigma_k^2 = \int x^2 k(x)dx$. Then the AMISE estimator of λ is

$$\lambda_{\text{AMISE}} = \frac{[R(k)]^{1/5}}{\sigma_k^{4/5}[R(f_t'')]^{1/5}N_t^{1/5}}. \tag{7.9}$$

Functions in the `kedd` package in `R` can be used to help with the bandwidth calculation. One can call the `h.amise` function through the following syntax:

```
h.amise(data, deriv.order = 0),
```

where the second argument, `deriv.order`, specifies the order of the derivative of a probability density to be estimated. For estimating the original density function, $f(\cdot)$, the `deriv.order` argument is set zero.

The kernel density estimation is considered a nonparametric approach not because the kernel function does not have parameters or not use a functional form; in fact it does have a parameter, which is the bandwidth. Being a nonparametric approach, the kernel function is different from the target density function that it aims to estimate. While the target density function functions may vary drastically from one application to another, or from one time instance to another instance in the same application, the kernel function used in the estimation remains more or less the same. The non-changing kernel function is able to adapt to the ever-changing density functions, as long as there are enough data. The parameter in the kernel function serves a role differing substantially from the parameters in a parametric density function, such as the mean and standard deviation parameters in a Gaussian distribution—while the parameters in a parametric density function have direct probability distribution linked interpretations, the bandwidth parameter in a kernel function does not.

7.2.3 *Penalized B-Splines*

Eilers and Marx (1996) propose another method for estimating smooth density functions. Their method uses the B-spline functions as the basic building block but adds a penalty term to enforce smoothness in the resulting density function estimation.

To describe the idea precisely, let $\eta_{it} = \log f_t(x_i)$ represent the log density evaluated at x_i. The B-spline approach models the log density as

$$\eta_{it} = \sum_{j=1}^{n} \alpha_{jt} B_j(x_i), \tag{7.10}$$

where n is the number of basis functions, $B_j(x)$ is the jth B-spline basis function, and α_{jt} is its coefficient at time t. Collectively, the coefficients can be expressed in a coefficient vector at time t, namely

$$\boldsymbol{\alpha}_t := [\alpha_{1t}, \alpha_{2t}, \cdots, \alpha_{nt}]^T.$$

Note that we slightly abuse the notation here, as $f_t(\cdot)$ above may differ from a true density function by a normalizing constant (which is needed to ensure the pdf integrable to one). We continue using the same notation for the un-normalized density function to maintain a notational simplicity. People can easily check the resulting function and can perform a post-modeling re-normalization, when necessary, to account for the normalizing constant.

It is assumed that the count variable Y_{it} follows a Poisson distribution with a density of $\exp(\eta_{it})$. The penalized B-spline method aims at maximizing the following penalized Poisson likelihood function, L_t, with $\boldsymbol{\alpha}_t$ as the decision variable:

$$L_t(\boldsymbol{\alpha}_t) = \sum_{i=1}^{m} Y_{it}\eta_{it} - \sum_{i=1}^{m} \exp(\eta_{it}) - \gamma \cdot \sum_{j=1}^{n-\kappa} \frac{(\Delta^\kappa \alpha_{jt})^2}{2}, \tag{7.11}$$

where γ is the penalty coefficient, as in Eq. (7.2), and Δ^κ is the κ-th order difference operator on index j. For instance, when $\kappa = 1$, $\Delta\alpha_{jt} = \alpha_{(j+1)t} - \alpha_{jt}$ and when $\kappa = 2$, $\Delta^2\alpha_{jt} = \alpha_{(j+2)t} - 2\alpha_{(j+1)t} + \alpha_{jt}$.

In Eq. (7.11), the first and second terms correspond to the Poisson likelihood, measuring the goodness-of-fit of the resulting density function to the observed histogram, whereas the third term is the penalty term to impose smoothness onto the resulting density function.

Eilers and Marx (1996) determine the optimal value of γ by minimizing an Akaike information criterion (AIC), defined as

$$\text{AIC}(\gamma) = \text{dev}(\mathbf{Y}_t; \gamma) + 2 \cdot \dim(\gamma). \tag{7.12}$$

where $\text{dev}(\mathbf{Y}_t; \gamma)$ is the deviance of the fitted curve and $\dim(\gamma)$ is the model's degree of freedom. For the density estimation, the deviance is calculated by

$$\mathrm{dev}(\mathbf{Y}_t;\gamma) = 2\sum_{i=1}^{m} Y_{it}\log\left(\frac{Y_{it}}{\hat{\lambda}_{it}}\right), \tag{7.13}$$

where λ_{it} is the Poisson parameter for interval i at time t, i.e., the average particle count and $\hat{\lambda}_{it}$ is the estimated average particle count at a given γ. The model's degree of freedom is

$$\dim(\gamma) = \mathrm{tr}(\mathbf{S}_\gamma),$$

where $\mathbf{S}_\gamma$ is the smoothing matrix of the penalized B-splines, estimated with a γ choice. Please refer to Eq. (26) in Eilers and Marx (1996) for the expression of $\mathrm{tr}(\mathbf{S})$.

7.3 Multiple Frames Methods

The single frame methods, processing the images one frame at a time, have the obvious disadvantage of not making full use of the temporally dependent information across the adjacent image frames. Recent research development produces two multi-frame methods: one is for retrospective (offline) analysis and the other is for prospective (online) analysis. A retrospective analysis assumes that the nanocrystal growth has reached an equilibrium and that researchers have collected all the observations from the process, whereas a prospective analysis is being conducted while the growth process is ongoing and researchers have only the observation data up to the current moment. This section presents the model suitable for retrospective analysis and Sect. 7.4 presents the state space model suitable for prospective analysis, i.e., online tracking, updating, and monitoring.

7.3.1 Retrospective Analysis

Qian et al. (2017) extend the single-frame penalized B-spline density estimation (Eilers and Marx 1996) to the multi-frame problems. Their basic action is to pool all image frames together and write an overall log-likelihood function. In order to make use of the temporal dependence information across adjacent image frames, a second penalty term is used, in addition to the penalty that is already there in Eq. (7.11). The second penalty is to ensure that the time-varying density functions change smoothly over time.

Recall the log-likelihood function in Eq. (7.11) for the single-frame penalized B-spline density estimation. Qian et al. (2017) rewrite it as:

$$L_t(\boldsymbol{\alpha}_t) = \sum_{i=1}^{m} Y_{it}\eta_{it} - \sum_{i=1}^{m} \exp(\eta_{it}) - \gamma_1 \sum_{j=1}^{n-1} \frac{(\Delta_1 a_{jt})^2}{2}, \tag{7.14}$$

where the subscript of "1" is added to both γ and Δ, for the purpose of differentiating this penalty term from a new penalty term to be added next. Eilers and Marx (1996) recommend that a penalty of order κ is to be used with B-splines of order $\kappa + 1$. Because Qian et al. (2017) use B-splines of order two, κ is then one here, meaning that a first order difference is used in Eq. (7.14).

When pooling the data of all image frames together, the new objective function becomes

$$L(\{\boldsymbol{\alpha}_1, \ldots, \boldsymbol{\alpha}_T\}) = \sum_{t=1}^{T} L_t(\boldsymbol{\alpha}_t) - \gamma_2 \sum_{t=1}^{T-1} \sum_{j=1}^{n} \frac{(\Delta_2 \alpha_{jt})^2}{2}, \tag{7.15}$$

where T is the number of image frames in the video, Δ_2 is a difference operator operating on index t, i.e., $\Delta_2 a_{jt} = a_{j(t+1)} - a_{jt}$, and γ_2 is the temporal smoothness penalty parameter. This new formulation enables borrowing information among different time frames and thus improves estimation efficacy, especially at those time frames having too few nanocrystals.

Recall that we use Δ^κ to denote a difference operator of order κ. We would like to stress that both Δ_1 and Δ_2 are the fist order operators, as their corresponding $\kappa = 1$.

Qian et al. (2017) maximize the penalized log-likelihood in Eq. (7.15) for estimating the spline coefficients associated with all density functions over the whole video duration. Apparently, the algorithm developed by Eilers and Marx (1996) does not apply since the new formulation has an extra index t and an extra penalty term. The main challenge is caused by the newly introduced second penalty term, which makes the objective function no longer separable with respect to t.

Here Qian et al. (2017) employ the alternating direction multiplier method (Boyd et al. 2011, ADMM) to decouple the relationship along the t index. Specifically, Qian et al. (2017) replace α_{jt}'s in the second penalty term by a set of new variables z_{jt}'s and solve the optimization problem under the constraint of $\alpha_{jt} = z_{jt}$.

A constrained optimization is performed through an augmented Lagrangian as follows:

$$\begin{aligned} L_\rho(\{\alpha_{jt}\}, \{z_{jt}\}, \{c_{jt}\}) = &\sum_{t=1}^{T} L_t(\boldsymbol{\alpha}_t) - \gamma_2 \sum_{t=1}^{T-1} \sum_{j=1}^{n} \frac{(\Delta_2 z_{jt})^2}{2} \\ &- \rho \sum_{t=1}^{T} \sum_{j=1}^{n} c_{jt}(\alpha_{jt} - z_{jt}) - \frac{\rho}{2} \sum_{t=1}^{T} \sum_{j=1}^{n} (\alpha_{jt} - z_{jt})^2, \end{aligned} \tag{7.16}$$

where c_{jt}'s are the Lagrangian multipliers and ρ is the penalty parameter of the augmented Lagrangian. Then the ADMM algorithm targets to find the saddle point of Eq. (7.16), defined by

$$(\{\hat{\alpha}_{jt}\}, \{\hat{z}_{jt}\}, \{\hat{c}_{jt}\}) = \arg \min_{\{c_{jt}\}} \max_{\{\alpha_{jt}\},\{z_{jt}\}} L_\rho(\{\alpha_{jt}\}, \{z_{jt}\}, \{c_{jt}\}), \tag{7.17}$$

where $\{\hat{\alpha}_{jt}\}$ is the optimized coefficients to be used in the resulting density estimation.

The saddle point is found by using the coordinate descent method (Luenberger 1973). The idea of the method is as follows. In the qth iteration of updating $\{\alpha_{jt}\}$, $\{z_{jt}\}$ and $\{c_{jt}\}$,

1. Firstly, apply Eliers and Marx's algorithm to find the optimal $\{\alpha_{jt}\}$, given $\{c_{jt}\}$ and $\{z_{jt}\}$ at their current values;
2. Then, fix $\{\alpha_{jt}\}$ and $\{c_{jt}\}$ and update $\{z_{jt}\}$ via the Lagrangian, which is a quadratic form in $\{z_{jt}\}$ and thus its optimization has a closed-form solution;
3. After that, the Lagrange multipliers, $\{c_{jt}\}$, are updated by a "price update" step:

$$c_{jt}^{(q+1)} = c_{jt}^{(q)} + (\alpha_{jt}^{(q)} - z_{jt}^{(q)}), \tag{7.18}$$

   where the superscript indicates that the value of c_{jt} at the $(q+1)$th iteration is updated using the values available at the qth iteration.

One continues the iteration until all these variables converge. At the convergence of the algorithm, one substitutes the convergent value of $\boldsymbol{\alpha}_t$ to Eq. (7.10) to get the estimate of NPSD, $\hat{f}_t(\cdot)$, for all the time frames. We present the detailed steps of the ADMM algorithm in Sect. 7.3.2.

To estimate the NPSD for Video 1, Qian et al. (2017) set $n = 10$ (the number of B-spline basis), $m = 50$ (the number of knots), and $T = 1,149$ (the number of frames in the video). The estimation is robust with respect to n and m. The parameter ρ only affects the convergence speed of ADMM, so that as long as the algorithm converges, there is no need to tune it aggressively. Qian et al. (2017) set ρ to 9.0. The remaining tuning parameters γ_1 and γ_2 are decided by using an AIC metric, like in Eilers and Marx (1996). For the learning formulation in Eq. (7.15), Qian et al. (2017) derive the AIC as:

$$\text{AIC}(\gamma_1, \gamma_2) = \text{dev}(\gamma_1, \gamma_2) + 2\text{dim}(\gamma_1, \gamma_2), \tag{7.19}$$

where $\text{dev}(\gamma_1, \gamma_2)$ is the deviance of the fitted curve, and $\text{dim}(\gamma_1, \gamma_2)$ is the effective dimension of parameters. The deviance is defined as

$$\text{dev}(\gamma_1, \gamma_2) = 2\sum_{t=1}^{T}\sum_{i=1}^{m} Y_{it} \log Y_{it} - 2\sum_{t=1}^{T}\sum_{j=1}^{n} \hat{\alpha}_{jt} B_{jt}^{+}, \tag{7.20}$$

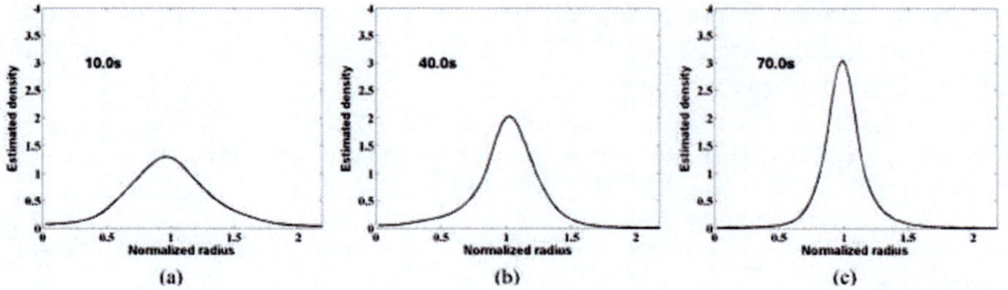

Fig. 7.3 The estimated NPSDs at 10, 40, and 70 s, all for Video 1. **(a)** NPSD at 10 s. **(b)** NPSD at 40 s. **(c)** NPSD at 70 s. (Reprinted with permission from Qian et al. 2017)

where $B_{jt}^{+} = \sum_{\ell=1}^{N_t} B_j(x_\ell)$. And Qian et al. (2017) derive the effective dimension of parameters as

$$\dim(\gamma_1, \gamma_2) = \text{tr}\{(\mathbf{B}^T\mathbf{B} + \gamma_1\mathbf{D}_1^T\mathbf{D}_1)^{-1}\mathbf{B}^T\mathbf{B}\}\text{tr}\{(\mathbf{I}_T + \gamma_2\mathbf{D}_2\mathbf{D}_2^T)^{-1}\}, \tag{7.21}$$

where $\mathbf{D}_1$ is an $n \times n$ matrix with $(\mathbf{D}_1)_{jj}$ as -1, $(\mathbf{D}_1)_{j(j-1)}$ as 1, for $j = 2, \ldots, n$, and all other elements as 0, $\mathbf{D}_2$ is a $T \times T$ matrix with $(\mathbf{D}_2)_{tt}$ as -1, $(\mathbf{D}_2)_{t(t+1)}$ as 1, for $t = 1, \ldots, T-1$, and all other elements as 0, $\mathbf{B}$ is the B-spline coefficient matrix, such that $(\mathbf{B})_{ij} = B_j(x_i)$, and $\text{tr}\{\cdot\}$ is the trace of the corresponding matrix.

By minimizing AIC(γ_1, γ_2), the two tuning parameters are chosen as $\gamma_1 = 0.5$ and $\gamma_2 = 1.5$ for Video 1. Figure 7.3 presents the NPSDs estimated at $t = 10$, 40, and 70 s for Video 1.

7.3.2 *Optimization for Density Estimation*

This section provides more details about the ADMM algorithm that solve the optimization problem in Eq. (7.16). Recall that the ADMM algorithm targets to find the saddle point defined in Eq. (7.17). The saddle point is found by using the coordinate decent method (Luenberger 1973).

Specifically, Qian et al. (2017) first change the min-max problem in Eq. (7.17) to a max-min one by adding a negative sign in Eq. (7.16) and rewrite it in a matrix form:

$$\begin{aligned} L'_\rho(\mathbf{A}, \mathbf{Z}, \mathbf{C}) = &-\sum_{t=1}^{T} L_t(\mathbf{A}) + (\gamma_2/2)\sum_{t=1}^{T-1}\sum_{j=1}^{n}(\Delta_2 z_{jt})^2 \\ &+ \rho\mathbf{C}^T(\mathbf{A} - \mathbf{Z}) + (\rho/2)\sum_{t=1}^{T}\|\mathbf{A} - \mathbf{Z}\|_2^2, \end{aligned} \tag{7.22}$$

where $(\mathbf{A})_{jt} = \alpha_{jt}$, $(\mathbf{Z})_{jt} = z_{jt}$, $(\mathbf{C})_{jt} = c_{jt}$, and $\|\cdot\|_2$ is the entrywise matrix 2-norm, or equivalently, the Frobenius norm. Then Qian et al. (2017) update **A**, **Z**, **C** iteratively. When updating one of the three variables, Qian et al. (2017) fix the other two. The values of variables in the qth iteration are signified via a "(q)" superscript.

To update **A**, Qian et al. (2017) solve:

$$\arg\min_{\mathbf{A}} -\sum_{t=1}^{T} L_t(\mathbf{A}) + (\rho/2)\|\mathbf{A} - \mathbf{Z}^{(q)} + \mathbf{C}^{(q)}\|_2^2. \tag{7.23}$$

The problem can be decomposed into T independent subproblems, each for a single image frame. Denote the t-th column of **A**, **Z** and **C** by $\boldsymbol{\alpha}_t$, $\mathbf{z}_t$ and $\mathbf{c}_t$, respectively. The difference operator Δ_1 can be rewritten as a matrix multiplication:

$$\sum_{j=1}^{n-1}(\Delta_1\alpha_{jt})^2 = \boldsymbol{\alpha}_t^T\mathbf{D}_1^T\mathbf{D}_1\boldsymbol{\alpha}_t. \tag{7.24}$$

As such, Qian et al. (2017) update $\boldsymbol{\alpha}_t$ by:

$$\begin{aligned}\boldsymbol{\alpha}_t^{(q+1)} &= \arg\min_{\boldsymbol{\alpha}_t}\Big\{-\sum_{i=1}^{m} Y_{it}\eta_{it} + \sum_{i=1}^{m}\exp(\eta_{it}) \\ &+ (\gamma_1/2)\boldsymbol{\alpha}_t^T\mathbf{D}_1^T\mathbf{D}_1\boldsymbol{\alpha}_t + (\rho/2)[\boldsymbol{\alpha}_t^T\boldsymbol{\alpha}_t - 2(\mathbf{z}_t^{(q)} - \mathbf{c}_t^{(q)})^T\boldsymbol{\alpha}_t]\Big\}.\end{aligned} \tag{7.25}$$

The solution approach of the above minimization problem could follow that in Eilers and Marx (1996), as they solved a similar problem.

According to Eilers and Marx (1996), the solution of $\boldsymbol{\alpha}_t$ is to set the first derivative of the above objective function to zero. That leads to the following equation:

$$\mathbf{B}_t^+ - \mathbf{B}^T\exp(\mathbf{B}\boldsymbol{\alpha}_t) = \gamma_1\mathbf{D}_1^t\mathbf{D}_1\boldsymbol{\alpha}_t + \rho[\boldsymbol{\alpha}_t - (\mathbf{z}_t^{(q)} - \mathbf{c}_t^{(q)})], \tag{7.26}$$

where $\mathbf{B}_t^+ = [B_{1t}^+, \cdots, B_{nt}^+]^T$. The above equation does not have a closed-form solution for $\boldsymbol{\alpha}_t$, so one has to solve it through an iterative procedure. The following equation uses a first order Taylor expansion of the exponential term, so that $\boldsymbol{\alpha}_t$ can be solved through a weighted linear regression:

$$\begin{aligned}&\mathbf{B}_t^+ - \mathbf{B}^T\exp(\mathbf{B}\hat{\boldsymbol{\alpha}}_t) + \mathbf{B}^T\mathbf{B}\hat{\boldsymbol{\alpha}}_t + \rho(\mathbf{z}_t^{(q)} - \mathbf{c}_t^{(q)}) \\ &= [\mathbf{B}^T\mathbf{B} + \gamma_1\mathbf{D}_1^T\mathbf{D}_1 + \rho\mathbf{I}_n]\boldsymbol{\alpha}_t,\end{aligned} \tag{7.27}$$

where $\mathbf{I}_n$ is the $n \times n$ identity matrix and $\hat{\boldsymbol{\alpha}}_t$ is the estimate from the previous iteration, whose initial value is set to be $\boldsymbol{\alpha}_t^{(q)}$ (from the q-th step). Once the numerical iterative procedure converges, the resulting $\boldsymbol{\alpha}_t$ is treated as $\boldsymbol{\alpha}_t^{(q+1)}$.

To update **Z**, Qian et al. (2017) solve:

$$\mathbf{Z}^{(q+1)} = \arg\min_{\mathbf{Z}} \left\{ \gamma_2 \sum_{t=1}^{T-1} \sum_{j=1}^{n} (\Delta_2 z_{jt})^2 + (\rho/2)\|\mathbf{A}^{(q+1)} - \mathbf{Z} + \mathbf{C}^{(q)}\|_2^2 \right\}. \tag{7.28}$$

The terms in the large bracket can be rewritten as:

$$\|\mathbf{A}^{(q+1)} - \mathbf{Z} + \mathbf{C}^{(q)}\|_2^2 + \frac{\gamma_2}{\rho} \sum_{t=1}^{T-1} \sum_{j=1}^{n} (\Delta_2 z_{jt})^2. \tag{7.29}$$

The second term can be transformed into a matrix form, i.e.,

$$\sum_{t=1}^{T-1} \sum_{j=1}^{n} (\Delta_2 z_{it})^2 = \|\mathbf{Z}\mathbf{D}_2\|_2^2. \tag{7.30}$$

where $\|\cdot\|_2$ is still the entrywise matrix 2-norm. The minimization problem in Eq. (7.28) has a closed-form solution for $\mathbf{Z}^{(q+1)}$, which is

$$\mathbf{Z}^{(q+1)} = \left[\mathbf{A}^{(q+1)} + \mathbf{C}^{(q)}\right] \left(\mathbf{I}_T + \frac{\gamma_2}{\rho}\mathbf{D}_2\mathbf{D}_2^T\right)^{-1}. \tag{7.31}$$

Lastly, Qian et al. (2017) update the Lagrangian multipliers, **C**, using the "price update" step shown in Eq. (7.18), which is rewritten in a matrix form as:

$$\mathbf{C}^{(q+1)} = \mathbf{C}^{(q)} + (\mathbf{A}^{(q+1)} - \mathbf{Z}^{(q+1)}). \tag{7.32}$$

The algorithm continues the iteration until all these variables converge. The solution algorithm is summarized in Algorithm 7.1.

7.4 State Space Modeling for Online Analysis

The method developed in Sect. 7.3 is a retrospective approach conducted offline. It uses the observations from all video frames and optimizes a likelihood function embodying all available data. Should this retrospective approach be applied to an online application, it ought to solve the optimization problem whenever a new image frame (or a couple of new images) comes. Doing so takes more time than allowed in a real-time processing. For instance, it takes a couple of minutes to solve the optimization problem using the ADMM algorithm for Video 1, not a big deal for offline analysis but not practical for online applications. The retrospective approach is hence inefficient and ill-suited for online applications. The approach

Algorithm 7.1 ADMM algorithm for solving the revised penalized B-spline density estimation

1. Set tuning parameters m, n, γ_1, γ_2 and ρ.
2. Construct the B-spline basis function $B_j(x_i)$ according to the knots x_i for $i = 1, \ldots, m$ and $j = 1, \ldots, n$. Then calculate $\mathbf{B}_t^+ = [B_{1t}^+, \ldots, B_{nt}^+]^T$ and $B_{jt}^+ = \sum_i^m B_j(x_i)$.
3. Initialize $\mathbf{A}^{(0)}$, $\mathbf{Z}^{(0)}$ and $\mathbf{C}^{(0)}$. Qian et al. (2017) recommend setting $(\mathbf{H}^{(0)})_{it} = \log(Y_{it})$, $\mathbf{A}^{(0)} = \mathbf{Z}^{(0)} = \mathbf{B}^{-1}\mathbf{H}^{(0)}$ and $\mathbf{C}^{(0)} = \mathbf{0}$.
4. Set $q = 0$.
5. Update $\mathbf{A}^{(q+1)}$. For $t = 1, \ldots, T$, Qian et al. (2017) update each column of $\mathbf{A}^{(q+1)}$ as follows:
   a. Set $\hat{\boldsymbol{\alpha}}_t = \boldsymbol{\alpha}_t^{(q)}$.
   b. Solve Eq. (7.27) and use the solution of $\boldsymbol{\alpha}_t$ to update $\hat{\boldsymbol{\alpha}}_t$.
   c. Repeat the previous step until $\hat{\boldsymbol{\alpha}}_t$ converges, then let $\boldsymbol{\alpha}_t^{(q+1)} = \hat{\boldsymbol{\alpha}}_t$.
6. Update $\mathbf{Z}^{(q+1)}$ by solving Eq. (7.31)
7. Update $C^{(q+1)}$ by solving Eq. (7.32), and then let $q = q + 1$.
8. Repeat Steps 5 to 7 until **A**, **Z** and **C** all converge.

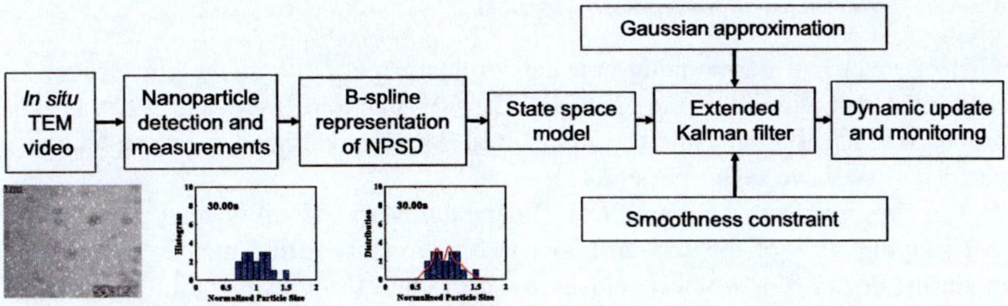

Fig. 7.4 The framework of a prospective analysis of in situ TEM videos. (Reprinted with permission from Qian et al. 2019)

presented next is a computationally efficient state space modeling approach that allows researchers to update the density estimation at time $t + 1$ using the density estimation at time t and the new data observed at time $t + 1$. Figure 7.4 presents an overview of the online, prospective analysis.

7.4.1 State Space Model for NPSD

As Y_{it} is the count of observations falling in $[x_i - \delta/2, x_i + \delta/2]$, it is rather standard to assume that Y_{it} follows the Poisson distribution with expectation of λ_{it} (Bishop et al. 1975), the same treatment as done in Sect. 7.2.3. This means

$$Y_{it} \sim \text{Poisson}\{\lambda_{it}\}, \; i = 1, \ldots, m. \tag{7.33}$$

Rewriting Eq. (7.10) using λ_{it}, one has

$$\log \lambda_{it} = \sum_{j=1}^{n} \alpha_{jt} B_j(x_i). \tag{7.34}$$

Using the log link function can guarantee a positive λ_{it}. Using the matrix expression of $\boldsymbol{\alpha}_t$ and $\mathbf{B}$, the Poisson model of Y_{it} is then expressed as:

$$Y_{it} \sim \text{Poisson}\{(\exp[\mathbf{B}\boldsymbol{\alpha}_t])_i\},\ i = 1, \ldots, m. \tag{7.35}$$

The maximum likelihood method can be used to estimate the spline coefficient vector $\boldsymbol{\alpha}_t$. Once $\boldsymbol{\alpha}_t$ is estimated, one can further obtain the NPSD, $f_t(x)$, as follows. As $\lambda_{it} \propto \int_{x_i-\delta/2}^{x_i+\delta/2} f_t(x)dx$, $\lambda_{it} \propto f_t(x_i)$ when δ is small, so that $f_t(x)$ can be estimated by the continuous form of Eq. (7.34) as

$$f_t(x) = \frac{1}{C_t(\boldsymbol{\alpha}_t)} \exp\left[\sum_{j=1}^{n} \alpha_{jt} B_j(x)\right], \tag{7.36}$$

where $C_t(\boldsymbol{\alpha}_t)$ is a normalizing constant to guarantee $f_t(x)$ integrable to one. To ensure the smoothness of the estimated density function, Eilers and Marx (1996) proposed to use a small δ and use a penalized Poisson likelihood with a smoothness penalty, as we have seen in Eq. (7.11).

The above approach adopted from Eilers and Marx (1996) is an approximation and simplification of the standard approach of directly fitting the raw data to the model in Eq. (7.36), which could be done through the maximization of the log-likelihood of

$$\sum_{\ell=1}^{N_t} \log f_t(x^{\ell}(t)) = \sum_{\ell=1}^{N_t}\sum_{j=1}^{n} \alpha_{jt} B_j(x^{\ell}(t)) - N_t \log C_t(\boldsymbol{\alpha}_t).$$

By applying the penalized Poisson likelihood to the binned data, the approach of Eilers and Marx (1996) can avoid computing the integral in the normalizing constant,

$$C_t(\boldsymbol{\alpha}_t) = \int_x \exp\left[\sum_{j=1}^{n} \alpha_{jt} B_j(x)\right] dx. \tag{7.37}$$

This simplification turns out to be helpful while developing an online updating algorithm. When a small δ is used, the binning action does not cause much information loss as compared to using the original observations of $x^{\ell}(t)$, and the inclusion of the smoothness penalty helps prevent overfitting.

With the B-spline representation in place, Qian et al. (2019) use the B-spline coefficient vector, $\boldsymbol{\alpha}_t$, as the state vector in the state space model, because the change in $\boldsymbol{\alpha}_t$ indicates the change of the underlying NPSD, $f_t(x)$. Previous studies (Lifshitz and Slyozov 1961; Aldous 1999) show that $f_t(x)$ undergoes small fluctuations during a growth stage in which the commanding physical growth mechanism remains the same, while $f_t(x)$ will see a much greater change as a different growth mechanism takes over. Based on this understanding, Qian et al. (2019) assume that the state vector, $\boldsymbol{\alpha}_t$, follows a random walk model

$$\boldsymbol{\alpha}_t = \boldsymbol{\alpha}_{t-1} + \mathbf{w}_t, \tag{7.38}$$

where the innovation, $\mathbf{w}_t$, is the disturbance vector of the state and assumed to follow the distribution of normal($\mathbf{0}$, $\mathbf{Q}$). The covariance matrix $\mathbf{Q}$ will be treated as a constant matrix throughout the process. The state updating equation, Eq. (7.38), and the observation equation, Eq. (7.35), constitute the state space model, which leads to the development of an online updating algorithm in the form of an extended Kalman filter.

The use of the random walk model is based on a trade-off between flexibility and tractability. Qian et al. (2019) state that they would like to choose a model that is flexibility enough but is still tractable. The main concern for tractability is the fast computation that allows real time online processing. Although using a stationary time series process in the state space model such as the autoregressive moving average (ARMA) approach is possible, real time updating of both the ARMA parameters and the state variables is a daunting task for online processing.

The simple random walk model can cope well with the nanocrystal growth process. Within a specific stage of nanocrystal growth, the normalized particle size distribution undergoes small fluctuations, and this is well captured by the random walk model. In a transition period moving from one growth stage to another, the random walk model serves as a prior distribution on the state variables, and the Kalman filter is able to adapt to structure changes by calculating the posterior after receiving new observations. When the Kalman filter outputs a large value in the innovation series, it signals possible growth mechanism changes, providing a simple way of identifying change points; see an example presented later in Fig. 7.8.

7.4.2 *Online Updating of State $\boldsymbol{\alpha}_t$*

Updating the estimation of NPSD is thus equivalent to updating the state vector in the state space model. In the dynamic systems and control theory, the Kalman filter (Kalman 1960) is arguably the most popular method used for conducting such update. For linear state space models with Gaussian observations, a Kalman filter uses the posterior mean, $\mathbb{E}(\boldsymbol{\alpha}_t|\mathbf{Y}_1, \cdots, \cdots, \mathbf{Y}_t)$, denoted as $\hat{\boldsymbol{\alpha}}_t$, to iteratively estimate $\boldsymbol{\alpha}_t$. There are two main steps in a Kalman filter. The *prediction step* is to predict, based on the observations received up to time $t-1$, the prior mean, $\hat{\boldsymbol{\alpha}}_t^-$, and the prior covariance matrix, $\mathbf{P}_t^-$, of the state at time t. When the new observation of $\mathbf{Y}_t$ arrives, the Kalman filter undertakes a *correction step*, also known as the *update step*, to obtain the posterior mean, $\hat{\boldsymbol{\alpha}}_t$, and the posterior covariance matrix, $\mathbf{P}_t$. For a Gaussian system, the Kalman filter has a closed-form solution for both prediction and update steps and can thus run very efficiently.

The state space model of the time-varying NPSD is not a Gaussian system since $\mathbf{Y}_t$ follows a Poisson distribution with an exponential link function in Eq. (7.35). To solve for the posterior mean, $\mathbb{E}(\boldsymbol{\alpha}_t|\mathbf{Y}_1, \cdots, \mathbf{Y}_t)$, one possible approach is to use sampling methods, such as particle filtering (Doucet et al. 2001; Ma et al. 2008; Zhang et al. 2017), to simulate the posterior distribution of the state of $\boldsymbol{\alpha}_t$. But the sampling approach is not ideal for online estimation because the approach's computational speed can hardly meet the online updating requirement. After knowing the model set up, the shortcoming of the sampling approach is even more obvious. To estimate the NPSD accurately, both $\mathbf{Y}_t$ and $\boldsymbol{\alpha}_t$ should have a moderate to high dimension; for instance, $m \geq 10$ and $n \geq 10$. To sample from a space of such dimension for approximating a posterior distribution, the sample size ought to be rather large, and hence, making it computationally efficient is not easy.

Qian et al. (2019) choose to extend the Kalman filter by adopting Durbin and Koopman (1997)'s method that finds a Gaussian approximation of the Poisson observation. A Gaussian distribution can have a similar shape as the Poisson within the neighborhood of $\hat{\boldsymbol{\alpha}}_t$. Based on this understanding, rather than approximating the Poisson distribution globally, Qian et al. (2019) derive a good local Gaussian approximation of the Poisson observation in Eq. (7.35) around the state's current estimation, $\hat{\boldsymbol{\alpha}}_t$. As the Kalman filter usually updates $\boldsymbol{\alpha}_t$ near its current estimation, such a local approximation can lead to an efficient and accurate estimation.

When used in the context of nano applications, the above thought suggests the following approximation:

$$\mathbf{Y}_t \sim \text{normal}(\mathbf{B}\boldsymbol{\alpha}_t + \boldsymbol{\mu}_t, \mathbf{H}_t), \tag{7.39}$$

so that the probability density functions of Eqs. (7.35) and (7.39) have the same first and second derivatives with respect to $\boldsymbol{\alpha}_t$ near $\hat{\boldsymbol{\alpha}}_t$. Following this thought, Qian et al. (2019) derive the following expressions for the mean vector $\boldsymbol{\mu}_t$ and the covariance matrix $\mathbf{H}_t$ (we defer the derivation details to Sect. 7.4.3), as

$$\begin{aligned} \boldsymbol{\mu}_t &= \mathbf{Y}_t - \mathbf{B}\hat{\boldsymbol{\alpha}}_t - \exp(-\mathbf{B}\hat{\boldsymbol{\alpha}}_t)[\mathbf{Y}_t - \exp(\mathbf{B}\hat{\boldsymbol{\alpha}}_t)], \quad \text{and} \\ \mathbf{H}_t &= \text{diag}[\exp(-\mathbf{B}\hat{\boldsymbol{\alpha}}_t)]. \end{aligned} \tag{7.40}$$

As such, the original state space model is converted into an approximated Gaussian state space model, now constituting of Eqs. (7.38) and (7.39), rewritten below:

$$\begin{aligned} \text{state equation:} \quad & \boldsymbol{\alpha}_{t+1} = \boldsymbol{\alpha}_t + \mathbf{w}_t, \ \ \mathbf{w}_t \sim \text{normal}(\mathbf{0}, \mathbf{Q}), \\ \text{observation equation:} \quad & \mathbf{Y}_t \sim \text{normal}(\mathbf{B}\boldsymbol{\alpha}_t + \boldsymbol{\mu}_t, \mathbf{H}_t). \end{aligned} \tag{7.41}$$

Technically, a standard Kalman filter can then be devised and applied. A remaining problem is that $\hat{\boldsymbol{\alpha}}_t$ is unknown when one calculates $\boldsymbol{\mu}_t$ and $\mathbf{H}_t$ in Eq. (7.40). To address that issue, Qian et al. (2019) use an iterative strategy to find $\hat{\boldsymbol{\alpha}}_t$, i.e., first use the prior estimator $\hat{\boldsymbol{\alpha}}_t^-$ to calculate $\boldsymbol{\mu}_t$ and $\mathbf{H}_t$; then update $\hat{\boldsymbol{\alpha}}_t$ by the Kalman filter; and after that, update $\boldsymbol{\mu}_t$ and $\mathbf{H}_t$ using the newly estimated $\hat{\boldsymbol{\alpha}}_t$. Repeat this process until $\hat{\boldsymbol{\alpha}}_t$ converges. According to both Durbin and Koopman (1997) and Qian et al. (2019), this process routinely converges in fewer than five steps.

This Kalman filter works iteratively from $t = 1$ to time T as follows. First, one predicts $\boldsymbol{\alpha}_t$ and its covariance matrix, according to the estimation of the previous step:

$$\begin{aligned} \hat{\boldsymbol{\alpha}}_t^- &= \hat{\boldsymbol{\alpha}}_{t-1}, \\ \mathbf{P}_t^- &= \mathbf{P}_{t-1} + \mathbf{Q}. \end{aligned} \tag{7.42}$$

When a new $\mathbf{Y}_t$ arrives, one estimates the innovation by $\boldsymbol{\nu}_t$ and its covariance matrix, according to the previous prediction $\hat{\boldsymbol{\alpha}}_t^-$ and the new input $\mathbf{Y}_t$:

$$\begin{aligned} \boldsymbol{\nu}_t &= \mathbf{Y}_t - \mathbf{B}\hat{\boldsymbol{\alpha}}_t^- - \boldsymbol{\mu}_t; \\ \mathbf{F}_t &= \mathbf{B}\mathbf{P}_t^-\mathbf{B}^T + \mathbf{H}_t. \end{aligned} \tag{7.43}$$

Then the Kalman gain is calculated by

$$\mathbf{K}_t = \mathbf{P}_t^-\mathbf{B}^T\mathbf{F}_t^{-1}. \tag{7.44}$$

After that, one updates the estimate of state $\boldsymbol{\alpha}_t$ and its covariance matrix through

$$\begin{aligned} \hat{\boldsymbol{\alpha}}_t &= \hat{\boldsymbol{\alpha}}_t^- + \mathbf{K}_t\boldsymbol{\nu}_t, \\ \mathbf{P}_t &= \mathbf{P}_t^-(\mathbf{I} - \mathbf{K}_t\mathbf{B})^T, \end{aligned} \tag{7.45}$$

where $\hat{\boldsymbol{\alpha}}_t$ is called the posterior estimator and $\mathbf{P}_t$ is the posterior covariance matrix. Algorithm 7.2 summarizes the estimation and updating process.

After one attains the posterior estimation of the state, $\hat{\boldsymbol{\alpha}}_t$, the corresponding NPSD, $\hat{f}_t(x)$, can be obtained through

$$\hat{f}_t(x) = \frac{1}{C_t(\hat{\boldsymbol{\alpha}}_t)} \exp\left[\sum_{j=1}^{n} \hat{\alpha}_{jt} B_j(x)\right]. \tag{7.46}$$

The normalizing constant, $C_t(\hat{\boldsymbol{\alpha}}_t)$, is computed by numerical integration after Algorithm 7.2 converges. Figure 7.5 highlights the main online updating steps for tracking the time-varying NPSD.

Algorithm 7.2 Kalman filter online updating method of the state space model

1. Set $t = 1$ and initialize $\hat{\boldsymbol{\alpha}}_0$ and $\mathbf{P}_0$.
2. Predict the prior estimator of the state as: $\hat{\boldsymbol{\alpha}}_t^- = \hat{\boldsymbol{\alpha}}_{t-1}$.
3. Predict the prior covariance matrix as: $\mathbf{P}_t^- = \mathbf{P}_{t-1} + \mathbf{Q}$.
4. Set $\hat{\boldsymbol{\alpha}}_t = \hat{\boldsymbol{\alpha}}_t^-$.
5. Calculate $\boldsymbol{\mu}_t$ and $\mathbf{H}_t$ as:

$$\boldsymbol{\mu}_t = \mathbf{Y}_t - \mathbf{B}\hat{\boldsymbol{\alpha}}_t - \exp(-\mathbf{B}\hat{\boldsymbol{\alpha}}_t)[\mathbf{Y}_t - \exp(\mathbf{B}\hat{\boldsymbol{\alpha}}_t)],$$
$$\mathbf{H}_t = \mathrm{diag}[\exp(-\mathbf{B}\hat{\boldsymbol{\alpha}}_t)].$$

6. Estimate the innovation and its covariance matrix:

$$\boldsymbol{\nu}_t = \mathbf{Y}_t - \mathbf{B}\hat{\boldsymbol{\alpha}}_t^- - \boldsymbol{\mu}_t; \quad \mathbf{F}_t = \mathbf{B}\mathbf{P}_t^-\mathbf{B}^T + \mathbf{H}_t.$$

7. Compute the Kalman gain as: $\mathbf{K}_t = \mathbf{P}_t^-\mathbf{B}^T\mathbf{F}_t^{-1}$.
8. Update the posterior estimator: $\hat{\boldsymbol{\alpha}}_t = \hat{\boldsymbol{\alpha}}_t^- + \mathbf{K}_t\boldsymbol{\nu}_t$.
9. Repeat Steps 5 to 8 until $\hat{\alpha}_t$ converges.
10. Update the posterior covariance matrix as: $\mathbf{P}_t = \mathbf{P}_t^-(\mathbf{I} - \mathbf{K}_t\mathbf{B})^T$.
11. Set $t = t + 1$, repeat from Step 2 until the process ends.

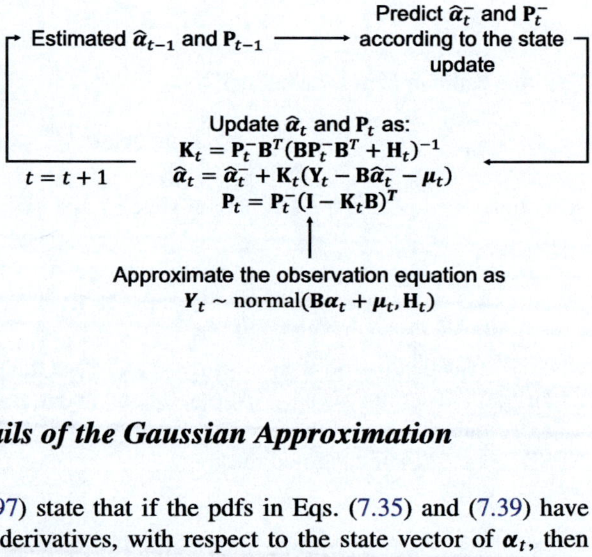

Fig. 7.5 Illustration of main steps in the Kalman filter's online updating. (Reprinted with permission from Qian et al. 2019)

7.4.3 *Technical Details of the Gaussian Approximation*

Durbin and Koopman (1997) state that if the pdfs in Eqs. (7.35) and (7.39) have the same first and second derivatives, with respect to the state vector of $\boldsymbol{\alpha}_t$, then Eq. (7.39) is considered a good approximation of Eq. (7.35) in updating the state space model. Qian et al. (2019) use this idea to calculate $\boldsymbol{\mu}_t$ and $\mathbf{H}_t$ in Eq. (7.39). To simplify the derivation, Qian et al. (2019) use $\mathbf{B}\boldsymbol{\alpha}_t$ instead of $\boldsymbol{\alpha}_t$ as the variable to calculate those derivatives.

The logarithm of the pdfs in Eqs. (7.35) and (7.39), as a function of $\mathbf{B}\boldsymbol{\alpha}_t$, can be expressed, respectively, as

$$\log f_{\mathrm{poi}}((\mathbf{B}\boldsymbol{\alpha}_t)_i) = Y_{it}(\mathbf{B}\boldsymbol{\alpha}_t)_i - \exp\{(\mathbf{B}\boldsymbol{\alpha}_t)_i\}, i = 1, \ldots, m, \tag{7.47}$$

and

$$\log f_{\text{nor}}(\mathbf{B}\boldsymbol{\alpha}_t) = -\frac{1}{2}(\mathbf{Y}_t - \mathbf{B}\boldsymbol{\alpha}_t - \boldsymbol{\mu}_t)^T \mathbf{H}_t^{-1}(\mathbf{Y}_t - \mathbf{B}\boldsymbol{\alpha}_t - \boldsymbol{\mu}_t) + \text{const}, \quad (7.48)$$

where 'const' is a term unrelated to $\boldsymbol{\alpha}_t$, and f_{poi} and f_{nor} denote the Poisson and normal density functions, respectively. Recall that we use $(\cdot)_i$ as an indexing notation; for example, $(\mathbf{B}\boldsymbol{\alpha}_t)_i$ denotes the ith element of the vector of $\mathbf{B}\boldsymbol{\alpha}_t$.

In Eq. (7.47), the pdf of each coordinate of $\mathbf{B}\boldsymbol{\alpha}_t$ is independent to each other, and Eq. (7.48) should have the same property, meaning that $\mathbf{H}_t$ should be a diagonal matrix. Qian et al. (2019) then rewrite Eq. (7.48) as:

$$\log f_{\text{nor}}((\mathbf{B}\boldsymbol{\alpha}_t)_i) = -\frac{1}{2(\mathbf{H}_t)_{ii}}(Y_{it} - (\mathbf{B}\boldsymbol{\alpha}_t)_i - (\boldsymbol{\mu}_t)_i)^2 + \text{const}, \quad (7.49)$$

Next, Qian et al. (2019) calculate the first and second derivatives of Eqs. (7.47) and (7.49) with respect to $(\mathbf{B}\boldsymbol{\alpha}_t)_i$, and equate them at $\hat{\boldsymbol{\alpha}}_t$. That leads to the following two equations:

$$Y_{it} - \exp\{(\mathbf{B}\hat{\boldsymbol{\alpha}}_t)_i\} = \frac{1}{(\mathbf{H}_t)_{ii}}(Y_{it} - (\mathbf{B}\hat{\boldsymbol{\alpha}}_t)_i - (\boldsymbol{\mu}_t)_i), \quad (7.50)$$

$$\exp\{(\mathbf{B}\hat{\boldsymbol{\alpha}}_t)_i\} = \frac{1}{(\mathbf{H}_t)_{ii}}. \quad (7.51)$$

The two equations further yield:

$$(\mathbf{H}_t)_{ii} = \frac{1}{\exp\{(\mathbf{B}\hat{\boldsymbol{\alpha}}_t)_i\}} = \exp\{(-\mathbf{B}\hat{\boldsymbol{\alpha}}_t)_i\}, \quad (7.52)$$

$$(\boldsymbol{\mu}_t)_i = Y_{it} - (\mathbf{B}\hat{\boldsymbol{\alpha}}_t)_i - \exp\{(-\mathbf{B}\hat{\boldsymbol{\alpha}}_t)_i\}(Y_{it} - \exp\{(\mathbf{B}\hat{\boldsymbol{\alpha}}_t)_i\}). \quad (7.53)$$

Rewriting Eqs. (7.52) and (7.53) in a matrix form, one obtains $\boldsymbol{\mu}_t$ and $\mathbf{H}$ as:

$$\begin{aligned} \boldsymbol{\mu}_t &= \mathbf{Y} - \mathbf{B}\hat{\boldsymbol{\alpha}}_t - \exp(-\mathbf{B}\hat{\boldsymbol{\alpha}}_t)[\mathbf{Y} - \exp(\mathbf{B}\hat{\boldsymbol{\alpha}}_t)], \\ \mathbf{H} &= \text{diag}[\exp(-\mathbf{B}\hat{\boldsymbol{\alpha}}_t)]. \end{aligned} \quad (7.54)$$

This concludes the derivation.

There are other approximation approaches. One of them is known as the root-unroot approach. The idea can be understood as follows. Because $Y \sim \text{Poisson}(\lambda)$, $\sqrt{Y + \frac{1}{4}}$ is approximately $N(\sqrt{\lambda}, \frac{1}{4})$ (Anscombe 1948). Based on this understanding, Brown et al. (2010) studied a root-unroot nonparametric density estimation method, which suggests a potentially simple solution to the online density updating problem, as follows—One first models the square-root of the density by a B-spline, i.e., $\sqrt{\lambda_t} = \mathbf{B}\boldsymbol{\alpha}_t$, applies the standard Kalman filter to the square-root transformed bin counts, and then unroots the estimator to obtain a proper density function.

Qian et al. (2019) find that this root-unroot approach does not work well, for two reasons: (1) The approximation of using the square root transformation has a large bias when λ is close to 0. Both Anscombe (1948) and Brown et al. (2010) note that the bias is proportional to $\lambda^{-2/3}$. In the context of probability density estimation using binned data, since there are many bins with zero counts (corresponding to small λ's), this square root approximation is too crude. (2) If applying the standard Kalman filter to update $\sqrt{\lambda_t} = \mathbf{B}\boldsymbol{\alpha}_t$, one may end up with a negative value for $\sqrt{\lambda_t}$. Although its square still gives a positive density estimation, there is an upward bias. In fact, when λ_t is near zero, updating the state variables to reduce the value of λ_t may yield a negative updated value of $\mathbf{B}\boldsymbol{\alpha}_t$ and thus can lead to a larger absolute value of λ_t, contrary to the purpose of reducing λ_t, creating more bias.

As a comparison, the quality of the local Gaussian approximation does not depend on the value of λ. Moreover, it is used in an iterative manner, so that the approximation improves along the iterations. Numerical results show that the local Gaussian approximation works rather well in the presence of empty bins, especially once the smoothness constraint is imposed.

7.4.4 Curve Smoothness for Distribution Estimation

While Algorithm 7.2 can provide an online estimation and updating of the time-varying NPSD, it does not impose any requirement on the smoothness of the estimated density function. Without a proper smoothness constraint, the resulting density estimation could be sensitive to choices like the number of intervals in the histogram, $\mathbf{Y}_t$, and the number of B-spline basis functions, and could become considerably inaccurate in the cases that some middle intervals in the input histogram turn out empty. The goal here is to incorporate the curve smoothness constraint and make it work with the state space model.

Qian et al. (2019) impose the curve smoothness constraint for the B-spline density estimation by penalizing the squared norm of the second order difference of the spline coefficient vector $\boldsymbol{\alpha}_t$, which is denoted as an $n-2$ dimension vector $\Delta^2\boldsymbol{\alpha}_t$ and defined as

$$\Delta^2\boldsymbol{\alpha}_t = \begin{bmatrix} -\alpha_{1t} + 2\alpha_{2t} - \alpha_{3t} \\ -\alpha_{2t} + 2\alpha_{3t} - \alpha_{4t} \\ \cdots \\ -\alpha_{(n-2)t} + 2\alpha_{(n-1)t} - \alpha_{nt} \end{bmatrix} = \begin{bmatrix} -1 & 2 & -1 & \cdots & 0 & 0 & 0 \\ 0 & -1 & 2 & \cdots & 0 & 0 & 0 \\ \vdots & \vdots & \vdots & \ddots & \vdots & \vdots & \vdots \\ 0 & 0 & 0 & \cdots & -1 & 2 & -1 \end{bmatrix} \boldsymbol{\alpha}_t. \tag{7.55}$$

In the smoothing spline literature, the commonly used penalty to enforce smoothness is the squared second derivative penalty $\int_x (f''(x))^2 dx$ (Wahba 1990). Section 3 of Eilers and Marx (1996) points out that the squared second difference penalty, $\|\Delta^2\boldsymbol{\alpha}_t\|^2$, can be viewed as a convenient approximation with the B-spline representation. The precise relation of the two penalties is given in Eq. 11 of Eilers and Marx (1996).

To put a constraint on $\|\Delta^2\boldsymbol{\alpha}_t\|^2$, Qian et al. (2019) transform linearly the original state, $\boldsymbol{\alpha}_t$, into another state, $\boldsymbol{\gamma}_t$. The new state, $\boldsymbol{\gamma}_t$, includes $\Delta^2\boldsymbol{\alpha}_t$ but should ideally have the same dimension as $\boldsymbol{\alpha}_t$. Qian et al. (2019) therefore make the last $n-2$ elements of $\boldsymbol{\gamma}_t$ equal to $\Delta^2\boldsymbol{\alpha}_t$ and then add something else as its first two elements for the purpose of making the new vector an n dimensional vector.

A straightforward choice for the first two elements is to let γ_{1t} be the summation of all the even numbered coordinates of $\boldsymbol{\alpha}_t$ and γ_{2t} be the summation of all the odd numbered coordinates of $\boldsymbol{\alpha}_t$. This choice ensures these two elements are orthogonal and have similar magnitudes. Because of the later property, Qian et al. (2019) set the variances of the corresponding innovations at the same value. Consequently, Qian et al. (2019) obtain an invertible transformation from $\boldsymbol{\alpha}_t$ to $\boldsymbol{\gamma}_t$, such that

$$\gamma_{1t} = \sum_{j=1}^{[n/2]} \alpha_{(2j)t}, \; \gamma_{2t} = \sum_{j=1}^{[n/2]} \alpha_{(2j-1)t}, \; \boldsymbol{\gamma}_{(3:n)t} = \Delta^2\boldsymbol{\alpha}_t, \tag{7.56}$$

where $[n/2]$ is the largest integer smaller than or equal to $n/2$. Qian et al. (2019) furthermore write this linear transform in a matrix format, such that $\boldsymbol{\alpha}_t = \mathbf{C}\boldsymbol{\gamma}_t$, where

$$\begin{aligned} &\mathbf{C}^{-1}_{1(2j)} = 1, \; \mathbf{C}^{-1}_{2(2j-1)} = 1, \; j = 1, \cdots, [n/2]; \\ &\mathbf{C}^{-1}_{j(j-2)} = -1, \mathbf{C}^{-1}_{j(j-1)} = 2, \mathbf{C}^{-1}_{jj} = -1, \; j = 2, \cdots, n; \end{aligned} \tag{7.57}$$

and other elements of $\mathbf{C}^{-1}$ are equal to 0.

By using the new state, $\boldsymbol{\gamma}_t$, Qian et al. (2019) express the state space model as:

$$\begin{aligned} Y_{it} &\sim \text{Poisson}\{(\exp[\mathbf{B}\mathbf{C}\boldsymbol{\gamma}_t])_i\}, \\ \boldsymbol{\gamma}_t &= \boldsymbol{\gamma}_{t-1} + \mathbf{w}_t, \end{aligned} \tag{7.58}$$

where $\mathbf{w}_t \sim \text{normal}(\mathbf{0}, \mathbf{Q})$. Here the notations are slightly abused—$\mathbf{w}_t$ and $\mathbf{Q}$ are used here again but they are of different values from those in Eq. (7.41).

This transformation in the state vector allows researchers to use the structure of $\mathbf{Q}$ to add the smoothness constraint on the estimated density $\hat{f}_t(x)$. Aware that $[\gamma_{1t}, \gamma_{2t}]$ are the summations of the even and odd terms of $\boldsymbol{\alpha}_t$, respectively, and $[\gamma_{3t}, \cdots, \gamma_{nt}]$ are the second differences of $\boldsymbol{\alpha}_t$, Qian et al. (2019) assume that their innovations are independent to each other, making $\mathbf{Q}$ a diagonal matrix, denoted as $\text{diag}(\sigma_1^2, \sigma_2^2, \cdots, \sigma_n^2)$, in which σ_1^2, σ_2^2 are the variances of w_{1t}, w_{2t}, and $\sigma_3^2, \cdots, \sigma_n^2$ are the variances of $w_{3t}, \cdots, w_{nt}$. For simplicity, Qian et al. (2019) further assume that σ_1^2 and σ_2^2 have the same value, denoted as σ_α^2, and all the remaining $\sigma_3^2, \cdots, \sigma_n^2$ are equal, their value denoted as σ_ϵ^2. Requiring $\sigma_\alpha^2 \gg \sigma_\epsilon^2$ for the new state vector is effectively forcing the second order difference of $\boldsymbol{\alpha}_t$ to be small and thus resulting the smoothness constraint imposed onto the estimated density curves. One can still use Algorithm 7.2 to update $\hat{\boldsymbol{\gamma}}_t$ after replacing $\mathbf{B}$ and $\boldsymbol{\alpha}_t$ with $\mathbf{BC}$ and $\boldsymbol{\gamma}_t$, respectively, so that the fast computation and online estimation/updating capability are retained.

7.5 Parameter Estimation

In order for the prospective analysis to work, one does need a short starting up period, which is to gather a limited amount of training video data to initialize the parameters in the model. Qian et al. (2019) use the first few hundreds of frames for parameter estimation, equivalent to the first 15–20 s of the process.

In the state space model, there are two parameters σ_α^2 and σ_ϵ^2 that need to be estimated using the training data from the short starting up period, from $t = 1$ until time T. The two parameters determine the covariance matrix of $\mathbf{w}_t$: σ_α^2 represents the degree of variability of the underlying state $\boldsymbol{\gamma}_t$, whereas σ_ϵ^2 controls its second order, indicating the smoothness of the estimated density curve. It is not convenient to find the values of σ_α^2 and σ_ϵ^2 through MLE, as the calculation of the likelihood of such a mixed system needs complicated process like importance sampling (Durbin and Koopman 1997) or simulation smoothing (de Jong and Shephard 1995), let alone to optimize the likelihood for estimating its parameters. Instead, Qian et al. (2019) adopt a Bayesian approach to estimate the two parameters in the covariance matrix.

7.5.1 Bayesian Modeling

Qian et al. (2019) regard σ_α^2 and σ_ϵ^2 as latent random variables and choose their prior distributions first. Then, Qian et al. (2019) obtain their posterior distribution through a sampling method and use the corresponding posterior means as the estimate of the parameters.

Since σ_α^2 and σ_ϵ^2 define the covariance matrix of $\mathbf{w}_t$, which is assumed to follow a normal distribution, Qian et al. (2019) choose the corresponding conjugate prior as an inverse-gamma distribution, making the posterior distribution in the same family. Qian et al. (2019) write the hierarchical structure of the Bayesian model as:

$$\begin{aligned}
&Y_{it} \sim \text{Poisson}\{(\exp[\mathbf{BC}\boldsymbol{\gamma}_t])_i\},\\
&\boldsymbol{\gamma}_t - \boldsymbol{\gamma}_{t-1} = \mathbf{w}_t \sim \text{normal}(\mathbf{0}, \mathbf{Q}),\ \mathbf{Q} = \text{diag}(\sigma_\alpha^2, \sigma_\alpha^2, \sigma_\epsilon^2, \cdots, \sigma_\epsilon^2),\\
&\sigma_\alpha^2 \sim \text{inverse-gamma}(a_1, b_1),\ \sigma_\epsilon^2 \sim \text{inverse-gamma}(a_2, b_2),
\end{aligned} \tag{7.59}$$

where the initial state $\boldsymbol{\gamma}_0$ is set as $[-2, -2, \cdots, -2]$ so that $f(x)$ will evolve from a zero function.

To determine the hyper-parameters in the inverse-gamma distributions of σ_α^2 and σ_ϵ^2, Qian et al. (2019) choose the non-informative prior (Spiegelhalter et al. 1996) as $a_1 = 1.0$ and $b_1 = 1.0$ for σ_α^2. To make sure $\sigma_\alpha^2 \gg \sigma_\epsilon^2$, Qian et al. (2019) choose the same shape parameter, i.e., $a_2 = 1.0$, but a much smaller scale parameter b_2 for σ_ϵ^2 (the mean of the inverse-gamma distribution is proportional to the scale parameter). Qian et al. (2019) recommend choosing b_2 such that $b_1/b_2 = 100$, and a robustness check of this choice is discussed in Sect. 7.5.3.

Compared to the original state space model, the hierarchical model adds another layer associated with the prior distributions of σ_α^2 and σ_ϵ^2. Once observing $\mathbf{Y}_1, \cdots, \mathbf{Y}_T$ in the starting up period, one can employ a Markov chain Monte Carlo (MCMC) sampling method to update the posterior distributions of σ_α^2 and σ_ϵ^2, and then use the posterior means as the estimate of the two parameters.

7.5.2 MCMC Sampling

This subsection presents the posterior distribution of σ_α^2 and σ_ϵ^2 in the Bayesian model, Eq. (7.59). Since $\mathbf{Q}$ is a covariance matrix, one can rewrite the second line of the model in Eq. (7.59) as:

$$\gamma_{jt} - \gamma_{j(t-1)} \sim \text{normal}(0, \sigma_\alpha^2), \ \ j = 1, 2, \ \ t = 2, \cdots, T, \tag{7.60}$$

and

$$\gamma_{jt} - \gamma_{j(t-1)} \sim \text{normal}(0, \sigma_\epsilon^2), \ \ j = 3, \cdots, n, \ \ t = 2, \cdots, T. \tag{7.61}$$

For $j = 1$ and 2, $\gamma_{jt} - \gamma_{j(t-1)}$ are regarded as $2(T-1)$ i.i.d variables following normal$(0, \sigma_\alpha^2)$. Since the conjugate prior $\sigma_\alpha^2 \sim$ inverse-gamma(a_1, b_1) is chosen, its posterior distribution has the same formation as inverse-gamma$(a_1^{\text{post}}, b_1^{\text{post}})$. As derived in Bolstad and Curran (2016), a_1^{post} and b_1^{post} are calculated by

$$a_1^{\text{post}} = a_1 + \frac{1}{2} 2(T-1) = a_1 + (T-1), \tag{7.62}$$

and

$$b_1^{\text{post}} = b_1 + \frac{1}{2} \sum_{j=1}^{2} \sum_{t=2}^{T} [\gamma_{jt} - \gamma_{j(t-1)}]^2. \tag{7.63}$$

For $j = 3, \cdots, n$, $\gamma_{jt} - \gamma_{j(t-1)}$ are regarded as $(n-2)(T-1)$ i.i.d variables following normal$(0, \sigma_\epsilon^2)$. Following the same derivation, the posterior distribution of σ_ϵ^2 is written as inverse-gamma$(a_2^{\text{post}}, b_2^{\text{post}})$ with

$$a_2^{\text{post}} = a_2 + \frac{1}{2}(n-2)(T-1), \tag{7.64}$$

and

$$b_2^{\text{post}} = b_2 + \frac{1}{2}\sum_{j=3}^{n}\sum_{t=2}^{T}[\gamma_{jt} - \gamma_{j(t-1)}]^2. \tag{7.65}$$

Denote the values in the kth iteration of MCMC by $\boldsymbol{\gamma}_1^{(k)}, \cdots, \boldsymbol{\gamma}_T^{(k)}$, $(\sigma_\alpha^2)^{(k)}$ and $(\sigma_\epsilon^2)^{(k)}$. After the initialization, one draws samples of $(\sigma_\alpha^2)^{(k)}$ and $(\sigma_\epsilon^2)^{(k)}$ through the Gibbs sampling method, given $\boldsymbol{\gamma}_1^{(k-1)}, \cdots, \boldsymbol{\gamma}_T^{(k-1)}$. Since the conjugate priors are used, the posterior distributions are still inverse-gamma as:

$$(\sigma_\alpha^2)^{(k)} \sim \text{inverse-gamma}(a_1^{\text{post}}, b_1^{\text{post}}),\ (\sigma_\epsilon^2)^{(k)} \sim \text{inverse-gamma}(a_2^{\text{post}}, b_2^{\text{post}}), \tag{7.66}$$

where a_1^{post}, b_1^{post}, a_2^{post} and b_2^{post} are determined by a_1, b_1, a_2, b_2, and the sampled $\boldsymbol{\gamma}_t^{(k-1)}$.

Then, one samples $\boldsymbol{\gamma}_1^{(k)}, \cdots, \boldsymbol{\gamma}_T^{(k)}$, given $\mathbf{Q}^{(k)} = \text{diag}[(\sigma_\alpha^2)^{(k)}, (\sigma_\alpha^2)^{(k)}, (\sigma_\epsilon^2)^{(k)}, \ldots, (\sigma_\epsilon^2)^{(k)}]$ and the observations, $\mathbf{Y}_1, \cdots, \mathbf{Y}_T$. Unfortunately, the posterior distributions of $\boldsymbol{\gamma}_t^{(k)}$ are not of a standard type. Qian et al. (2019) therefore implement a Metropolis-Hastings algorithm to sample $\boldsymbol{\gamma}_t^{(k)}$ from $t = 1$ to T. For each individual t, Qian et al. (2019) first draw $\boldsymbol{\gamma}_t^{(k)}$ from the following proposal distribution:

$$\boldsymbol{\gamma}_t^{(k)} \sim \text{normal}(\boldsymbol{\gamma}_t^{(k-1)}, \mathbf{R}), \tag{7.67}$$

where $\mathbf{R} = \text{diag}(\sigma_1^2, \sigma_1^2, \sigma_2^2, \cdots, \sigma_2^2)$ shares a similar structure as $\mathbf{Q}$. The acceptance ratio of a newly sampled $\boldsymbol{\gamma}_t^{(k)}$, r, is defined in a standard way, as the ratio of the conditional pdf given the current $\boldsymbol{\gamma}_t^{(k)}$ to that given the previous $\boldsymbol{\gamma}_t^{(k-1)}$. After getting r, one compares it with a uniform random variable, u, in [0, 1], to determine whether to accept the new $\boldsymbol{\gamma}_t^{(k)}$ or not.

After repeating the above sampling iterations K times, the posterior means can be obtained by:

$$\hat{\sigma}_\alpha^2 = \frac{1}{K - K_B}\sum_{k=K_B+1}^{K}(\sigma_\alpha^2)^{(k)},\ \hat{\sigma}_\epsilon^2 = \frac{1}{K - K_B}\sum_{k=K_B+1}^{K}(\sigma_\epsilon^2)^{(k)}, \tag{7.68}$$

where K_B is the amount of the burn-in steps.

We summarize the main steps in Algorithm 7.3.

Algorithm 7.3 Parameter estimation through Bayesian sampling

1. Initialize $\boldsymbol{\gamma}_1^{(0)}, \cdots, \boldsymbol{\gamma}_T^{(0)}, (\sigma_\alpha^2)^{(0)}$ and $(\sigma_\epsilon^2)^{(0)}$.
2. Set $k = 1$, then sample $(\sigma_\alpha^2)^{(k)}$ and $(\sigma_\epsilon^2)^{(k)}$ as: $(\sigma_\alpha^2)^{(k)} \sim$ inverse-gamma$(a_1^{\text{post}}, b_1^{\text{post}})$ and $(\sigma_\epsilon^2)^{(k)} \sim$ inverse-gamma$(a_2^{\text{post}}, b_2^{\text{post}})$, where
$$a_1^{\text{post}} = a_1 + (T-1), b_1^{\text{post}} = b_1 + \tfrac{1}{2}\textstyle\sum_{j=1}^2 \sum_{t=2}^T [\gamma_{jt}^{(k-1)} - \gamma_{j(t-1)}^{(k-1)}]^2,$$
$$a_2^{\text{post}} = a_2 + \tfrac{n-2}{2}(T-1), b_2^{\text{post}} = b_2 + \tfrac{1}{2}\textstyle\sum_{j=3}^n \sum_{t=2}^T [\gamma_{jt}^{(k-1)} - \gamma_{j(t-1)}^{(k-1)}]^2.$$
3. Let $\mathbf{Q}^{(k)} = \text{diag}[(\sigma_\alpha^2)^{(k)}, (\sigma_\alpha^2)^{(k)}, (\sigma_\epsilon^2)^{(k)}, \cdots, (\sigma_\epsilon^2)^{(k)}]$.
4. Set $t = 1$, sample $\boldsymbol{\gamma}_t^{(k)}$ from a proposal distribution: $\boldsymbol{\gamma}_t^{(k)} \sim$ normal$(\boldsymbol{\gamma}_t^{(k-1)}, \mathbf{R})$.
5. Calculate the acceptance rate r as:
$$r = \frac{\prod_{i=1}^m f_{\text{poi}}(Y_{it}|(\mathbf{BC}\boldsymbol{\gamma}_t^{(k)})_i) f_{\text{nor}}(\boldsymbol{\gamma}_t^{(k)}|\boldsymbol{\gamma}_{t-1}^{(k)}, \mathbf{Q}^{(k)}) f_{\text{nor}}(\boldsymbol{\gamma}_t^{(k)}|\boldsymbol{\gamma}_{t+1}^{(k-1)}, \mathbf{Q}^{(k)})}{\prod_{i=1}^m f_{\text{poi}}(Y_{it}|(\mathbf{BC}\boldsymbol{\gamma}_t^{(k-1)})_i) f_{\text{nor}}(\boldsymbol{\gamma}_t^{(k-1)}|\boldsymbol{\gamma}_{t-1}^{(k)}, \mathbf{Q}^{(k)}) f_{\text{nor}}(\boldsymbol{\gamma}_t^{(k-1)}|\boldsymbol{\gamma}_{t+1}^{(k-1)}, \mathbf{Q}^{(k)})},$$
where $f_{\text{poi}}(\cdot|\cdot)$ is the pdf of a Poisson distribution and $f_{\text{nor}}(\cdot|\cdot,\cdot)$ is the pdf of a multivariate normal distribution.
6. Generate a uniform random number, u, in [0, 1]. If $r > u$, accept $\boldsymbol{\gamma}_t^{(k)}$; otherwise set $\boldsymbol{\gamma}_t^{(k)} = \boldsymbol{\gamma}_t^{(k-1)}$.
7. Set $t = t + 1$, and repeat Steps 4 to 6 until $t = T$.
8. Set $k = k + 1$, and repeat Steps 2 to 7 until $k = K$.
9. Estimate σ_α^2 and σ_ϵ^2 as the posterior means:
$$\hat{\sigma}_\alpha^2 = \frac{1}{K - K_B} \sum_{k=K_B+1}^K (\sigma_\alpha^2)^{(k)}, \ \hat{\sigma}_\epsilon^2 = \frac{1}{K - K_B} \sum_{k=K_B+1}^K (\sigma_\epsilon^2)^{(k)}.$$

7.5.3 Select the Hyper-Parameters

The hyper-parameters in the Bayesian model, Eq. (7.59), and in the MCMC algorithm include: a_1, b_1, a_2 and b_2 in the prior distribution, the initial values of the MCMC sampling, $\boldsymbol{\gamma}_t^{(0)}$, $(\sigma_\alpha^2)^{(0)}$ and $(\sigma_\epsilon^2)^{(0)}$, and σ_1^2 and σ_2^2 in the covariance matrix $\mathbf{R}$ of the proposal distribution.

The parameters in the MCMC sampling matter less, as a long burn-in stage (namely a large enough K_B) makes the MCMC robust to initialization. As long as the MCMC has a good mixing, different proposal distributions give similar estimation outcomes. Qian et al. (2019) recommend setting those parameters in the following way: $(\sigma_\alpha^2)^{(0)} = 4 \times 10^{-2}$, $(\sigma_\epsilon^2)^{(0)} = 2 \times 10^{-3}$, run the extended Kalman filter in Algorithm 7.2 to obtain $\boldsymbol{\gamma}_t^{(0)}$, and let $\sigma_1^2 = 2 \times 10^{-2}$ and $\sigma_2^2 = 1 \times 10^{-3}$.

While a_1, b_1 and a_2 are specified in Sect. 7.5.1, one should run MCMC to find a suitable value for b_2. Qian et al. (2019) find that as long as b_1/b_2 is large enough, say, more than an order of magnitude, the estimation outcome appears robust. Table 7.1 presents the posterior means of the two parameters estimated from Video

Table 7.1 The parameters, $\hat{\sigma}_\alpha^2$ and $\hat{\sigma}_\epsilon^2$, and their 90% credible intervals, estimated using Video 1 data and under different b_2 values. In the table, $a_1 = a_2 = b_1 = 1.0$. (*Source*: Qian et al. 2019)

b_2	b_1/b_2	$\hat{\sigma}_\alpha^2$	$\hat{\sigma}_\epsilon^2$	$\hat{\sigma}_\alpha^2/\hat{\sigma}_\epsilon^2$
0.1	10	$5.94\ (4.44,\ 7.70) \times 10^{-2}$	$4.03\ (3.52,\ 4.64) \times 10^{-3}$	14.73
0.05	20	$6.47\ (4.89,\ 8.41) \times 10^{-2}$	$3.92\ (3.44.\ 4.54) \times 10^{-3}$	16.49
0.01	100	$6.39\ (4.46,\ 8.77) \times 10^{-2}$	$3.82\ (3.27,\ 4.35) \times 10^{-3}$	16.72
0.005	200	$6.39\ (4.80,\ 8.16) \times 10^{-2}$	$3.66\ (3.25,\ 4.18) \times 10^{-3}$	17.46

1, with a total of $K = 10^5$ iterations and $K_B = 4 \times 10^4$ burn-in steps. Although significantly different b_2's are used, the estimated results for other parameters stay similar. In practice, Qian et al. (2019) recommend using $b_2 = 0.01$ as the default setting. One should also check the convergence of the MCMC by plotting the chains of $(\sigma_\alpha^2)^{(k)}$ and $(\sigma_\epsilon^2)^{(k)}$ with multiple initial values and find that all the chains mix well after the burn-in stage.

7.6 Case Study

Qian et al. (2019) test the state space model and its online updating on the three clips of in situ TEM video described in Sect. 7.1.2. The number of the B-spline basis functions is fixed at 20 in all three cases. Because of incorporation of the smoothness constraint in the state space model, the final estimation of the NPSD is not sensitive to the choices of this parameter.

7.6.1 Analysis of the Three Videos

The first step is to find σ_α^2 and σ_ϵ^2 for each clip of videos. In Video 1, there are 1, 149 frames in total, with 15 frame per second (fps) imaging rate. Qian et al. (2019) choose the first 300 frames as the training set, corresponding to the first 20 s of the process. Using the Bayesian estimation method in described in Sect. 7.5 with the default parameter setting, the estimate of the two system parameters is $\hat{\sigma}_\alpha^2 = 6.39 \times 10^{-2}$ and $\hat{\sigma}_\epsilon^2 = 3.82 \times 10^{-3}$.

Qian et al. (2019) apply the updating method to the whole video. In this test, the TEM videos have already been fully recorded. Qian et al. (2019) mimic a prospective analysis, starting at the end of the initialization period. For the remaining 849 frames in Video 1, the total processing time of using the online algorithm is 1.23 s, or 1.5×10^{-4} s per frame, much faster than the frame rate of the video (which is 15 fps or 0.067 s per frame). Combined with the image processing time (0.04 s per frame), the overall model processing is still fast enough for online monitoring. Figure 7.6 illustrates the updating process running from 25.67 s through 28.33 s.

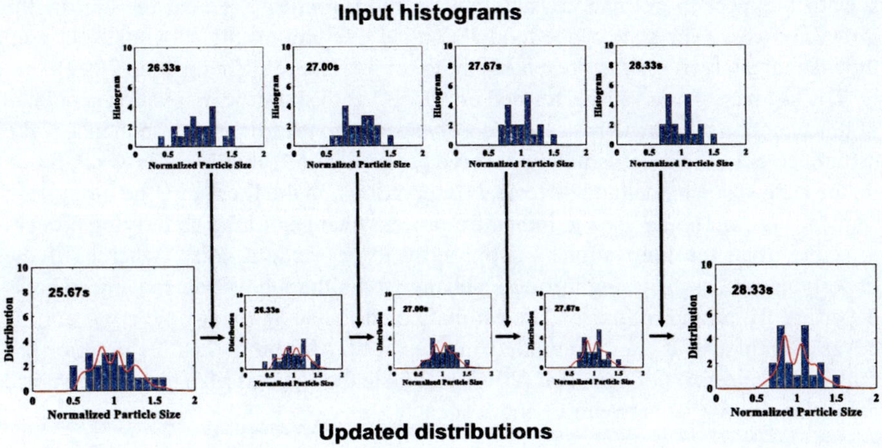

Fig. 7.6 Illustration of the updating outcomes of the state space model. (Reprinted with permission from Qian et al. 2019)

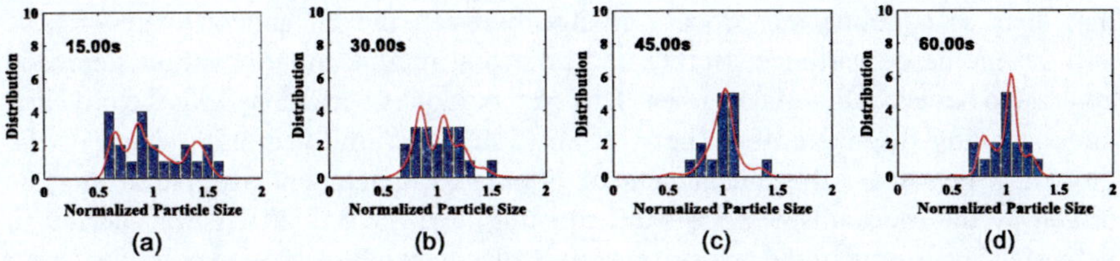

Fig. 7.7 The estimated NPSD of Video 1 at different growth stages. (Reprinted with permission from Qian et al. 2019)

The upper row shows the input histograms, whereas the lower row shows the updated NPSDs. To demonstrate the difference of the estimated distributions, the time difference between two consecutive images in that plot is set to be 10 frames.

Figure 7.7 presents the estimated NPSDs in different growth stages at 15, 30, 45, and 60 s, respectively. Figure 7.7a presents the NPSD at the beginning of the growth stage when the nanocrystals are initializing in the chemical solution. The variance of the particle sizes is large and the support of the distribution is broad. Figure 7.7b presents a NPSD at the orientated attachment (Aldous 1999) growth stage, at which time the smaller particles collide with each other and are merged into larger ones. The variance of the particle sizes is smaller than that of the first stage. There is a noticeable bimodal pattern in the NPSD, in which the two peaks correspond to the sizes of the smaller particles and the merged (larger) particles, respectively. The final two plots in Fig. 7.7c, d are in the final growth stage, known as the Ostwald ripening (Lifshitz and Slyozov 1961) stage. In that stage, the larger particles grow at the expense of dissolving smaller particles. The size distribution tends to get concentrated and become uni-modal. The variance continues to decrease. Material scientists expect to get nanocrystals having more uniform sizes at the end of the growth process. The state space model's online tracking results are consistent with the manual analysis results presented in the original report (Zheng et al. 2009).

The last part of analysis performed on Video 1 is to show the innovation sequence of this nanocrystal growth process. Loosely speaking, the innovation sequence is the difference between what is newly observed at time t and what is anticipated, based on the state space model and historical observations. In the literature, the innovation sequence is commonly used to indicate a process change—if the underlying process is stable, then the innovation is supposedly to be random noise, whereas if the underlying process is going through a change, then the innovation sequence shows departure from random noise. The estimate of the innovation at time t, $\boldsymbol{\nu}_t$, and its covariance matrix $\mathbf{F}_t$, are computed in Step 6 of Algorithm 7.2. To monitor the multivariate vector $\boldsymbol{\nu}_t$, Qian et al. (2019) calculate the squared Mahalanobis distance (Mahalanobis 1936) between $\boldsymbol{\nu}_t$ and $\mathbf{0}$ at each t, i.e.,

$$A_t = \boldsymbol{\nu}_t^T \mathbf{F}_t^- \boldsymbol{\nu}_t. \tag{7.69}$$

The sequence $\{A_1, A_2, \ldots,\}$ for Video 1 is plotted in Fig. 7.8. One can observe that there is a noticeable process change between the 20 and 40 s time marks with an increased variance. Before and after that period, the innovation sequence appears to have smaller magnitudes. This observation is consistent with the physical understanding discovered by Zheng et al. (2009), i.e., the beginning stage of the growth is driven by the mechanism of orientated attachment, the latter stage is driven by the mechanism of Ostwald ripening, and there is a transition period in between. The timing of the transition period, discovered by a change point detection method in the retrospective analysis (Qian et al. 2017, and also see Chapter 9), is between 25.8 and 39.9 s. The three-sigma control limits for the first and latter

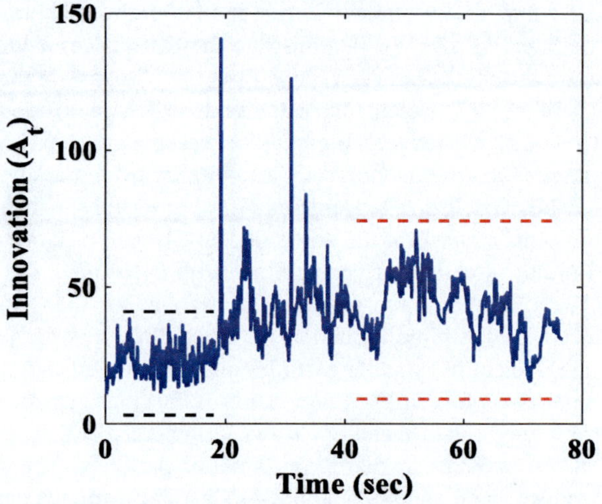

Fig. 7.8 Statistic, A_t, obtained from the innovation sequence of the Kalman filter. The dashed lines are the three-sigma control limits for the two respective growth stages. (Reprinted with permission from Qian et al. 2019)

stages are plotted in Fig. 7.8, where the peaks in the transition period is far greater than the upper control limits. The result in Fig. 7.8 shows that by tracking the innovation sequence of the state space model, it offers the opportunity to detect possible mechanism changes in the process.

Qian et al. (2019) next test the online algorithm on Video 2, which was published in the same paper as Video 1 (Zheng et al. 2009) and that captures a similar nanocrystal self-assembly growth process. There are total 637 frames in Video 2 with 15 fps imaging rate. Qian et al. (2019) again choose the first 300 frames to estimate the parameters. The Bayesian method produces the estimate of σ_α^2 as 7.24×10^{-2} and that of σ_ϵ^2 as 4.19×10^{-3}. Using these parameters, Qian et al. (2019) estimate the NPSD. The total updating time is 0.098 s or 1.54×10^{-4} s per frame; this computational performance is consistent with that for processing Video 1 (and the image processing also takes 0.04 s per frame). Video 2 is a shorter clip and contains fewer particles. The density plots are presented in Fig. 7.9.

Qian et al. (2019) further test the algorithm on Video 3. It was published in Woehl et al. (2013) and captures a different growth process than that in Videos 1 and 2. This process is of silver nanocrystal growth. There are only 112 frames in this video clip with one fps imaging rate, so Qian et al. (2019) pick the first 50 frames as the training set to estimate the parameters. For the process in Video 3, the parameters are accordingly estimated as $\sigma_\alpha^2 = 1.75 \times 10^{-1}$ and $\sigma_\epsilon^2 = 7.56 \times 10^{-3}$. Applying the online updating method to Video 3, the total run time is 0.02 s or 1.79×10^{-4} s per frame. The image processing time for Video 3 is 0.2 s per frame, so that the combined computation is again faster than the frame rate. Figure 7.10 presents the estimated NPSD of Video 3. In this process, the NPSD is always uni-modal and its variance gets larger in the process.

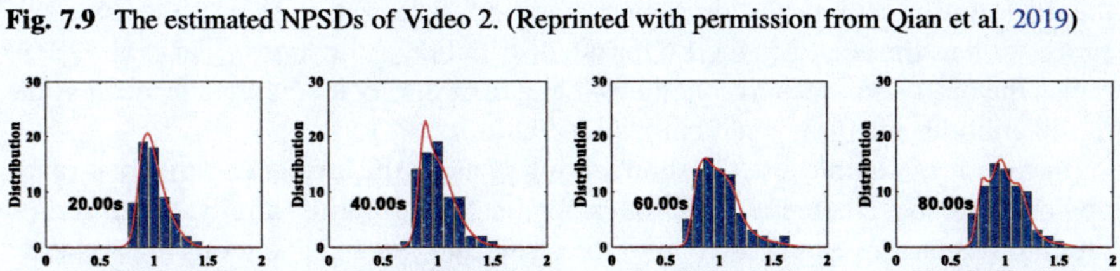

Fig. 7.9 The estimated NPSDs of Video 2. (Reprinted with permission from Qian et al. 2019)

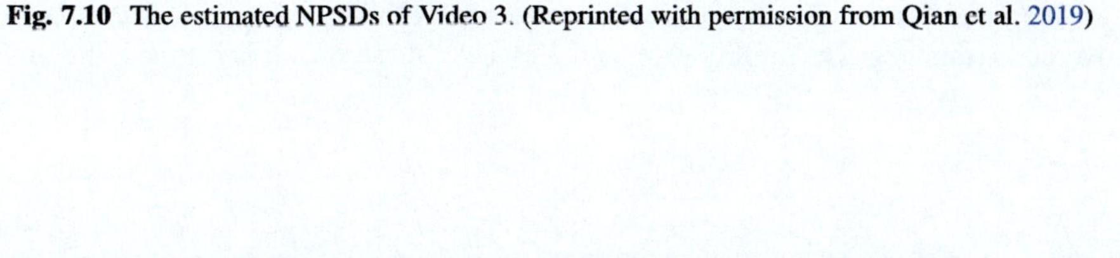

Fig. 7.10 The estimated NPSDs of Video 3. (Reprinted with permission from Qian et al. 2019)

7.6.2 Comparison with Alternative Methods

This subsection demonstrates the benefit of having both the curve smoothness and temporal continuity constraints. All comparison results are demonstrated using Video 1 but the same insight holds true for other videos. The state space method is not compared with the retrospective method in Sect. 7.3.1 because a retrospective (offline) method sees all data and has the luxury of time, whereas a prospective (online) method only sees a subset of the data, unless it reaches the very end of the video, and must be time conscious.

The first comparison is to conduct an out-of-sample quantitative test, comparing the state space method with the curve smoothness constraint with three types of alternative: the first type is a pure histogram-based treatment (no smoothness constraint at all), the second type is the single-frame methods presented in Sect. 7.2, and the third type is a state space model without the curve smoothness (i.e., with temporal continuity across frames but no curve smoothness within a frame). For the state space model without the curve smoothness, $\boldsymbol{\alpha}_t$, instead of $\boldsymbol{\gamma}_t$, is used as the state vector, and the covariance matrix $\mathbf{Q}$ is set as $\text{diag}(\sigma_\alpha^2, \sigma_\alpha^2, \cdots, \sigma_\alpha^2)$. The single parameter σ_α^2 can be estimated by a simplified Bayesian model, assuming $\sigma_\alpha^2 \sim$ inverse-gamma$(1, 1)$. The first 300 frames are used for the training purpose. The Bayesian estimate of σ_α is 5.9×10^{-2}, which is rather close to that estimated in Sect. 7.6.1.

The out-of-sample test calculates the log-likelihood of the estimated probability density functions based on a number of observed nanocrystals. Qian et al. (2019) randomly pick 90% the observed nanocrystals in each and every image frame and use them to establish the state space model and estimate the NPSD. Then, Qian et al. (2019) use the remaining 10% observed nanocrystals in each and every frame to calculate the log-likelihood. For a given testing nanocrystal observation having a normalized particle size x^ℓ at frame t, its log-likelihood is:

$$\log f_t(x^\ell) = \sum_{j=1}^{n} B_j(x^\ell)[\mathbf{C}\boldsymbol{\gamma}_t]_j - \log C_t(\mathbf{C}\boldsymbol{\gamma}_t). \tag{7.70}$$

Qian et al. (2019) calculate the summation of the log-likelihoods for all of the 10% out-of-sample testing nanocrystals at all time frames and then use this summation as the accuracy metric for the distribution estimation. Qian et al. (2019) repeat the out-of-sample test 500 times for each of the six methods. The mean of the log-likelihoods results are summarized in Table 7.2.

In the out-of-sample test, the shortcoming of using the histogram directly is made obvious—almost all the log-likelihoods obtained are negative infinity. When certain samples fall into an empty interval of the histogram (meaning that this interval does not have any training observations), the direct histogram method sets the likelihood of this testing sample as zero, causing the log-likelihood to be negative infinity.

The single-frame density estimation methods with the curve smoothness can overcome this negative infinity problem. However, these methods estimate the dis-

Table 7.2 Comparison results of the out-of-sample test among six approaches: using the observed histograms directly, three single-frame estimation methods, the state space method without the curve smoothness, and the state space method with the curve smoothness; all tested on Video 1. (*Source*: Qian et al. 2019)

Methods		Mean of log-likelihoods
Observed histograms (no constraint)		$-\infty$
Curve smoothness only	Smoothed histograms	−41.6
	Kernel estimation	−24.4
	Penalized B-splines	−46.7
State space model (with temporal continuity)	Without curve smoothness	129.8
	With curve smoothness	196.1

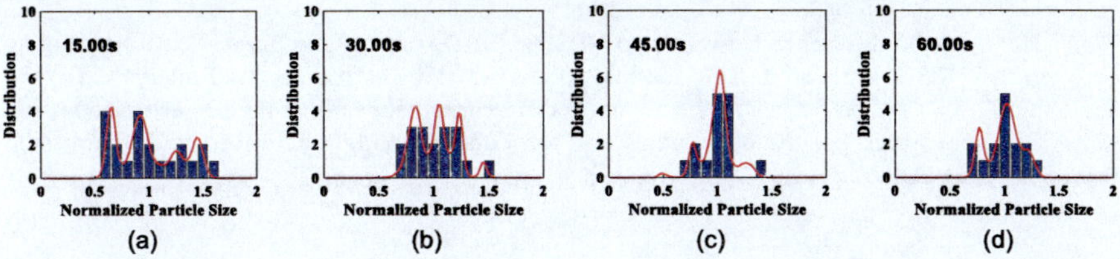

Fig. 7.11 The estimated NPSDs of Video 1 by the state space model without the curve smoothness. (Reprinted with permission from Qian et al. 2019)

tribution from each frame independently, lacking the ability to borrow information across time frames. When the number of observations at individual frames is not large enough, they fail to produce a quality estimate, as evident by the poor results in the out-of-sample test.

By using the state space transition equations, the two state space methods incorporate the temporal continuity, allowing the estimators to borrow information from other image frames and leading to much better performances than the other alternatives.

Between the state space models with and without the curve smoothness, the one with the curve smoothness produces a much higher log-likelihood measure. Qian et al. (2019) conduct a statistical testing and see whether the log-likelihood difference between the two approaches is significant. A one-way ANOVA, in which the null hypothesis is that the two log-likelihoods have the same mean, yields a p-value of 6×10^{-162}, which confirm that the difference is indeed significant.

Given the benefit of using the state space framework demonstrated above, the focus of the next two comparisons is set between the two state space models, with and without the curve smoothness.

The second comparison is to inspect the resulting NPSD obtained by the two state space models. Figure 7.11 presents the NPSDs at 15, 30, 45 and 60s, respectively, estimated by the state space model without the curve smoothness. Comparing the results in Fig. 7.7 obtained at the same time marks by the state space model with the curve smoothness, the estimated distributions in Fig. 7.11 are worse, as the

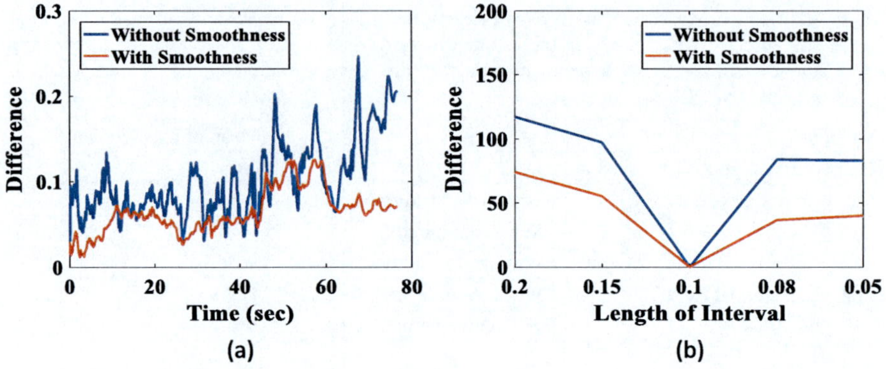

Fig. 7.12 L_2-norm difference between two NPSDs: **(a)** L_2-norm differences at each time frame between the NPSDs estimated using the histograms with 10 intervals and 20 intervals; **(b)** the summation over all time frames of the L_2-norm differences between the NPSDs estimated using histograms of various lengths of intervals and the default setting. (Reprinted with permission from Qian et al. 2019)

state space model without the curve smoothness apparently overfits the histogram, and consequently, it is sensitive to small changes in the number of particles in a bin. To see this point, consider the following observations. In Fig. 7.11b, while the orientated attachment growth mechanism suggests a bimodal distribution, the estimated distribution gives us three peaks. Between Fig. 7.11c, d, the variance is supposed to decrease, as this is in the Ostwald ripening growth stage, but the estimated distribution shows an increasing variance. When displaying the online distribution estimation frame by frame, it is obvious that the state space model without the curve smoothness produces a time-varying NPSD that is far more volatile and reacts too dramatically to noises and disturbances.

The third comparison is to check the robustness of the state space method to possible changes in the number of intervals in the input histograms. The default width of interval while constructing a histogram is 0.1, which gives 20 intervals in the histogram. Qian et al. (2019) test the cases by setting the length of interval to 0.2, 0.15, 0.08 and 0.05, respectively, and then estimate the corresponding NPSD, using the state space model with and without the curve smoothness. Qian et al. (2019) compare the resulting NPSDs with that obtained under the default setting, i.e., the length of interval 0.1 or 20 intervals in the histogram. The difference between the two NPSDs is measured by a L_2 norm of the two density function curves.

Figure 7.12a plots the L_2-norm differences at each time frame between the NPSDs estimated, respectively, using the binned data with 10 intervals (the length of an interval 0.2) and 20 intervals (the length of an interval 0.1). It is apparent that inclusion of the curve smoothness leads to an estimation less sensitive to the number of intervals, especially in the later stage of the process. Figure 7.12b presents the summation over all frames of the L_2-norm differences between the NPSDs estimated, respectively, using binned data of a various number of intervals and the default setting (i.e., 20 intervals or interval length 0.1). In the broad range of choices, the curve smoothness penalty generally decreases the L_2-norm differences by half. Those results show that imposing the curve smoothness constraint can alleviate the overfitting when using a small interval for binned data.

7.7 Future Research Need: Learning-on-the-Fly

The essence of the state space modeling is to provide a real-time, dynamic framework for active learning. The importance of the real-time learning is rooted in the need for process control and real-time decision making. For engineering systems, including the targeted nano material processes, to alter a process, a control action is typically needed before a change point in that process. After the change point, the process may no longer be reversible. In general, control options possible in the early growth stage may not be feasible any more in the later stages. For instance, adding certain surfactants can change the surface dynamics of nanocrystals, so as to affect their shape and size and control subsequent monomer addition, but the timing of such process interventions is critically important. For an online application, the learning action must be fast enough to capture the changes in the underlying density functions; how fast is enough is dictated by the underlying physical process and the imaging speed. For the applications at hand, the imaging speed is now 15 fps, leaving 67 ms between two images. But the imaging process can easily go orders of magnitude faster in the near future as technology upgrades, leaving a much shorter time interval for reaction.

Controlling crystal growth of ensembles of nanomaterials amounts to controlling the structuring of matter at the scale of individual atoms with high-precision repeatability across a large amount of individual objects. Phenomena occurring at a very small length scales also tend to operate on a short time scale—the nucleation and growth of nano meter scale objects is governed by molecular reactions with transient behavior on the scale of milliseconds or shorter (Park and Ding 2019). At this rate, even simply storing all the image data is a challenge, because when the imaging rate is microseconds, a single second of imaging could yield as much as eight tera bytes image data, and as such, the digital system will be overwhelmed by the fast arrival of data. Learning-on-the-fly is no longer a nicety but a necessity.

References

Aldous D (1999) Deterministic and stochastic models for coalescence (aggregation and coagulation): A review of the mean-field theory for probabilists. Bernoulli 5(1):3–48

Anscombe FJ (1948) The transformation of Poisson, binomial and negative-binomial data. Biometrika 35(3/4):246–254

Bishop Y, Fienberg S, Holland P (1975) Discrete Multivariate Analysis: Theory and Practice. Springer-Verlag, New York

Boal AK, Ilhan F, DeRouchey JE, Thurn-Albrecht T, Russell TP, Rotello VM (2000) Self-assembly of nanoparticles into structured spherical and network aggregates. Nature 404(6779):746–748

Bolstad WM, Curran JM (2016) Introduction to Bayesian Statistics. John Wiley & Sons, New York

Boyd S, Parikh N, Chu E, Peleato B, Eckstein J (2011) Distributed optimization and statistical learning via the alternating direction method of multipliers. Foundations and Trends in Machine Learning 3(1):1–122

Brown L, Cai T, Zhang R, Zhao L, Zhou H (2010) The root-unroot algorithm for density estimation as implemented via wavelet block thresholding. Probability Theory and Related Fields 146(3-4):401–433

Doucet A, Gordon NJ, Krishnamurthy V (2001) Particle filters for state estimation of jump Markov linear systems. IEEE Transactions on Signal Processing 49(3):613–624

Durbin J, Koopman SJ (1997) Monte Carlo maximum likelihood estimation for non-Gaussian state space models. Biometrika 84(3):669–684

Eilers PH, Marx BD (1996) Flexible smoothing with B-splines and penalties. Statistical Science 11(2):89–102

Grzelczak M, Vermant J, Furst EM, Liz-Marzán LM (2010) Directed self-assembly of nanoparticles. ACS Nano 4(7):3591–3605

de Jong P, Shephard N (1995) The simulation smoother for time series models. Biometrika 82(2):339–350

Kalman RE (1960) A new approach to linear filtering and prediction problems. Journal of Basic Engineering 82(1):35–45

Li M, Schnablegger H, Mann S (1999) Coupled synthesis and self-assembly of nanoparticles to give structures with controlled organization. Nature 402(6760):393–395

Lifshitz I, Slyozov V (1961) The kinetics of precipitation from supersaturated solid solutions. Journal of Physics and Chemistry of Solids 19(1):35–50

Luenberger DG (1973) Introduction to Linear and Nonlinear Programming. Addison-Wesley, Reading, MA

Ma J, Kockelman KM, Damien P (2008) A multivariate Poisson-lognormal regression model for prediction of crash counts by severity using Bayesian methods. Accident Analysis and Prevention 40(3):964–975

Mahalanobis PC (1936) On the generalized distance in statistics. Proceedings of the National Institute of Sciences of India 2(1):49–55

Park C, Ding Y (2019) Automating material image analysis for material discovery. MRS Communications 9(2):545–555

Park C, Woehl TJ, Evans JE, Browning ND (2015) Minimum cost multi-way data association for optimizing multitarget tracking of interacting objects. IEEE Transactions on Pattern Analysis and Machine Intelligence 37(3):611–624

Qian Y, Huang JZ, Li X, Ding Y (2016) Robust nanoparticles detection from noisy background by fusing complementary image information. IEEE Transactions on Image Processing 25(12):5713–5726

Qian Y, Huang JZ, Ding Y (2017) Identifying multi-stage nanocrystal growth using *in situ* TEM video data. IISE Transactions 49(5):532–543

Qian Y, Huang JZ, Park C, Ding Y (2019) Fast dynamic nonparametric distribution tracking in electron microscopic data. The Annals of Applied Statistics 13(3):1537–1563

Sheather SJ, Jones MC (1991) A reliable data-based bandwidth selection method for kernel density estimation. Journal of the Royal Statistical Society, Series B 53(3):683–690

Silverman BW (1986) Density Estimation for Statistics and Data Analysis. Chapman and Hall/CRC, London and New York

Simonoff JS (1983) A penalty function approach to smoothing large sparse contingency tables. The Annals of Statistics 11(1):208–218

Spiegelhalter D, Thomas A, Best N, Gilks W (1996) BUGS 0.5: Bayesian inference using Gibbs sampling manual (version ii). MRC Biostatistics Unit, Institute of Public Health, Cambridge, UK pp 1–59

Wahba G (1990) Spline Models for Observational Data. Society for Industrial and Applied Mathematics, Philadelphia, PA

Woehl TJ, Park C, Evans JE, Arslan I, Ristenpart WD, Browning ND (2013) Direct observation of aggregative nanoparticle growth: Kinetic modeling of the size distribution and growth rate. Nano Letters 14(1):373–378

Zhang C, Chen N, Li Z (2017) State space modeling of autocorrelated multivariate Poisson counts. IISE Transactions 49(5):518–531

Zheng H, Smith RK, Jun YW, Kisielowski C, Dahmen U, Alivisatos AP (2009) Observation of single colloidal platinum nanocrystal growth trajectories. Science 324(5932):1309–1312

Chapter 8
Dynamic Shape Modeling for Shape Changes

8.1 Problem of Shape Distribution Tracking

Many material processes take place in a liquid or gas environment, where a large number of material seeds are formed and evolved into the final material structures. Large degrees of uncertainties in the evolution process suggest that material objects can go through different evolution steps even under the same environment. Naturally, the sizes and shapes of materials objects follow a probability distribution at any given time, instead of being identical. Furthermore, the size and shape distributions change over time. Tracking these population distributions and their change over time is referred to as distribution tracking, as explained in Chap. 7. While Chap. 7 discusses the distribution tracking methods for size only changes, this chapter focuses on the shape distribution tracking, i.e., to track the temporal evolution of the shape distribution.

The shape distribution tracking is related to the mathematical problem of representing the probability distribution on a shape and modeling the temporal evolution of the distribution. We previously reviewed in Chap. 4 some statistical modeling approaches of modeling shape distributions, such as the complex Bingham distribution, offset normal distribution, and wrapped Gaussian distribution. In this section we explain how the modeling approaches are extended for time-varying shape distributions, which are referred to as dynamic shape distributions in the literature (Fontanella et al. 2019).

One natural way to model a dynamic shape distribution is to use a conventional (static) shape distribution as the baseline model and allow the distribution parameters to vary in time through a time-series modeling of the distribution parameters. For example, when the Kendall's shape representation for planar shapes is used, one can use a complex representation $z \in \mathbb{C}^K$ to represent a preshape as defined in Eq. (4.4), and use the complex Watson density to model the probability distribution of the preshape z, such as

C. Park, Y. Ding, *Data Science for Nano Image Analysis*, International Series in Operations Research & Management Science 308,
https://doi.org/10.1007/978-3-030-72822-9_8

$$p_w(z; \boldsymbol{\mu}_t, \kappa) = c_1(\kappa)^{-1} \exp\{\kappa |z^* \boldsymbol{\mu}_t|^2\}, \tag{8.1}$$

where $*$ denotes the complex conjugate transpose, $\boldsymbol{\mu}_t \in \mathbb{C}^K$ is the mean vector that varies with the observation time t, $\kappa > 0$ is a concentration parameter, and $c_1(\kappa)$ is a normalizing constant depending on κ.

Apparently the dynamic probability distribution of a preshape depends on the dynamics of its mean vector $\boldsymbol{\mu}_t$. But modeling the time-series of $\boldsymbol{\mu}_t$ is not straightforward, because modeling $\boldsymbol{\mu}_t$ is to attempt a time series model on a manifold, and doing so is complicated. Recall that the mean parameter belongs to a shape space and all of the shape spaces that we described in Sect. 4.1 are manifolds, not Euclidean spaces.

When the landmark-based shape representation is used, the approach to model a dynamic shape distribution is to use an offset normal distribution. Specifically, the centered configuration matrix of observed landmarks is assumed to follow a multivariate normal distribution. Then the corresponding probability distribution of the shape features extracted from the configuration matrix can be thus induced. For example, when the Bookstein coordinates of the configuration matrix is used as the shape features, the induced distribution is the Dryden-Mardia offset normal distribution (Dryden and Mardia 1991) as described in Sect. 4.1.1.3. Fontanella et al. (2019) extended the approach to a dynamic offset normal distribution. We will describe this approach in more details in Sect. 8.2

The offset normal model has its density function in a complex form. Because of this, the necessary numerical steps for statistical inference can be very complicated. A simpler approach is to pre-align the configuration matrix $\tilde{z}$ to a reference shape for removing the scale, rotation, and translation effect of it and then treat the aligned data as Euclidean features. After this pre-aligning treatment, many conventional statistical modeling approaches, developed for Euclidean features, can be used to model the distribution. The alignment is often performed by the general Procrustes analysis (Gower 1975). As we described in Sect. 4.1.1, the Procrustes mean vector can be estimated in the alignment procedure and set to the reference shape, and the configuration matrix is aligned to the reference shape. The coordinates in the aligned configuration matrix are referred to as the Procrustes tangent coordinates, which are treated as the Euclidean features. The temporal changes of the tangent coordinates are then modeled as a conventional time series modeling (Kent et al. 2000, 2001),

$$\tilde{z}_t = \tilde{z}_0 + \boldsymbol{F}(\boldsymbol{\mu}) \boldsymbol{A} \boldsymbol{\phi}_t + \boldsymbol{\epsilon}_t,$$

where $\tilde{z}_0 \in \mathbb{C}^K$ is the baseline, $\boldsymbol{F}(\boldsymbol{\mu})$ is a $K \times K$ matrix of the thin-plate spline loadings that define the shape deviation from the reference shape $\boldsymbol{\mu}$, $\boldsymbol{\phi}_t$ defines the temporal variation, and $\boldsymbol{\epsilon}_t$ defines the random variation. The unknown model parameters consist of the baseline $\tilde{z}_0$ and the matrix $\boldsymbol{A}$. They can be estimated from a sample of the observed shapes. We will describe this approach in more details in Sect. 8.3.

Instead of pre-aligning the configuration matrix, Dryden et al. (2019) took the centered configuration matrix of landmark data and modeled its linkage to unknown size-and-shape features using a rotation parameter. The size-and-shape features are regressed over an input space, e.g., time, such as,

$$\tilde{z}_t = \boldsymbol{f}(t)\boldsymbol{O}_t + \boldsymbol{\epsilon}_t,$$

where $\boldsymbol{f}(t)$ is a linear regression model that relates time t to the size-and-shape features, $\boldsymbol{O}_t$ is a rotation matrix, and $\boldsymbol{\epsilon}_t$ is an independent noise. After assuming the prior models for the regression parameters and rotation parameters, the posterior inference of the regression model is conducted through an Markov chain Monte Carlo algorithm. We will describe the details of this approach in Sect. 8.4.

A shape of an object can also be represented by a parametric curve that describes the outline of the object; see Sect. 4.1.2. A probability distribution of the shape can then be defined as the probability distribution on the parametric curve. The temporal evolution of the probability distribution is modeled as a time-series of the parametric curve and its random variation. The Gaussian radial growth model (Jónsdóttir and Jensen 2011) formulated the stochastic growth of a featureless star-shaped object as a random expansion of the object's outline over time. A Gaussian distribution fits the degree of the outline expansion occurring during a certain time period. One advantage of this approach is that the random expansion explains the growth in the object better than simple increments in a one-dimensional size coordinate of the landmark data. Park (2014) extended the random expansion method and used it for modeling the temporal evolution of a non-Gaussian shape distribution. Section 8.5 will provide detailed description of this modeling approach.

8.2 Dynamic Shape Distribution with Bookstein Shape Coordinates

Fontanella et al. (2019) used an offset normal distribution to describe a dynamic shape distribution with a temporal sequence of landmark data in $\mathbb{R}^2$. Suppose that landmark data are available at time $t \in \{1, 2, \ldots, N\}$. Let $\tilde{\boldsymbol{X}}_t$ denote the configuration matrix of K landmark coordinates observed at t, i.e.,

$$\tilde{\boldsymbol{X}}_t = \begin{bmatrix} \tilde{X}_{11t} & \tilde{X}_{12t} \\ \tilde{X}_{21t} & \tilde{X}_{22t} \\ \vdots & \vdots \\ \tilde{X}_{K1t} & \tilde{X}_{K2t} \end{bmatrix}. \tag{8.2}$$

To make $\tilde{\boldsymbol{X}}_t$ invariant to translation, rotation, and scaling of an object, Fontanella et al. (2019) converted the configuration matrix to the Bookstein coordinates. Following the description of the Bookstein coordinates in Sect. 4.1.1.3, the centered configuration matrix is first taken as

$$X_t = L(\tilde{X}_t - \mathbf{1}_K c_t^T),$$

where $L = [\mathbf{0}_{K-2}, \mathbf{0}_{K-2}, I_{K-2}]$ and $c_t = (\tilde{X}_{11t}, \tilde{X}_{12t})^T$ represents the first landmark coordinate. As previously defined, $\mathbf{0}_{K-2}$ is a $(K-2)$-dimensional column vector of zeros, and I_{K-2} is a $(K-2) \times (K-2)$ identity matrix.

Let X_{kmt} denote the (k, m)th element of X_t. Then, the corresponding Bookstein coordinate can be defined as $U_t = X_t R_t$, where

$$R_t = \frac{1}{X_{11t}^2 + X_{12t}^2} \begin{bmatrix} X_{11t} & -X_{12t} \\ X_{12t} & X_{11t} \end{bmatrix}.$$

The inverse relation can be achieved as

$$X_t = U_t \begin{bmatrix} X_{11t} & X_{12t} \\ -X_{12t} & X_{11t} \end{bmatrix}.$$

Its vectorization version is written as

$$\text{vec}(X_t) = (I_2 \otimes U_t) \begin{bmatrix} 1 & 0 \\ 0 & -1 \\ 0 & 1 \\ 1 & 0 \end{bmatrix} h_t, \tag{8.3}$$

where $\text{vec}(\cdot)$ is the vectorization operator and $h_t = (X_{11t}, X_{12t})^T$.

8.2.1 Joint Estimation of Dynamic Shape Distribution

The N centered configurations observed at N different time points are concatenated into a $(K-2) \times N$ matrix, $X = (X_1, \ldots, X_N)$. The data matrix X is assumed to be Gaussian, that is,

$$\text{vec}(X) \sim \mathcal{N}(\mu, \Sigma). \tag{8.4}$$

Let us rewrite Eq. (8.3) as

$$\text{vec}(X) = Wh, \tag{8.5}$$

where $h = (h_1^T, \ldots, h_N^T)^T$, and W is a blocked diagonal matrix of N blocks with its tth block given by

$$\boldsymbol{W}_t = (\boldsymbol{I}_2 \otimes \boldsymbol{U}_t) \begin{bmatrix} 1 & 0 \\ 0 & -1 \\ 0 & 1 \\ 1 & 0 \end{bmatrix}.$$

In Eq. (8.5), the first term $\boldsymbol{W}$ depends only on the Bookstein coordinates. In this way, this equation shows how the centered configuration can be decomposed into the shape coordinates and extra information uninterested.

By a change of variables in the density function of Eq. (8.4), the joint density function of $\boldsymbol{U}$ and $\boldsymbol{h}$ can be derived as

$$f(\boldsymbol{U}, \boldsymbol{h}; \boldsymbol{\mu}, \boldsymbol{\Sigma}) = \frac{1}{(2\pi)^{(K-1)N}|\boldsymbol{\Sigma}|^{1/2}} \exp\left\{-\frac{1}{2}\psi(\boldsymbol{U}, \boldsymbol{h}; \boldsymbol{\mu}, \boldsymbol{\Sigma})\right\} \prod_{t=1}^{N} ||\boldsymbol{h}_t||^{2(K-2)}, \tag{8.6}$$

where $\psi(\boldsymbol{U}, \boldsymbol{h}; \boldsymbol{\mu}, \boldsymbol{\Sigma}) = (\boldsymbol{W}\boldsymbol{h} - \boldsymbol{h})^T \boldsymbol{\Sigma}^{-1}(\boldsymbol{W}\boldsymbol{h} - \boldsymbol{h})$. Since the quadratic term $\psi(\boldsymbol{U}, \boldsymbol{h}; \boldsymbol{\mu}, \boldsymbol{\Sigma})$ can be decomposed into

$$\psi(\boldsymbol{U}, \boldsymbol{h}; \boldsymbol{\mu}, \boldsymbol{\Sigma}) = g(\boldsymbol{U}; \boldsymbol{\mu}, \boldsymbol{\Sigma}) + (\boldsymbol{h} - \boldsymbol{\eta})^T \boldsymbol{\Gamma}^{-1}(\boldsymbol{h} - \boldsymbol{\eta}),$$

where $\boldsymbol{\Gamma} = \boldsymbol{W}^T \boldsymbol{\Sigma}^{-1} \boldsymbol{W}$, $\boldsymbol{\eta} = \boldsymbol{\Gamma}^T \boldsymbol{W}^T \boldsymbol{\Sigma}^{-1} \boldsymbol{\mu}$, and $g(\boldsymbol{U}; \boldsymbol{\mu}, \boldsymbol{\Sigma}) = \boldsymbol{\mu}^T \boldsymbol{\Sigma}^{-1} \boldsymbol{\mu} - \boldsymbol{\eta}^T \boldsymbol{\Gamma}^{-1} \boldsymbol{\eta}$, the joint density function can be rewritten as

$$f(\boldsymbol{U}, \boldsymbol{h}; \boldsymbol{\mu}, \boldsymbol{\Sigma}) = \frac{\exp\{-g(\boldsymbol{U}; \boldsymbol{\mu}, \boldsymbol{\Sigma})/2\}; \boldsymbol{\mu}, \boldsymbol{\Sigma})}{(2\pi)^{(K-1)N}|\boldsymbol{\Sigma}|^{1/2}} \exp\left\{-\frac{1}{2}(\boldsymbol{h} - \boldsymbol{\eta})^T \boldsymbol{\Gamma}^{-1}(\boldsymbol{h} - \boldsymbol{\eta})\right\} \prod_{t=1}^{N} ||\boldsymbol{h}_t||^{2(K-2)}.$$

By integrating the density function with respect to $\boldsymbol{h}$, we can obtain the marginal distribution of only the shape feature $\boldsymbol{U}$,

$$f(\boldsymbol{U}; \boldsymbol{\mu}, \boldsymbol{\Sigma}) = \frac{|\boldsymbol{\Gamma}|^{1/2} \exp\{-g(\boldsymbol{U}; \boldsymbol{\mu}, \boldsymbol{\Sigma})/2\}; \boldsymbol{\mu}, \boldsymbol{\Sigma})}{(2\pi)^{(K-2)N}|\boldsymbol{\Sigma}|^{1/2}} \int \prod_{t=1}^{N} ||\boldsymbol{h}_t||^{2(K-2)} \phi(\boldsymbol{h}; \boldsymbol{\eta}, \boldsymbol{\Gamma}) d\boldsymbol{h},$$

where ϕ is the normal density with mean $\boldsymbol{\eta}$ and covariance $\boldsymbol{\Gamma}$. Fontanella et al. (2019) described the maximum likelihood estimation of $\boldsymbol{\mu}$ and $\boldsymbol{\Sigma}$ using a set of observed landmark data.

So far the approach does not process images and corresponding shapes piece-by-piece in a serial fashion. The autoregressive model presented in Sect. 8.2.2 is suitable for serial and real-time data processing.

8.2.2 Autoregressive Model

The temporal sequence of $\boldsymbol{X}_t$ can also be modeled by an autoregressive model. Consider a simple AR(p) process of $\boldsymbol{X}_t$, i.e.,

$$\mathrm{vec}(\boldsymbol{X}_t) = \boldsymbol{\mu}_t + \sum_{j=1}^{p} \phi_j(\mathrm{vec}(\boldsymbol{X}_{t-j}) - \boldsymbol{\mu}_{t-j}) + \boldsymbol{e}_t,$$

where $\boldsymbol{\mu}_t$ is a $2(K-2)$-dimensional column vector representing the mean of $\mathrm{vec}(\boldsymbol{X}_t)$, and $\boldsymbol{e}_t \sim \mathcal{N}(\boldsymbol{0}, \boldsymbol{\Sigma}_S)$ is a $2(K-2)$-dimensional random vector representing the white noise. A constant mean is assumed, i.e., $\boldsymbol{\mu}_t = \boldsymbol{\mu}_0$ and $\boldsymbol{\Sigma}_S = \sigma^2(\boldsymbol{I}_2 \otimes \boldsymbol{L})(\boldsymbol{I}_2 \otimes \boldsymbol{L}^T)$. The AR model is rewritten as

$$\mathrm{vec}(\boldsymbol{X}_t) = (1 - \phi_1 - \ldots - \phi_p)\boldsymbol{\mu}_0 + \mathrm{vec}(\boldsymbol{X}_{t-1}) + \ldots + \mathrm{vec}(\boldsymbol{X}_{t-p}) + \boldsymbol{e}_t. \tag{8.7}$$

The conditional distribution of $\mathrm{vec}(\boldsymbol{X}_t)$, conditioned on all p previous configuration matrices, is

$$\mathrm{vec}(\boldsymbol{X}_t)|\,\mathrm{vec}(\boldsymbol{X}_{t-1}), \ldots, \mathrm{vec}(\boldsymbol{X}_{t-p}), \boldsymbol{\mu}_0, \sigma^2 \sim \mathcal{N}(\boldsymbol{\mu}_{t|t-p}, \boldsymbol{\Sigma}_S), \tag{8.8}$$

where the conditional mean is

$$\boldsymbol{\mu}_{t|t-p} = (1 - \phi_1 - \ldots - \phi_p)\boldsymbol{\mu}_0 + \mathrm{vec}(\boldsymbol{X}_{t-1}) + \ldots + \mathrm{vec}(\boldsymbol{X}_{t-p}).$$

Accordingly, the conditional distribution of the corresponding shape feature $\boldsymbol{U}_t$ can be derived in the form of an offset normal distribution, so that the maximum likelihood estimation of the AR parameters can be carried out (Fontanella et al. 2019).

8.3 Dynamic Shape Distribution with Procrustes Tangent Coordinates

Suppose that J different individual objects are observed at N different time points. Each of the objects can be represented by K landmark coordinates in $\mathbb{R}^2$. Let $\boldsymbol{v}_{jt} \in \mathbb{C}^K$ denote the partial Procrustes tangent coordinates, which can be achieved from the raw landmark data following the description in Sect. 4.1.1.2. Let $\boldsymbol{\mu} \in \mathbb{C}^K$ denote the Procrustes tangent mean of the $J \times N$ landmark data, which is just an arithmetic mean of the scaled $\boldsymbol{v}_{jt}$'s, i.e.,

$$\boldsymbol{\mu} = \frac{1}{N}\frac{1}{J}\sum_{j=1}^{J}\sum_{t=1}^{N}\frac{\boldsymbol{v}_{jt}}{||\boldsymbol{v}_{jt}||}.$$

It is assumed that the landmark $\boldsymbol{v}_{jt}$ is a variant of the mean shape $\boldsymbol{\mu}$. A thin-plate spline model can thus be used to map the deformation of $\boldsymbol{\mu}$ to $\boldsymbol{v}_{jt}$.

Let $f(s)$ denote the vector of p thin-plate spline basis functions defined on $\mathbb{C}$, $\mu_k \in \mathbb{C}$ denote the kth element of $\boldsymbol{\mu}$, $k = 1, \ldots, K$, and

$$\boldsymbol{F}(\boldsymbol{\mu}) = \begin{bmatrix} f(\mu_1)^T \\ f(\mu_2)^T \\ \vdots \\ f(\mu_K)^T \end{bmatrix}$$

denote the $K \times p$ matrix of the thin-plate spline function values evaluated at the K coordinates of $\boldsymbol{\mu}$. Then, the individual tangent coordinate $\boldsymbol{v}_{jt}$ is represented as a spatiotemporal variation of $\boldsymbol{\mu}$ by

$$\boldsymbol{v}_{jt} = z_0 + \boldsymbol{F}(\boldsymbol{\mu})\boldsymbol{\beta}_t + \boldsymbol{\epsilon}_{jt}, \tag{8.9}$$

where z_0 is the intercept, $\boldsymbol{\beta}_t$ is a $p \times 1$ vector of the model parameters representing the temporal variation, and the $K \times 1$ vector $\boldsymbol{\epsilon}_{jt}$ represents random variations of the K landmark coordinates in $\boldsymbol{v}_{jt}$. The model parameter $\boldsymbol{\beta}_t$ is time-varying, modeled as

$$\boldsymbol{\beta}_t = \boldsymbol{A}\boldsymbol{g}(t), \tag{8.10}$$

where $\boldsymbol{g}(t)$ is a q-dimensional vector of the time basis functions, e.g., some polynomial functions of t, and $\boldsymbol{A}$ is a $p \times q$ matrix of unknown model parameters. Combining Eq. (8.9) and Eq. (8.10), we have

$$\boldsymbol{v}_{jt} = z_0 + \boldsymbol{F}(\boldsymbol{\mu})\boldsymbol{A}\boldsymbol{g}(t) + \boldsymbol{\epsilon}_{jt}. \tag{8.11}$$

The random variation $\boldsymbol{\epsilon}_{jt}$ can be modeled as a multivariate complex normal random vector with zero mean, i.e.,

$$\boldsymbol{\epsilon}_{jt} \sim \mathcal{CN}(\boldsymbol{0}, \boldsymbol{\Gamma}_1, \boldsymbol{\Gamma}_2),$$

where $\boldsymbol{\Gamma}_1 = E[\boldsymbol{\epsilon}_{jt}\boldsymbol{\epsilon}_{jt}^*]$, $\boldsymbol{\Gamma}_2 = E[\boldsymbol{\epsilon}_{jt}\boldsymbol{\epsilon}_{jt}^T]$, $*$ is the complex conjugate transpose operator and T is the matrix transpose operator. For white noises, $\boldsymbol{\Gamma}_1 = \sigma^2\boldsymbol{I}, \sigma^2 > 0$, and $\boldsymbol{\Gamma}_2$ is a zero matrix. Given this time-series modeling, the dynamic shape distribution of $\boldsymbol{v}_{jt}$ can be induced as follows:

$$\boldsymbol{v}_{jt} \sim \mathcal{CN}(z_0 + \boldsymbol{F}(\boldsymbol{\mu})\boldsymbol{A}\boldsymbol{g}(t), \boldsymbol{\Gamma}_1, \boldsymbol{\Gamma}_2).$$

This model is adequate for all landmark data having only a small variation from the overall mean $\boldsymbol{\mu}$. The distribution has two parameters, z_0 and $\boldsymbol{A}$. The maximum likelihood estimate of z_0 and $\boldsymbol{A}$ can be obtained by an alternating algorithm: (a) given $\boldsymbol{A}$, estimate z_0 by averaging the residual of Eq. (8.11) over j and t, and (b) given z_0, estimate $\boldsymbol{A}$ by solving the following maximization for $\boldsymbol{A}$,

$$\max \sum_{j=1}^{J} \sum_{t=1}^{N} ||\boldsymbol{v}_{jt} - z_0 - \boldsymbol{F}(\boldsymbol{\mu})\boldsymbol{A}\boldsymbol{g}(t)||^2.$$

8.4 Bayesian Linear Regression Model for Size and Shape

Dryden et al. (2019) proposed a Bayesian linear regression for modeling the size and shape of landmark data in $\mathbb{R}^3$. Consider the configuration matrix of landmark coordinates in $\mathbb{R}^3$ in the form of Eq. (4.2). Suppose that there are N such landmark data observed. Let $\boldsymbol{X}_n$ denote the configuration matrix of the nth landmark data and t_n denote the observed time of the landmark data. The configuration matrix is assumed to be centered so that the sum of each column of $\boldsymbol{X}_n$ is zero.

The centered configuration matrix is invariant to the translation of the corresponding landmark coordinates. Dryden et al. (2019) used the centered configuration matrix as the preshape to study the size and shape of landmark data and their temporal evolution. Dryden et al. (2019) first modeled the conditional distribution of $\boldsymbol{X}_n$ with the mean function $\boldsymbol{\mu}(t)$ at time t_n subject to an unknown rotation transformation $\tilde{\boldsymbol{O}}_n \in SO(3)$. The conditional mean is then

$$E[\boldsymbol{X}_n | t_n, \tilde{\boldsymbol{O}}_n] = \boldsymbol{\mu}(t_n)\tilde{\boldsymbol{O}}_n.$$

The corresponding regression model for $\boldsymbol{X}_n$ with a noise term is

$$\boldsymbol{X}_n = \boldsymbol{\mu}(t_n)\tilde{\boldsymbol{O}}_n + \boldsymbol{\epsilon}_n, \tag{8.12}$$

where $\boldsymbol{\epsilon}_n \sim \mathcal{MN}_{K\times 3}(\boldsymbol{0}, \sigma^2 \boldsymbol{I}_M, \boldsymbol{I}_K)$ is assumed to be an i.i.d. $K \times M$ multivariate random variable. Here, $\mathcal{MN}$ represents the matrix normal distribution. Dryden et al. (2019) used a linear model for the conditional mean, that is,

$$\boldsymbol{\mu}(t_n) = \boldsymbol{A}_0 + \boldsymbol{A}_1 t_n,$$

where $\boldsymbol{A}_0, \boldsymbol{A}_1 \in \mathbb{R}^{K\times 3}$.

The model in Eq. (8.12) is unidentifiable, because $\boldsymbol{\mu}(t_n)\boldsymbol{O}_n^T$ and $\boldsymbol{O}_n\tilde{\boldsymbol{O}}_n$ would give the same model for an arbitrary choice of $\boldsymbol{O}_n \in SO(3)$. To make the model identifiable, LQ decompositions ought to be taken for $\boldsymbol{A}_0 = \boldsymbol{B}_0\boldsymbol{Q}_n$ and $\boldsymbol{A}_1 = \boldsymbol{B}_1\boldsymbol{Q}_n$, where $\boldsymbol{B}_0$ and $\boldsymbol{B}_1$ are $K \times 3$ lower triangular matrices and $\boldsymbol{Q}_n \in SO(3)$. With the LQ decomposition, Eq. (8.12) is rewritten as

$$\begin{aligned} \boldsymbol{X}_n &= (\boldsymbol{B}_0 \boldsymbol{Q}_n + \boldsymbol{B}_1 \boldsymbol{Q}_n t_n)\tilde{\boldsymbol{O}}_n + \boldsymbol{\epsilon}_n, \\ &= (\boldsymbol{B}_0 + \boldsymbol{B}_1 t_n)\boldsymbol{O}_n + \boldsymbol{\epsilon}_n \end{aligned} \tag{8.13}$$

where $\boldsymbol{O}_n = \boldsymbol{Q}_n \tilde{\boldsymbol{O}}_n$ belongs to $SO(3)$, and $\boldsymbol{B}_0$ and $\boldsymbol{B}_1$ are all restricted to $K \times 3$ lower triangular matrices.

An inverse gamma prior was assumed for σ^2. The rotation matrix $\boldsymbol{O}_n$ is parameterized by three angles $(\theta_{n1}, \theta_{n2}, \theta_{n3})$ using the ZXZ convention (Landau and Lifschitz 1976),

$$\boldsymbol{O}_n = \begin{bmatrix} \cos(\theta_{n3}) & \sin(\theta_{n3}) & 0 \\ -\sin(\theta_{n3}) & \cos(\theta_{n3}) & 0 \\ 0 & 0 & 1 \end{bmatrix} \begin{bmatrix} 1 & 0 & 0 \\ 0 & \cos(\theta_{n3}) & \sin(\theta_{n3}) \\ 0 & -\sin(\theta_{n3}) & \cos(\theta_{n3}) \end{bmatrix} \begin{bmatrix} \cos(\theta_{n1}) & \sin(\theta_{n1}) & 0 \\ -\sin(\theta_{n1}) & \cos(\theta_{n1}) & 0 \\ 0 & 0 & 1 \end{bmatrix},$$

where $\theta_{n1}, \theta_{n3} \in [0, 2\pi]$ and $\theta_{n2} \in [0, \pi]$. A matrix Fisher prior is assumed for $\boldsymbol{O}_n$,

$$\boldsymbol{O}_n \sim \text{matrixFisher}(\boldsymbol{F}_0),$$

where $\boldsymbol{F}_0$ is a 3×3 parameter matrix, and the matrix Fisher distribution has its density proportional to $\exp(\text{trace}(\boldsymbol{F}_0^T \boldsymbol{O}_n))$. Moreover, a uniform prior is assumed for each of the model parameters $\boldsymbol{B}_0$ and $\boldsymbol{B}_1$. Given the prior models, the posterior inference of the model parameters can be performed using the Gibbs sampling and Metropolis Hastings algorithm within the Gibbs steps (Dryden et al. 2019).

8.5 Dynamic Shape Distribution with Parametric Curves

Park (2014) proposed a dynamic shape distribution model based on a parametric curve representation of shapes. In such model, a parametric curve is used to depict the outline of each individual object. The parametric curve for an object is transformed to shape features after removing uninterested factors such as the orientation and location of the object. In particular, Park (2014) transformed the parametric curves to the centroid distance functions, described previously in Sect. 4.2.1. Here we give a brief recap.

Suppose that J_t different individual objects and their outlines in $\mathbb{R}^2$ are taken for each time $t \in \{1, \ldots, L\}$. The jth outline taken at time t is represented by a parametric curve $z_{tj} : \mathbb{S}^1 \mapsto \mathbb{C}$ in the form of Eq. (4.14). The parametric curve $z_{tj}(s)$ is then transformed to the centroid distance function,

$$r_{tj}(s) = |z_{tj}(s) - c_{tj}|, s \in \mathbb{S}^1,$$

where $c_{tj} = \int_{\mathbb{S}^1} z_{tj}(s) ds$ represents the centroid of z_{tj}. Please refer to Fig. 8.1 for illustration of the notations. The location effect is removed in the centroid distance function.

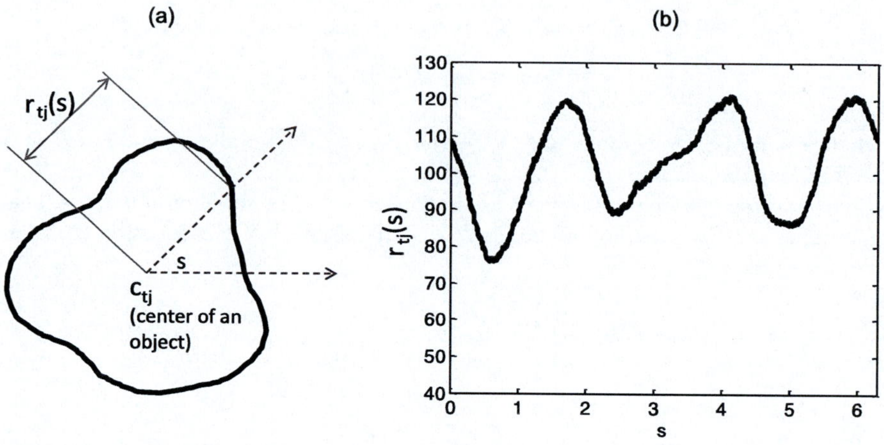

Fig. 8.1 Centroid distance function representation of an object's outline. Reprinted with permission from Park (2014)

In digital images, the outline is only observed at a finite number of locations $s \in S_{tj}$. The available data are thus the set of the discrete observations, namely

$$\tilde{r}_{tj} := \{r_{tj}(s), s \in S_{tj}\}.$$

The centroid distance function is still influenced by the orientation of an object. In order to remove the orientation, Park (2014) rotationally aligned the centroid distance functions using a general Procrustes analysis. When an object is rotated by $\theta_{tj} \in [0, 2\pi]$, the centroid distance function is circularly shifted as described below:

$$r_{tj} \circ h_{\theta_{tj}}(s),$$

where $h_\theta(s) := (s+\theta)$ modulo 2π denotes the circular shift by θ. For aligning $\{z_{tj}\}$ completely, the general Procrustes analysis iteratively updates the centroid distance functions through the following steps:

A. Initialize $\theta_{tj} = 0$ for $t \in \{1, \ldots, L\}$ and $j \in \{1, \ldots, J_t\}$.
B. Updated the mean: Estimate a mean radius function in the form of $\mu(s) = \sum_n \beta_n g_n(s)$, where g_n's are periodic spline basis functions. The unknown coefficients $\{\beta_n\}$ are estimated by minimizing

$$\sum_{t=1}^{L} \sum_{j=1}^{J_t} \sum_{s \in S_{jt}} (r_{tj} \circ h_{\theta_{tj}}(s) - \mu(s))^2.$$

C. Update θ_{tj} by minimizing the following objective function, while keeping $\mu(\cdot)$ unchanged:

$$\sum_{s \in S_{jt}} (r_{tj} \circ h_{\theta_{tj}}(s) - \mu(s))^2.$$

A Newton Raphson algorithm is used for the minimization.

D. Repeat Step B and Step C until convergence.

After the convergence, r_{tj} is updated with $r_{tj} \circ h_{\theta_{tj}}(s)$

$$\boldsymbol{r}_{tj} := \{r_{tj} \circ h_{\theta_{tj}}(s), s \in S_{tj}\}.$$

The aligned data will be analyzed to estimate the dynamic shape distribution. For the rest of the chapter, we introduce the notation,

$$\mathcal{X} = \{\boldsymbol{r}_{tj}; t = 1, \dots, L, j = 1, \dots, J_t\}, \tag{8.14}$$

to denote all the available shape data.

After this preprocessing step and given the availability of $\mathcal{X}$, Park (2014) presents both the parametric and nonparametric versions of his dynamic shape distribution model. Section 8.5.1 describes a Gaussian regression modeling for r_{tj} and induces the Gaussian dynamic shape distribution from the regression modeling. Section 8.5.2 extends the model to a mixture of Gaussian regression models, which is the nonparametric version of the dynamic shape distribution.

8.5.1 Bayesian Regression Modeling for Dynamic Shape Distribution

The data $\mathcal{X}$ are analyzed to track the size and shape information of the outlines. The underlying evolution process of the size and shape is modeled through a general basis expansion model,

$$r_t(s) = \sum_{m=1}^{M} \sum_{n=1}^{N} \alpha_{m,n} \phi_m(t) \gamma_n(s), \ t \geq 0 \text{ and } s \in [0, 2\pi), \tag{8.15}$$

where $\{\phi_m(t)\}$ are the polynomial basis functions, $\{\gamma_n(s)\}$ are the uniform B-spline basis functions, and $\{\alpha_{m,n}\}$ are the corresponding basis coefficients. The observed centroid distance data are the noisy observations of the underlying process,

$$r_{tj}(s) = r_t(s) + \epsilon_{tjs}, \tag{8.16}$$

where ϵ_{tjs} is a zero-mean Gaussian random variable with variance σ^2.

In a Bayesian treatment, one needs to specify the prior distributions for the two modeling parameters, σ^2 and $\{\alpha_{m,n}\}$. Park (2014) suggested the standard conjugate prior for the noise variance σ^2, which is an inverse gamma distribution, i.e., $\sigma^2 \sim IG(\nu_0, \lambda_0)$.

The prior for $\{\alpha_{m,n}\}$ needs to be so chosen that it reflects the characteristics of the underlying evolution process because the basis coefficients, $\{\alpha_{m,n}\}$, determine the dynamics of the underlying process. Park (2014) intended to model the dynamic shape distribution of nanocrystals, where nanocrystals are growing in time. For the nanocrystal growth process, the centroid distance function must be monotonically increasing in time. Park (2014) therefore stated a sufficient condition for $r_t(s)$, which is that $r_t(s)$ is an increasing function of t, i.e.,

$$\alpha_{m+1,n} \geq \alpha_{m,n}, \quad m = 1, \ldots, M-1 \text{ and } n = 1, \ldots, N. \tag{8.17}$$

Please note that the sufficient conditions are all linear.

Let $Q := \{\boldsymbol{\alpha}_n : \alpha_{m+1,n} \geq \alpha_{m,n}, m = 1, \ldots, M-1, n = 1, \ldots, N\}$ denote a collection of the linear constraints. A natural choice to define a probability distribution over constrained variables is a truncated distribution with the support of the distribution being restricted to Q. Park (2014) proposed a truncated normal prior for $\boldsymbol{\alpha} = (\alpha_{1,1}, \ldots, \alpha_{M,1}, \alpha_{1,2}, \ldots, \alpha_{M,N})^T$,

$$\boldsymbol{\alpha} \sim N_Q(\mathbf{0}, \sigma_0^2 \boldsymbol{I}), \tag{8.18}$$

where $Q := \{\boldsymbol{\alpha}_n : \alpha_{m+1,n} \geq \alpha_{m,n}, m = 1, \ldots, M-1, n = 1, \ldots, N\}$ is a collection of random coefficients satisfying the sufficient condition in Eq. (8.17). Since the support of the truncated normal distribution is restricted to satisfy the sufficient condition in Eq. (8.17), the centroid distance functions, $\{r_t(s)\}$, with $\boldsymbol{\alpha}$ randomly sampled from the truncated distribution, would always have monotonically increasing sample paths in time t.

The conditional distribution of $\boldsymbol{r}_{tj}$, conditioned on $\boldsymbol{\alpha}$ and σ^2, has the density of

$$p(\boldsymbol{r}_{tj}|\boldsymbol{\alpha}, \sigma^2) = f(\boldsymbol{r}_{tj}; \boldsymbol{\Phi}_{tj}\boldsymbol{\alpha}, \sigma^2 \boldsymbol{I}), \tag{8.19}$$

where $f(\boldsymbol{x}; \boldsymbol{\mu}, \boldsymbol{\Sigma})$ is a multivariate normal density function with mean $\boldsymbol{\mu}$ and covariance $\boldsymbol{\Sigma}$, and $\boldsymbol{\Phi}_{tj}$ is an $|S_{tj}| \times MN$ matrix of the B-spline basis function values, $\{\phi_m(t)\gamma_n(s); s \in S_{tj}, m = 1, \ldots, M, n = 1, \ldots, N\}$. One can show that the conditional posterior distribution of $\boldsymbol{\alpha}$, given σ^2, follows the truncated Gaussian distribution of

$$\boldsymbol{\alpha}|X, \sigma^2 \sim N_Q(\boldsymbol{\alpha}; \boldsymbol{\mu}, \boldsymbol{\Sigma}), \tag{8.20}$$

where

$$\boldsymbol{\mu} = \frac{\sigma_0^2}{\sigma^2}\left(\boldsymbol{I} - \sum_{t,j} \boldsymbol{\Phi}_{tj}^T \boldsymbol{\Phi}_{tj} \left(\frac{\sigma^2}{\sigma_0^2}\boldsymbol{I} + \sum_{t,j} \boldsymbol{\Phi}_{tj}^T \boldsymbol{\Phi}_{tj}\right)^{-1}\right) \sum_{t,j} \boldsymbol{\Phi}_{tj} \boldsymbol{r}_{tj}, \text{ and}$$

$$\boldsymbol{\Sigma} = \frac{1}{\sigma_0^2}\boldsymbol{I} + \frac{1}{\sigma^2}\sum_{t,j} \boldsymbol{\Phi}_{tj}^T \boldsymbol{\Phi}_{tj}.$$

By conditional conjugacy, the conditional posterior distribution of σ^2, given $\boldsymbol{\alpha}$, is

$$\sigma^2|X, \boldsymbol{\alpha} \sim IG(\sigma^2; \nu_0 + \sum_{t,j} |S_{tj}|/2,$$
$$\lambda_0 + \sum_{t,j} (\boldsymbol{\Phi}_{tj}\boldsymbol{\alpha} - \boldsymbol{r}_{tj})^T (\boldsymbol{\Phi}_{tj}\boldsymbol{\alpha} - \boldsymbol{r}_{tj})/2). \tag{8.21}$$

Posterior sampling of $\boldsymbol{\alpha}$ and σ^2 can be efficiently performed by a Gibbs sampler, which samples, iteratively, $\boldsymbol{\alpha}$ from Eq. (8.20) and σ^2 from Eq. (8.21). Sampling $\boldsymbol{\alpha}$ from Eq. (8.20) requires sampling from a multivariate normal distribution under the linear constraints $\boldsymbol{Q}$, which can be achieved using the Gibbs sampler proposed by Geweke (1991). Once the posterior samples of $\boldsymbol{\alpha}$ and σ^2 are obtained, the posterior samples of the centroid distance function r_{tj} can be taken from the density in Eq. (8.19). Then, the empirical distribution formed by the posterior samples presents an estimate of the probability distribution of r_{tj}.

8.5.2 *Mixture of Regression Models for Nonparametric Dynamic Shape Distribution*

In the Bayesian regression model in Eq. (8.16), the probability density function of the random coefficients $\boldsymbol{\alpha}$ fully characterizes the random behavior of the centroid distance function r_{tj} and accordingly determines the corresponding dynamic shape distribution. When the prior distribution of $\boldsymbol{\alpha}$ was assumed a truncated normal distribution, $\mathcal{N}_Q(\boldsymbol{0}, \sigma_0^2 \boldsymbol{I})$, the corresponding posterior distribution is then also a truncated normal distribution, $\mathcal{N}_Q(\boldsymbol{\mu}, \boldsymbol{\Sigma})$. The truncated distribution has a single mode, i.e., its density function has one peak, because the normal density is strictly quasiconcave, and every strictly quasiconcave function has a unique maxima on a convex polytope feasible region such as Q (Boyd and Vandenberghe 2004, Section 3.4). Consequently, the posterior distribution of $\boldsymbol{r}_{tj}$ is unimodal.

However, a dynamic shape distribution could be multi-modal in many applications. The nanocrystal growth is one such process. As shown in Fig. 8.2, the nanocrystals are almost equally sized and shaped at the beginning of $t = 30$, but they evolve into different shapes at the end. The observed shapes can hardly be fit to a unimodal distribution. Because of these complexities, a nonparametric approach to model the shape distribution is necessary and more practical to meet the requirements in various applications.

Park (2014) proposed a mixture distribution to model the nonparametric dynamic shape distribution. The proposed approach assumes the Dirichlet mixture prior for $\boldsymbol{\alpha}$, where each mixture component is the truncated normal in the form of Eq. (8.18). The number of the mixture components determines the number of the modes of the mixture distribution. The number is automatically estimated using the data extracted from nanocrystal images.

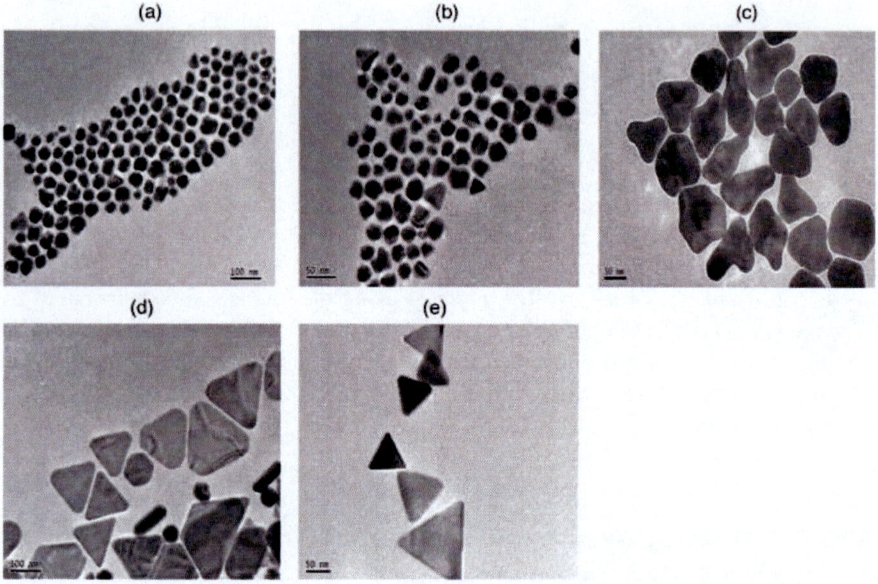

Fig. 8.2 Images taken from a nanocrystal growth process. (**a**) Time = 30. (**b**) Time = 45. (**c**) Time = 60. (**d**) Time = 90. (**e**) Time = 90^+. Reprinted with permission from Park (2014)

The Dirichlet mixture prior used by Park (2014) is an infinite mixture model (Ferguson 1983; Sethuraman 1994; Escobar and West 1995). The infinite mixture size could certainly create an over-complicated model. So the goal of model fitting is to reduce the mixture size to a finite number by eliminating unnecessary mixture components through the use of a sparsity principle. During the iterative model fitting process, the mixture size may increase when the current mixture model is insufficient to represent data or shrink when some mixture components are too similar to each other. This flexibility is an advantage of the Dirichlet mixture prior.

Specifically, Park (2014)'s Dirichlet infinite mixture prior distribution of $\boldsymbol{\alpha}$ is expressed as

$$\boldsymbol{\alpha} \sim \sum_{k=1}^{\infty} \omega_k \delta_{\boldsymbol{\alpha}_k},$$

where $\omega_k = v_k \prod_{k'<k}(1 - v_{k'})$ with $v_k \sim Beta(1, \eta_0)$ independently, $\boldsymbol{\alpha}_k \sim \mathcal{N}_Q(\mathbf{0}, \sigma_0^2 \boldsymbol{I})$, and $\delta_{\boldsymbol{\alpha}_k}$ is a Dirac delta function of $\boldsymbol{\alpha}$; its value is 1 if $\boldsymbol{\alpha} = \boldsymbol{\alpha}_k$ and 0 otherwise. The mixture prior distribution is equivalent to putting a Dirichlet process prior DP($\eta_0 P_0$) on $\boldsymbol{\alpha}$, where η_0 is the Dirichlet process precision parameter and P_0 is a base probability measure which is here defined by $\mathcal{N}_Q(\mathbf{0}, \sigma^2 \boldsymbol{I})$. Intuitively, in the mixture model, the mixture component parameter $\boldsymbol{\alpha}_k$ characterizes one of many unimodal dynamic shape distributions potentially existing, and the

mixing proportion parameter ω_k represents the relative frequency of the unimodal shape distribution. Given the mixing proportion and mixture component parameters $\{\omega_k, \boldsymbol{\alpha}_k\}$, the data likelihood function becomes an infinite location mixture model:

$$L(\boldsymbol{r}_{tj}|\{\omega_k, \boldsymbol{\alpha}_k\}, \sigma^2) = \sum_{k=1}^{\infty} \omega_k f(\boldsymbol{r}_{tj}; \boldsymbol{\Phi}_{tj}\boldsymbol{\alpha}_k, \sigma^2 \boldsymbol{I}). \tag{8.22}$$

Let us introduce a hidden mixture component index variable, k_{tj}, for $\boldsymbol{r}_{tj}$, where $k_{tj} = k$ indicates that $\boldsymbol{r}_{tj}$ is sampled from the kth dynamic shape distribution. The mixing proportion of the kth mixture component is ω_k, i.e., $Pr(k_{tj} = k) = \omega_k$, so that the joint likelihood of $\boldsymbol{r}_{tj}$ and the hidden variable k_{tj} is

$$L(\boldsymbol{r}_{tj}, k_{tj}|\{\omega_k, \boldsymbol{\alpha}_k\}, \sigma^2) = \omega_{k_{tj}} f(\boldsymbol{r}_{tj}; \boldsymbol{\Phi}_{tj}\boldsymbol{\alpha}_{k_{tj}}, \sigma^2 \boldsymbol{I}).$$

The ultimate goal of the model specification is to estimate the posterior predictive distribution of the centroid distance function $r_t(s)$ at time t. Towards that goal, the posterior distributions of the model parameters $\{\boldsymbol{\alpha}_k\}$ and $\{\omega_k\}$ need to be estimated, and the predictive distribution is then computed. The posterior estimation can be performed by the Gibbs sampling algorithms developed for solving general Dirichlet process mixture models. Park (2014) followed the exact block Gibbs sampler (Papaspiliopoulos 2008), which combines the advantages of the Dirichlet process mixture slice sampler (Walker 2007) and the retrospective MCMC sampler (Papaspiliopoulos and Roberts 2008). With a sufficiently large number of Gibbs iterations, the samples of $\{\boldsymbol{\alpha}_k\}$ and $\{\omega_k\}$ from the exact Gibbs sampler follow the true posterior distribution of $\{\boldsymbol{\alpha}_k\}$ and $\{\omega_k\}$.

For the exact Gibbs sampler, Park (2014) introduced slice sampling variables $\boldsymbol{u} = \{u_{tj}\}$ with $u_{tj} \sim \text{Uniform}(0, w_{k_{tj}})$. The reason is the following. Sampling parameters of the infinite mixture model in Eq. (8.22) is very difficult because the number of parameters is infinite. Using slice sampling provides an efficient way to sample parameters from the infinite mixture model; see Walker (2007) for details. Park (2014) performed sequential sampling from the conditional posterior distribution of $\{k_{tj}\}$, $\{\boldsymbol{\alpha}_k\}$, σ^2, and $\{\omega_k\}$ using Algorithm 8.1.

In Algorithm 8.1, the Dirichlet process precision parameter η_0 can be fixed to a constant (a popular choice is $\eta_0 = 1$), or its posterior sampling can be incorporated to the exact block Gibbs sampler, following some existing trials (West 1992; Escobar and West 1995). The posterior sampling for η_0 is performed as follows. We first assume a Gamma prior $\eta_0 \sim \text{Gamma}(\nu_1, \lambda_1)$ with $\nu_1 > 0$ and $\lambda_1 > 0$. For simpler posterior sampling of η_0, let us introduce a continuous random variable κ, such that $\kappa|\eta_0 \sim \text{Beta}(\eta_0 + 1, J)$, where $J = \sum_{t=1}^{L} |J_t|$. We can sample η_0 from its conditional posterior distribution, which is a mixture of two Gamma distributions, i.e.,

$$\eta_0|\kappa, S \sim \pi_\kappa \text{Gamma}(\nu_1 + S, \lambda_1 - \log(\kappa)) + (1 - \pi_\kappa)\text{Gamma}(\nu_1 + S - 1, \lambda_1 - \log(\kappa)), \tag{8.23}$$

where π_κ is a constant satisfying $\pi_\kappa/(1+\pi_\kappa) = (\nu_1 + S - 1)/(J(\lambda_1 - \log(\kappa)))$.

Algorithm 8.1 The exact block Gibbs sampler for sampling $\{\boldsymbol{\alpha}_k\}$ and $\{\omega_k\}$

A. Sample $u_{tj} \sim \text{Uniform}(0, \omega_{k_{tj}})$.

B. Sample k_{tj}: Let V be the smallest index satisfying

$$\sum_{v=1}^{V} \omega_v > 1 - \min\{u_{tj}; t = 1, \ldots, L, j = 1, \ldots, J_t\}.$$

Let $K = \max\{k_{tj} : t = 1, \ldots, L, j = 1, \ldots, J_t\}$. If $K < V$, for $k > K$, sample $v_k \sim Beta(1, \eta_0)$ to get $\omega_k = v_k(1 - \sum_{k'=1}^{k-1} \omega_{k'})$, and sample $\boldsymbol{\alpha}_k \sim \mathcal{N}_Q(0, \sigma_0^2 \boldsymbol{I})$.
Let $A(u) := \{v : \omega_v \geq u, v = 1, \ldots, V\}$. Sample the mixture component indicator variable k_{tj} from

$$Pr(k_{tj} = v | X, \{\boldsymbol{\alpha}_k\}, \sigma^2, \{\omega_k\}) \propto 1(v \in A(u_{tj})) f(\boldsymbol{r}_{tj}; \boldsymbol{\Phi}_{tj}\boldsymbol{\alpha}_v, \sigma^2 \boldsymbol{I}).$$

C. Sample $\boldsymbol{\alpha}_k$ from the conditional posterior:

$$\boldsymbol{\alpha}_k | X, \{k_{tj}\}, \sigma^2, \{\omega_k\} \sim \mathcal{N}_Q(\boldsymbol{\alpha}_k; \boldsymbol{\mu}_k, \boldsymbol{\Sigma}_k),$$

where

$$\boldsymbol{\mu}_k = \frac{\sigma_0^2}{\sigma^2} \left(\boldsymbol{I} - \sum_{k_{tj}=k} \boldsymbol{\Phi}_{tj}^T \boldsymbol{\Phi}_{tj} \left(\frac{\sigma^2}{\sigma_0^2} \boldsymbol{I} + \sum_{t,j} \boldsymbol{\Phi}_{tj}^T \boldsymbol{\Phi}_{tj} \right)^{-1} \right) \sum_{t,j} \boldsymbol{\Phi}_{tj} \boldsymbol{r}_{tj}, \text{ and}$$

$$\boldsymbol{\Sigma}_k = \frac{1}{\sigma_0^2} \boldsymbol{I} + \frac{1}{\sigma^2} \sum_{k_{tj}=k} \boldsymbol{\Phi}_{tj}^T \boldsymbol{\Phi}_{tj}.$$

Park (2014) used the Gibbs sampler to sample from the truncated normal distribution (Geweke 1991).

D. Sample σ^2 from

$$\sigma^2 | X, \{k_{tj}\}, \{\omega_k\}, \{\boldsymbol{\alpha}_k\} \sim IG(\sigma^2; \nu_0 + \frac{1}{2} \sum_{t,j} |S_{tj}|,$$

$$\lambda_0 + \frac{1}{2} \sum_{k=1}^{K} \sum_{k_{tj}=k} (\boldsymbol{\Phi}_{tj}\boldsymbol{\alpha}_k - \boldsymbol{r}_{tj})^T (\boldsymbol{\Phi}_{tj}\boldsymbol{\alpha}_k - \boldsymbol{r}_{tj})).$$

E. Sample the stick-breaking random variables v_k's, as follows,

$$v_k \sim Beta\left(1 + \sum_{t,j} 1(k_{tj} = k), \eta_0 + \sum_{t,j} 1(k_{tj} > k)\right),$$

and then update $\omega_k = v_k \prod_{k'<k} (1 - v_{k'})$.

F. Repeat Steps A–E to iterate.

Source: Park (2014) with permission

Let $\{\boldsymbol{\alpha}_k^{(q)}, \omega_k^{(q)}; k = 1, \ldots, K\}$ be the posterior samples of $\boldsymbol{\alpha}_k$ and ω_k taken at the qth iteration of the exact block Gibbs sampler. Given the posterior samples, we are interested in estimating the posterior mean and variance of the mixture component parameters. In the estimation of the mixture component parameters, label switching can be a serious issue. We first applied the relabeling algorithm (Stephens 1997) to minimize the label switching problem; the algorithm conceptually does re-indexing $\{\boldsymbol{\alpha}_k^{(q)}, \omega_k^{(q)}; k = 1, \ldots, K\}$ to make sure that $\boldsymbol{\alpha}_k^{(q)}$'s and $\omega_k^{(q)}$'s are sampled from the same mixture component. After the relabeling, the posterior mean estimates of $\boldsymbol{\alpha}_k$ and ω_k are simply the sample averages of the posterior samples, $\{\boldsymbol{\alpha}_k^{(q)}\}$ and $\{\omega_k^{(q)}\}$, respectively. The posterior variance estimates can be likewise obtained by the sample variances of the posterior samples.

Given the posterior samples $\{\boldsymbol{\alpha}_k^{(q)}, \omega_k^{(q)}; k = 1, \ldots, K, q = 1, \ldots, R\}$, we would like to estimate the posterior distribution of $r_t(s)$ for each evaluation locations $s \in S$ and each test time t. Let $\boldsymbol{\Phi}_t$ is an $|S| \times MN$ matrix of the B-spline basis function values evaluated at the test locations, $\{\phi_m(t)\gamma_n(s); s \in S, m = 1, \ldots, M, n = 1, \ldots, N\}$ and let $\boldsymbol{r}_t = (r_t(s) : s \in S)$. The posterior samples of $\boldsymbol{r}_t$ can be sampled as follows:

- For each q, for $q = 1, \ldots, R$
  - Draw B samples from a mixture distribution, $\sum_{k=1}^{K} \omega_k^{(q)} N(\boldsymbol{\Phi}_t^T \boldsymbol{\alpha}_k^{(q)}, \sigma^2)$. The bth sample is denoted by $\boldsymbol{r}_k^{(q,b)}(t)$ for $b = 1, \ldots, B$.
- The resulting RB samples are analyzed to estimate the mean and quantiles of $\boldsymbol{r}_t$.

8.6 Case Study: Dynamic Shape Distribution Tracking with Ex Situ Measurements

We applied the method described in Sect. 8.5.2 to material images shown in Fig. 8.2. The five TEM images of nanocrystals were taken from a nanocrystal synthesis process at five different times. The measurement times are 30, 45, 60, 90, and 90^+ s, respectively, where 90^+ implies that the sample was taken at the 90 s after the chemical reaction started but it went through some aging steps before being imaged. The aging steps do not change nanocrystal morphologies significantly (Kundu and Liang 2010), so that the image at 90^+ s was considered as an image taken at 90 s, i.e., $t = 90^+ = 90$. In other words, the nanocrystal estimates obtained for 90 s come from both these images.

All the images were taken in an ex situ fashion, meaning that different nanocrystal samples were taken and imaged at these five times. Because of this, the nanocrystals observed in an earlier image do not correspond to any nanocrystals in the later images, meaning that the images and the nanocrystal outlines extracted thereof are not longitudinal data. Without the longitudinal data, one cannot track how individual nanocrystals evolve, but instead can only sense how the nanocrystals collectively change their morphology through tracking the evolution of the shape distribution over time.

In this numerical study, the image segmentation algorithm developed by Park et al. (2013) is applied for extracting the outlines of nanocrystals from the five images. The following is a tally of the nanocrystal outlines obtained at each time: 134 at $t = 30$, 86 at $t = 45$, 25 at $t = 60$, and 26 at $t = 90$ and $t = 90^+$ combined, of which 20 are taken at $t = 90$ and six at $t = 90^+$. The outlines are converted into the corresponding centroid distance functions, which are then aligned by the general Procrustes analysis as described in Sect. 8.5.

We ran the Gibbs sampling steps in Algorithm 8.1 using the aligned data. In total, 50,000 Gibbs samples were sampled with the first 10,000 discarded as a burn-in. We used the Raftery and Lewis's diagnostic (Raftery and Lewis 1992) to estimate the necessary burn-in period, which was estimated to be 594. In addition, we graphically diagnosed the convergence of the Gibbs samplers taken after the burn-in period by checking whether the Gibbs samples exhibit good mixing after burn-in.

We set the hyperparameters $\nu_0 = \lambda_0 = 0.1$ for the inverse Gamma prior of σ^2. The choice of ν_0 and λ_0 has a negligible effect on the posterior sampling of σ^2. This can be understood by observing the posterior distribution σ^2 at Step D in Algorithm 8.1, which is an inverse gamma distribution having $\nu_0 + \frac{1}{2}N$ and $\lambda_0 + \frac{1}{2}MSE$ as its parameters, where $N = \sum_{t,j} |S_{tj}|$ is proportional to the total number of nanocrystals and $MSE = \sum_{k=1}^{K} \sum_{k_{tj}=k} (\boldsymbol{\Phi}_{tj}\boldsymbol{\alpha}_k - \boldsymbol{r}_{tj})^T (\boldsymbol{\Phi}_{tj}\boldsymbol{\alpha}_s - \boldsymbol{r}_{tj})$ is the mean squared distance between the model prediction and observations. Since ν_0 and λ_0 are far smaller than N and MSE, respectively, the adjustment made by λ_0 and ν_0 does not affect the parameters in the inverse gamma distribution much nor the final result.

We also set the hyperparameters $\nu_1 = \lambda_1 = 0.1$ as the prior of the concentration parameter η_0. The choice of ν_1 and λ_1 does affect the sampling of the Dirichlet process concentration parameter η_0, as

$$\eta_0|\kappa, S \sim \pi_\kappa \text{Gamma}(\nu_1+S, \lambda_1-\log(\kappa))+(1-\pi_\kappa)\text{Gamma}(\nu_1+S-1, \lambda_1-\log(\kappa)).$$

But the effect of ν_1 and λ_1 is also limited due to the existence of S and $\log(\kappa)$ in the gamma distribution parameters. Generally speaking, the expected value of the gamma random variable increases as either ν_1 or λ_1 increases.

The posterior mode of the Gibbs samples of K informs us about the number of mixture components in the dynamic shape distribution. For the ex situ images processed, the number of mixture components is 13; see Fig. 8.3 for the trace plot of the posterior samples of K.

Each of the 13 mixture components represents a mode of the dynamic shape distribution. Plotting the centroid distance functions for each mixture component over time shows how the shapes evolve in the mixture component. Figure 8.4 presents the eight shape modes that have the eight largest mixing proportion values (ω_s). The summation of the eight largest mixing proportion values is more than 90%, meaning that these eight modes explain the vast majority of the morphological

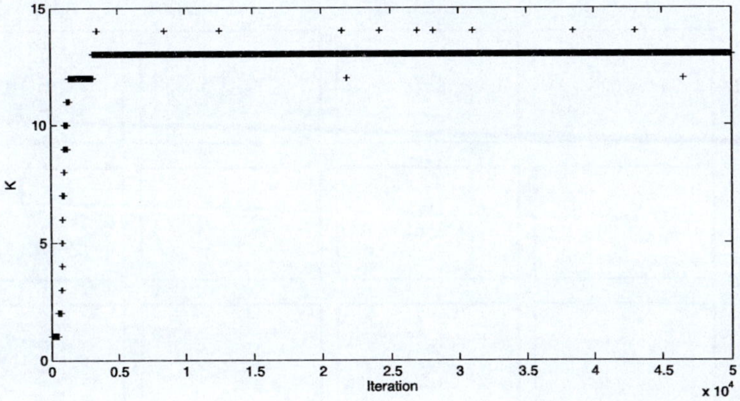

Fig. 8.3 The trace plot of K. The posterior mode of K is 13. Reprinted with permission from Park (2014)

information in the image data. What is shown in the figure is the estimated mean of the centroid distance functions at different time points with the corresponding 95% credible intervals. Furthermore, the observed nanocrystal outlines are overlaid on top of the estimated outlines, for those times when the nanocrystals of a corresponding shape mode were observed. Among the eight shape modes, five of them, corresponding to $k = 1, 2, 4, 6, 7$, explain the shape variation of different sizes for triangular nanocrystals, whereas the other two shape modes, i.e., $k = 3, 5$, explain the shape variation of spherical nanocrystals. The eighth shape mode, explaining the shape evolution to rod-shaped nanocrystals, is relatively minor and appeared much less frequently.

An important motivation driving the nanocrystal morphology research is to produce nanocrystals with precisely controlled sizes and shapes. Achieving the minimal size and shape variations around the designed or targeted size and shape is desirable. The dynamic shape distribution models provide a monitoring tool to quantify the size and shape variations at different times or stages in a nanocrystal synthesis process. Towards this purpose, in addition to the credible intervals of the centroid distance functions, depicting how each growth trajectory evolve as in Fig. 8.4, another useful tool is the density map of the posterior samples of the centroid distance functions.

Figure 8.5 presents the density plots of the posterior samples of $r_t(s)$, which are the samples of the centroid distance functions from different mixture components. The areas where the sampled centroid distance functions densely locate look brighter, whereas the areas where the sampled centroid distance functions do not appear often look darker. Comparing the density plots obtained at different time points, we cannot help noticing that the lower bounds of the dense regions increase slowly over time, while the upper bounds of the dense regions increase more noticeably. This is due to the difference in the growth speeds of nanocrystals of

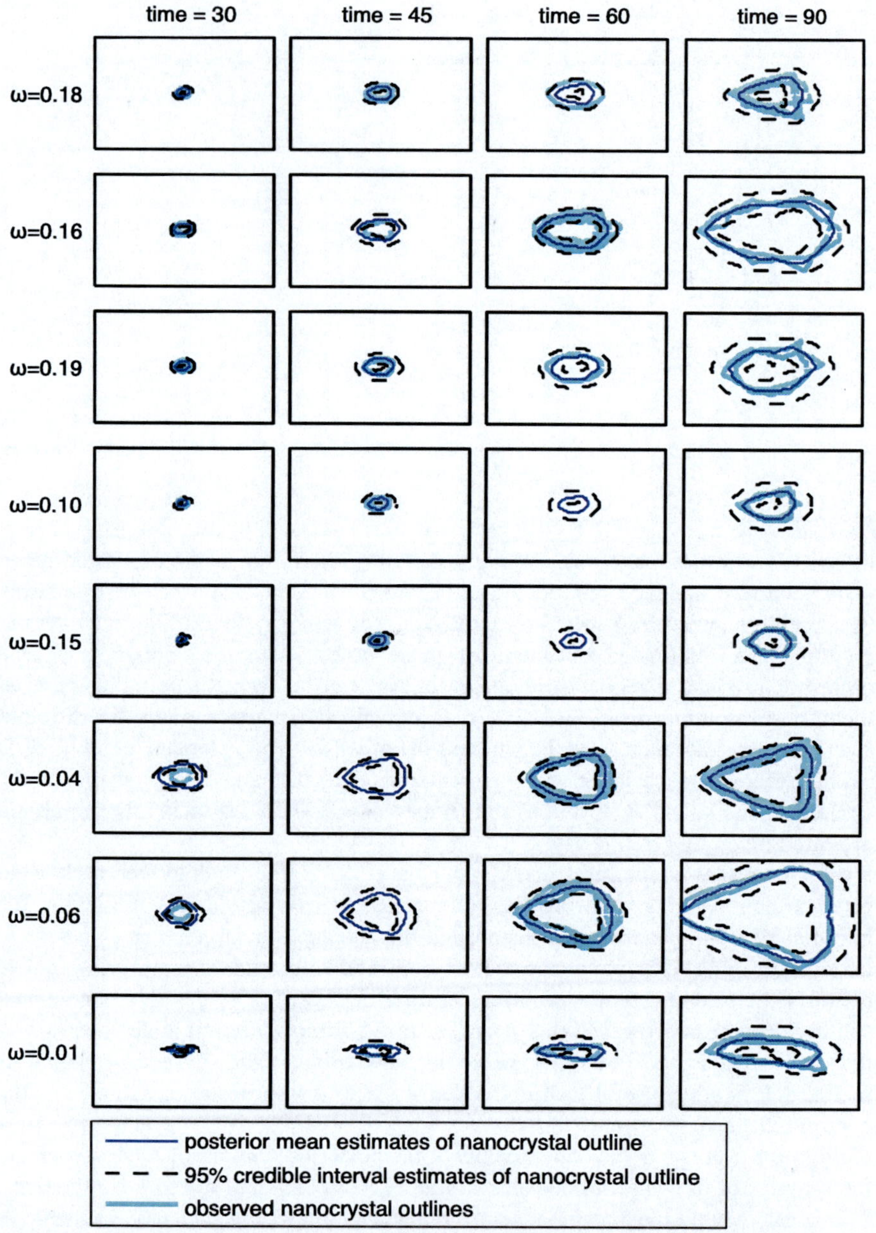

Fig. 8.4 Major modes of shapes of nanocrystals and their evolution, identified from the ex situ images. The estimated nanocrystal outlines are overlaid with the observed nanocrystal outlines. Each row shows one shape evolution. The values of ω to the left of each row represent the mixing proportion of the corresponding shape evolution. In some of the plots, there is no observed nanocrystal outline due to the absence of nanocrystals of the corresponding shape mode at that specific time. Reprinted with permission from Park (2014)

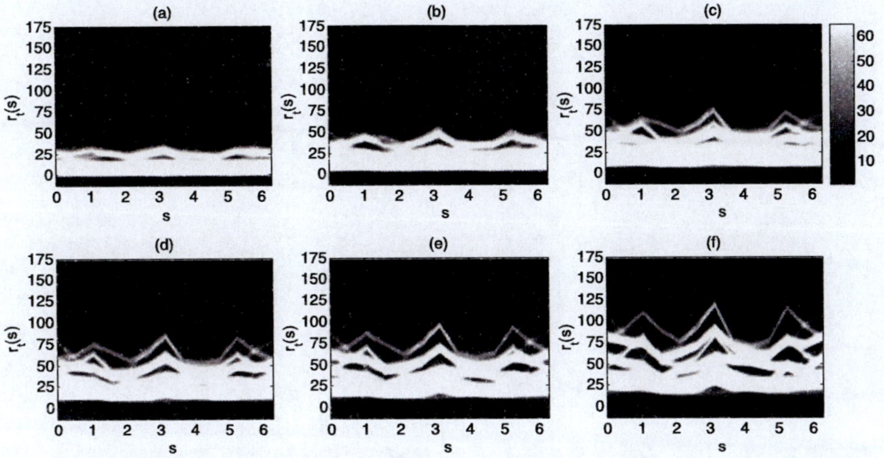

Fig. 8.5 Posterior density maps of radius functions in time. The high density regions look brighter. **(a)** t=30. **(b)** t=45. **(c)** t=60. **(d)** t=67. **(e)** t=75. **(f)** t=90. Reprinted with permission from Park (2014)

different sizes. The smaller nanocrystals grow slower and the larger ones grow faster, causing the variations in nanocrystal's sizes and shapes to increase over time. This observation appears to be consistent with the understanding coming from the *Ostwald ripening process* in the particle physics, in which 'larger crystals grow at the expense of smaller ones' (Cheon et al. 2004). Moreover, we also observed different growth speeds for different s's. There are three clear peaks in the radius functions at $t = 90$, suggesting an anisotropic growth of nanocrystals towards triangle-like shapes.

8.7 Case Study: Dynamic Shape Distribution Tracking with In Situ Measurements

Unlike in Sect. 8.6 where ex situ images were used, we in this study use an in situ video, capturing a nanocrystal growth process and providing a longitudinal image dataset. The original video, as an online supplement of Zheng et al. (2009), is available at http://www.sciencemag.org/cgi/content/full/324/5932/1309/DC1.

The longitudinal dataset is in the form of a 72-s long MPEG video file that contains the time-sequenced images of platinum nanocrystals taken every second during their growth processes. Analyzing the longitudinal data, we can obtain the temporal sequences of individual nanocrystal shapes observed at a common set of times. On the other hand, we can extract image snapshots from the video and then fit the dynamic shape distribution model, as done to the ex situ images in Sect. 8.5.2. The temporal sequences of individual nanocrystals can be matched with

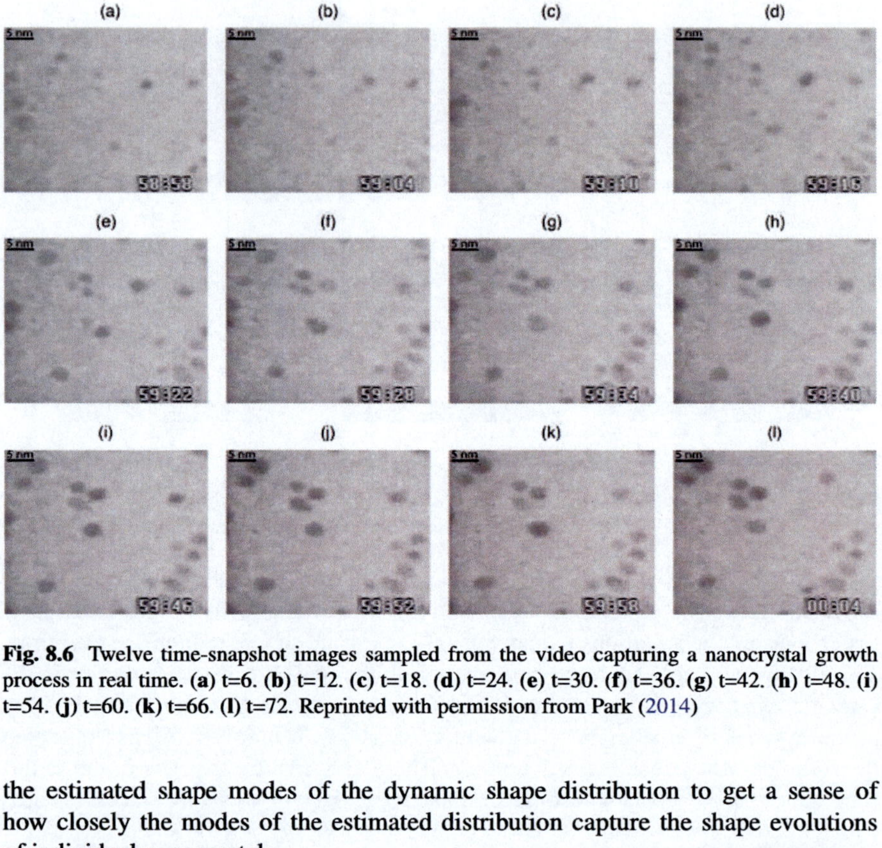

Fig. 8.6 Twelve time-snapshot images sampled from the video capturing a nanocrystal growth process in real time. **(a)** t=6. **(b)** t=12. **(c)** t=18. **(d)** t=24. **(e)** t=30. **(f)** t=36. **(g)** t=42. **(h)** t=48. **(i)** t=54. **(j)** t=60. **(k)** t=66. **(l)** t=72. Reprinted with permission from Park (2014)

the estimated shape modes of the dynamic shape distribution to get a sense of how closely the modes of the estimated distribution capture the shape evolutions of individual nanocrystals.

As displayed in Fig. 8.6, we sub-sampled twelve time snapshot images from the video with a time interval of 6 s, and used the snapshot images as if they were the ex situ image data for estimating the dynamic shape distribution model.

The dynamic shape distribution model was fit to the twelve snapshots in the same way as done in Sect. 8.6. The parameter settings used for learning the Bayesian growth model is the same as well, i.e., $\nu_0 = \lambda_0 = 0.1$ for the inverse Gamma prior of σ^2 and $\nu_1 = \lambda_1 = 0.1$ for the hyperprior of the Dirichlet concentration parameter η_0. A total of 60,000 Gibbs samples were obtained, with the first 15,000 discarded as burn-in.

The dynamic distribution model identified a total of eighteen different modes. Among them, we chose the seven modes that have the largest mixing proportion values. The summation of the seven largest mixing proportion values was more than 73%. Figure 8.7 shows the predictive mean estimates of the seven major modes at eight time points.

We manually extracted the temporal shape sequences of nine observed nanocrystals from the same video. Each temporal sequence is compared with one of the eighteen modes of the dynamic shape distribution model. Suppose that the

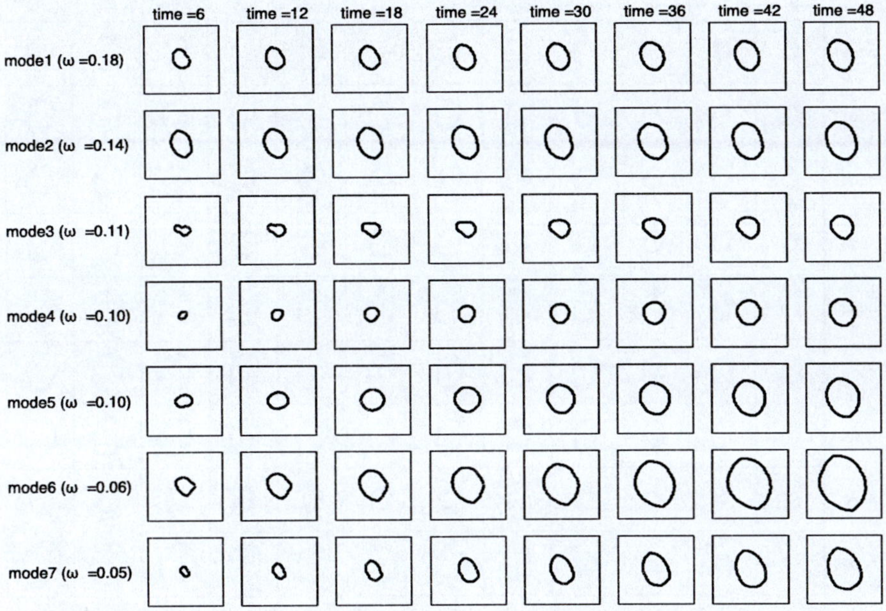

Fig. 8.7 Seven major modes of the nanocrystal's dynamic shape distribution, estimated using the twelve time-snapshot images taken from the video frame data. The solid lines represent the predictive mean estimates of the centroid distance functions for the sth mode. The value of ω to the left in each row represents the mixing proportion of the corresponding shape mode. Reprinted with permission from Park (2014)

nanocrystal outlines in the temporal sequence are denoted by $\{\boldsymbol{r}_t\}$. Then, the mode selected to match the temporal shape sequence is the one that maximizes the conditional probability of observing $\{\boldsymbol{r}_t\}$, given the posterior modes of $\boldsymbol{\alpha}_s$.

Figure 8.8 presents the observed temporal sequences versus the matched mode of the estimated dynamic shape distribution. The nth row of the figure shows the nth observed temporal sequence of nanocrystal shapes, overlaid with the mean function of the matched distribution mode. The caption to the left in a row indicates the index of the matched distribution mode. For instance, if the caption says 'mode 2,' it means that this temporal sequence is matched to the second row of Fig. 8.7.

All nine observed shape sequences found a good match with one of the five major distribution modes, i.e., mode 1 to mode 5, but some modes are matched with multiple observed sequences. For instance, nearly a half of the nine observed sequences fit well to mode 1, the mode with the largest mixing proportion, and three sequences are matched to mode 2. This does not come as a surprise, as the mode's respective mixing proportions do suggest a higher frequency for these sequences to be actually observed. A single sequence is matched to mode 3 and mode 5, respectively, while no sequence is matched to mode 4 among the nine chosen sequences.

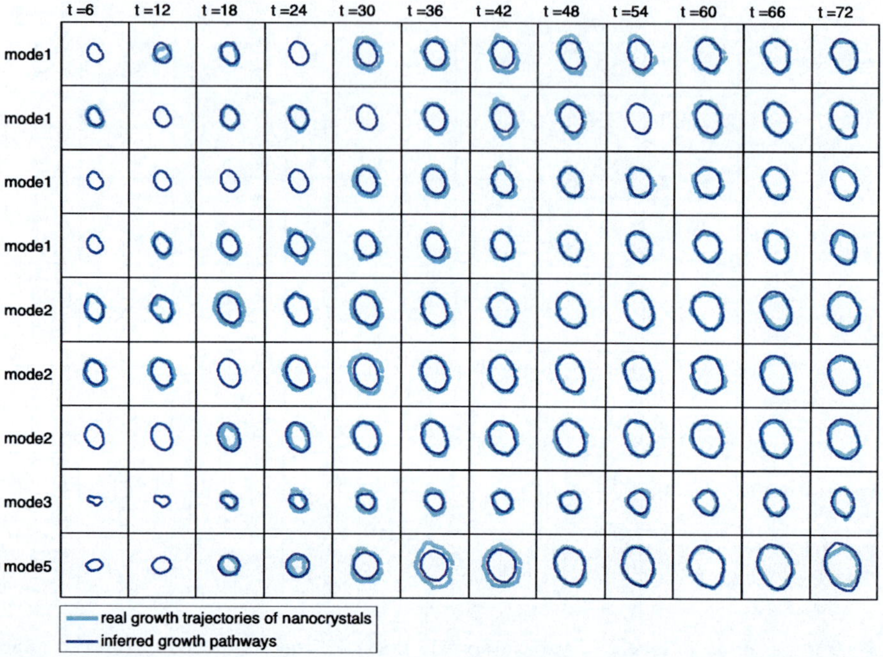

Fig. 8.8 Comparison between the observed shape sequences and the estimated shape distribution modes. The mode numbers on the left indicate which distribution mode in Fig. 8.7 were matched with the observed shape sequences in the corresponding row. Some panels do not have the observed nanocrystal outlines, because some nanocrystals look too faint in certain time frames for their outlines to be captured. Reprinted with permission from Park (2014)

References

Boyd S, Vandenberghe L (2004) Convex Optimization. Cambridge University Press, New York, NY, USA

Cheon J, Kang NJ, Lee SM, Lee JH, Yoon JH, Oh SJ (2004) Shape evolution of single-crystalline iron oxide nanocrystals. Journal of the American Chemical Society 126(7):1950–1951

Dryden IL, Mardia KV (1991) General shape distributions in a plane. Advances in Applied Probability 23(2):259–276

Dryden IL, Kim KR, Le H (2019) Bayesian linear size-and-shape regression with applications to face data. Sankhyā A 81(1):83–103

Escobar MD, West M (1995) Bayesian density estimation and inference using mixtures. Journal of the American Statistical Association 90(430):577–588

Ferguson TS (1983) Bayesian density estimation by mixtures of normal distributions. In: Rizvi MH, Rustagi JS, Siegmund D (eds) Recent Advances in Statistics, Academic Press, Cambridge, Massachusetts, pp 287–303

Fontanella L, Ippoliti L, Kume A (2019) The offset normal shape distribution for dynamic shape analysis. Journal of Computational and Graphical Statistics 28(2):374–385

Geweke J (1991) Efficient simulation from the multivariate normal and student-t distributions subject to linear constraints and the evaluation of constraint probabilities. In: Proceedings of the 23rd Symposium on the Interface between Computing Science and Statistics, pp 571–578

Gower JC (1975) Generalized Procrustes analysis. Psychometrika 40(1):33–51

Jónsdóttir K, Jensen E (2011) Gaussian radial growth. Image Analysis & Stereology 24(2):117–126

Kent JT, Mardia KV, Morris RJ, Aykroyd RG (2000) Procrustes growth models for shape. In: Proceedings of the First Joint Statistical Meeting, pp 236–238

Kent JT, Mardia KV, Morris RJ, Aykroyd RG (2001) Functional models of growth for landmark data. In: Proceedings of Functional and Spatial Data Analysis, pp 109–115

Kundu S, Liang H (2010) Shape-controlled synthesis of triangular gold nanoprisms using microwave irradiation. Journal of Nanoscience and Nanotechnology 10(2):746–754

Landau L, Lifschitz E (1976) Mechanics, 3rd Ed. Pergamon Press, Oxford

Papaspiliopoulos O (2008) A note on posterior sampling from Dirichlet mixture models. Unpublished Technical Report

Papaspiliopoulos O, Roberts G (2008) Retrospective Markov Chain Monte Carlo methods for Dirichlet process hierarchical model. Biometrika 95(1):169–186

Park C (2014) Estimating multiple pathways of object growth using nonlongitudinal image data. Technometrics 56(2):186–199

Park C, Huang J, Ji J, Ding Y (2013) Segmentation, inference and classification of partially overlapping nanoparticles. IEEE Transactions on Pattern Analysis and Machine Intelligence 35(3):669–681

Raftery AE, Lewis S (1992) How many iterations in the Gibbs sampler. Bayesian Statistics 4(2):763–773

Sethuraman J (1994) A constructive definition of Dirichlet priors. Statistics Sinica 4(2):639–650

Stephens M (1997) Bayesian methods for mixtures of normal distributions. PhD thesis, University of Oxford

Walker S (2007) Sampling the Dirichlet mixture model with slices. Communications in Statistics - Simulation and Computation 36(1):45–54

West M (1992) Hyperparameter estimation in Dirichlet process mixture models. Technical Report 92-A03, Institute of Statistics and Decision Sciences, Duke University, Durham, USA

Zheng H, Smith RK, Jun Yw, Kisielowski C, Dahmen U, Alivisatos AP (2009) Observation of single colloidal platinum nanocrystal growth trajectories. Science 324(5932):1309–1312

Chapter 9
Change Point Detection

9.1 Basics of Change Point Detection

The problem of change point detection has originated from the statistical quality control (SQC) or statistical process control (SPC) (Montgomery 2009). The objective of quality control is to maintain a stable production process that outputs nearly identical products. In other words, the key characteristics of the product are expected to stay more or less the same, subject to a small amount of fluctuation caused by many random factors in the production process, albeit none outstanding. When there are some significant changes to the process, the quality control methods or tools are supposed to flag the changes, so that remedial actions can be taken to adjust or correct the process back to where it is supposed to be.

The underlying physical process of interest here, whose observations form a time series, is stochastic in nature. What this means is that if one compares the current observation with its previous observations, one would find the process changing all the time. But not all those changes are of interest to decision makers who want to control the process; only interesting are the changes significantly greater than the level of natural random fluctuations. The methods in SQC are developed to detect these significant changes beyond natural random fluctuations.

The beginning of SQC is credited to the work by Walter A. Shewhart starting in 1920s. Shewhart distinguished the variations in a process into two categories: the chance cause variation and the assignable cause variation. The former is the natural fluctuation that one has to live with, whereas the latter are the "significant" changes that, once identified and after the assignable cause is removed through proper corrective actions, can be eliminated or substantially subdued.

The underlying process, being stochastic, is thus characterized by a probability density distribution (pdf). The process is considered unchanged if its pdf remains the same, or considered changed otherwise. A pdf is often represented in a parametric form, e.g., a Gaussian distribution. When one have a finite number of sample

C. Park, Y. Ding, *Data Science for Nano Image Analysis*, International Series in Operations Research & Management Science 308,
https://doi.org/10.1007/978-3-030-72822-9_9

observations from the pdf, the parameters of the underlying probability distribution is estimated using the samples. Consequently, the comparison to signify whether a pdf has changed from the baseline pdf is through a statistical comparison procedure, or more specifically, a hypothesis testing procedure, in which the null hypothesis, H_0, states that the underlying pdf does not change, while the alternative hypothesis, H_1, states that the underlying pdf has changed (two sided hypothesis) or the underlying pdf has changed to a specific direction (one sided hypothesis). In the early days of SQC, the distribution describing the stochastic behavior of the process is typically assumed a Gaussian distribution, which can be fully specified by its mean, μ, and standard deviation, σ. Under that setting, the hypotheses are stated as

$$\begin{aligned} H_0 &: \mu = \mu_0 \quad \text{and} \quad \sigma = \sigma_0 \quad \text{in presence of chance causes only, in-control,} \\ H_1 &: \mu \neq \mu_0 \quad \text{or} \quad \sigma \neq \sigma_0 \quad \text{in presence of assignable causes, out-of-control,} \end{aligned} \tag{9.1}$$

where μ_0 and σ_0 are the parameters of the Gaussian distribution while the process is in control. Walter Shewhart invented a practical and intuitive way for monitoring a running process, which is to plot certain test statistics against the time of observations on a run chart embodied with an upper and a lower control limit. When the observed test statistic exceeds the prescribed control limits, it triggers an alarm, signaling the rejection of the null hypothesis at a prescribed significance level and leading to the conclusion that a change has likely taken place. The run chart method is commonly referred to as the Shewhart control charts, named after its inventor.

Figure 9.1 presents control chart examples. The two control charts detect different types of changes in a process. The change in the left panel is a sustained shift in the process mean, taking place at time t_c. The random process fluctuates subsequently around the shifted mean until a corrective action is taken or until another change happens to the process. The change in the right panel illustrates an increase in process variance, while the mean of this second process stays unchanged. Of course, a process change could be a combination of the mean and variance

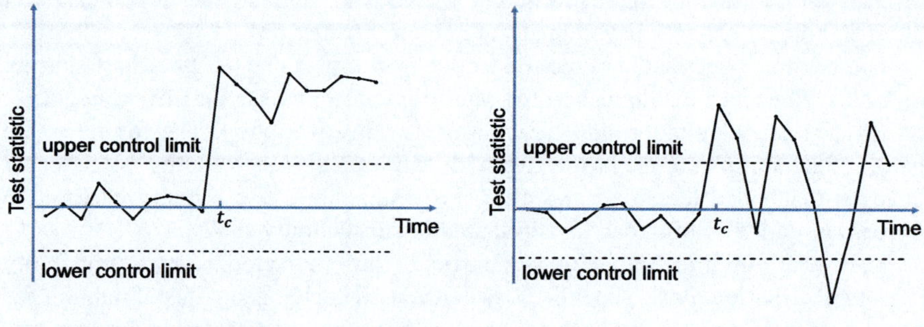

Fig. 9.1 Examples of control charts. Left panel: a sustained mean shift. Right panel: an increased process variance

changes, that is to say, both a mean shift and an variance increase could take place simultaneously.

9.1.1 Performance Metrics

Change detection methods are evaluated on two metrics—type-I and type-II errors. The type-I error, usually denoted by α (not to be confused with the $\boldsymbol{\alpha}$ vector, used as the B-splines coefficient vector), is the probability that when the process is in control (no change), the detection method mistakes it to be out of control (a change occurs). The type-I error is also known as false alarms or false positives. The type-II error, usually denoted by β, is the probability that when the process has undergone a change, the detection method fails to flag such change. The type-II error is also known as missed detections or false negatives. The type-II error can be reported as a delay in detection. It is well understood that a larger type-II error leads to a longer delay in detecting a sustained change. Practitioners would ideally like for a change detection method to have both errors as small as possible, or more intuitively, to have false alarms as infrequently as possible and have a genuine change detected as soon as possible.

In the current SQC literature, the control limits on a control chart are decided primarily based on a prescribed type-I error rate, α. When the test statistic exceeds the control limits, one say that the null hypothesis is rejected at the $100(1-\alpha)\%$ significance level. The smaller the α, the more certain the rejection. The broadly used three-sigma control limits in SQC correspond to $\alpha = 0.0027$ under the Gaussianity assumption. In the latter sections of this chapter, the type-I error rate is denoted by ν to avoid possible confusion with the notation for the B-splines coefficients.

9.1.2 Phase I Analysis Versus Phase II Analysis

There are two phases in the analysis for change point detection: Phase I versus Phase II analyses.

Phase I analysis is an offline, retrospective analysis. Researchers gather data and treat them as a training dataset, and then look back to see if there are any outliers or sustained changes. Outliers are eliminated, and the sustained change point are used to separate the training data into different segments. The purpose of Phase I analysis is to select one data segment that is supposed to be in control and representative of the normal operation state of the process. The selected data is then used as the benchmark baseline when it comes to monitoring the future, ongoing process. In the context of the hypothesis test setting in Eq. (9.1), the Phase I analysis is to produce a good estimate of μ_0 and σ_0 after analyzing and cleaning a training dataset.

Phase II analysis is the online, prospective analysis. It compares the ongoing process, based on real-time, in-process measurements, with the benchmark baseline, decided through the Phase I analysis, and flags a process change as it occurs. The Phase II detection of size-only changes in nanocrystals is demonstrated through the use of the innovation process in Fig. 7.8 associated with a state space model.

This chapter discusses both Phase I and Phase II analyses for shape change detection but focuses primarily on Phase I analysis for size-only change detection. We want to note that aside from process monitoring purposes, Phase I analysis can also aid discovery in the material exploration process. In the presence of dynamic video data, it is highly laborious to manually process thousands or more image frames for finding the change points in a nanoparticle growth process. Yet finding those change points is important, as they may provide clues to unearth the basic science behind nanoparticle growth processes. For that, the data science solutions can play a helping hand in automating change point detections and expedite material discovery.

9.1.3 Univariate Versus Multivariate Detection

The aforementioned examples use a single-dimensional response variable. Detection of changes in such process characteristics is a univariate detection problem. In many practical circumstances, one wants to detect changes in a vector of variables—such detection is a multivariate detection problem. The multivariate techniques are particularly relevant to the detection needs when the response data is a functional curve or a shape. The goal is then to compare a curve with other curves. Detecting changes in series of curves is a multivariate detection problem because a curve can be considered a collection of variables densely distributed along the curve (imagine that the curve is discretized in the form of a dense set of points.

The detection problem in the nanoparticle applications concerns with its morphological features, such as size and shape. Detecting changes in *shape* is naturally a multivariate detection problem. The *size* model, such as the normalized particle radius used in Chap. 7, is a scalar model. However, even for the size model, the detection problem eventually becomes multivariate because the density function representation of the normalized particle radius is not in a simple parametric form but modeled through a nonparametric B-splines model; see the arguments made in Sect. 7.1.1. Consequently, the detection problem is to compare two nonparametric pdf curves associated with the particle size, and doing so makes the detection problem multivariate. This is different from the circumstance under Gaussianity, where detecting the mean change in two Gaussian pdf curves can be reduced to the comparison of two means, and thus is a univariate detection problem.

It is understood that detection in high-dimensional spaces suffers from the "curse of dimensionality" and is prone to having poor performance. A preprocessing step popular in multivariate detection is the use of a dimension reduction technique for reducing the number of variables to be monitored and thus avoiding the curse in high dimensionality. Principal component analysis (PCA) is one of the widely used dimension reduction tools (Jolliffe 2002). PCA attempts to find a small number of significant projections of the original vector onto a lower-dimensional space, with the hope that the dimension-reduced vector still represents well the original vector.

The basic idea of PCA is to find a linear combination of the variables in the original vector, such that the first principal component has the largest variance, the second principal component has the largest variance in a subspace orthogonal to the first principal component, and the k-th principal component has the largest variance in a subspace orthogonal to the first $k - 1$ principal components. One can decide the principal components by computing the eigenvalue and eigenvector pairs of the covariance matrix associated with the original random vector. The first principal component is the one corresponding to the largest eigenvalue, the second principal component corresponding to the second largest eigenvalue, and so on. Because covariance matrices are symmetric with real number entries, its eigenvalues are all real. Also because the eigenvalues of a covariance matrix are variances, they are supposed to be nonnegative.

One can sort the eigenvalues in descent order and then select the largest few principal components to approximate the original vector; this is the idea behind the dimension reduction. The quality of such approximation can be quantified by the ratio of the summation of the eigenvalues of the selected principal components over the total summation of all eigenvalues. One can generate a Pareto plot of the eigenvalues and use it to aid visually the selection of the most significant principal components; one such plot is available in Fig. 9.3a.

Let us consider a two-dimensional example to help understand the concept, terminology, and steps in dimension reduction. There is rarely any practical need to reduce a two-dimensional data space further, but this is only for the illustration purpose. The concept and ideas illustrated here are applicable to truly high dimensional cases.

Suppose that a random vector, $\boldsymbol{x}$, follows a Gaussian distribution of $\mathcal{N}(\boldsymbol{\mu}_x, \boldsymbol{\Sigma}_x)$, where

$$\boldsymbol{\mu}_x = \begin{pmatrix} 0 \\ 0 \end{pmatrix},$$

$$\boldsymbol{\Sigma}_x = \begin{pmatrix} 0.25 & 0.45 \\ 0.45 & 1 \end{pmatrix}.$$

Figure 9.2a presents the scatter plot of data sampled from this Gaussian distribution. Apparently the largest variance direction associated with this dataset is along the long axis of the data cloud. The direction of the short axis is the largest variance direction perpendicular to the long axis. PCA is to find these principal variance directions through computing the eigenvalue and eigenvector pairs of $\boldsymbol{\Sigma}_x$, which are

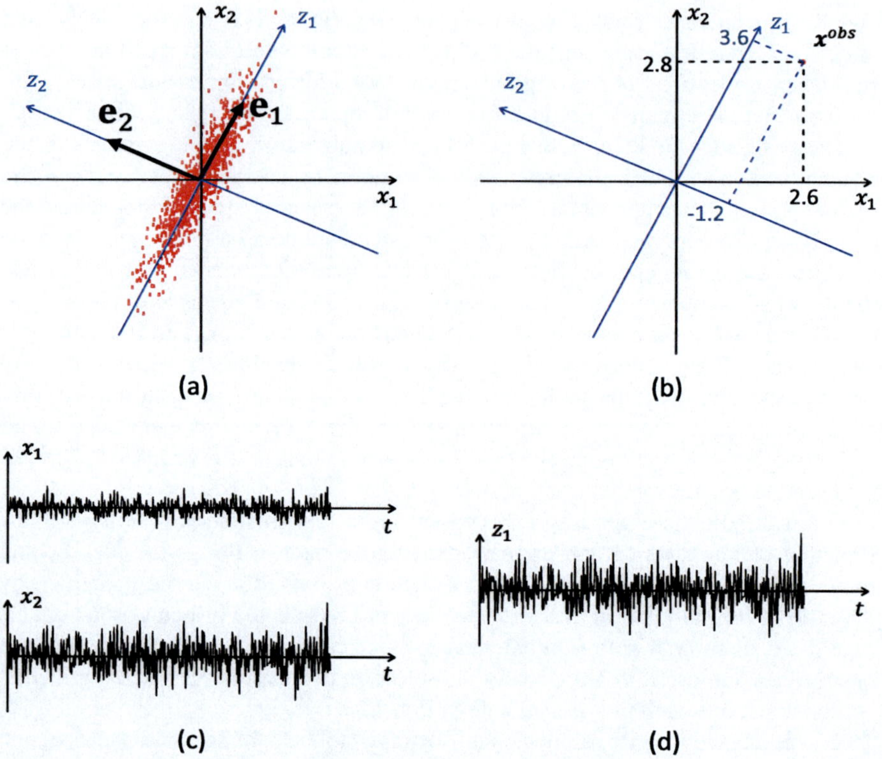

Fig. 9.2 Illustration of principal component analysis

$$\text{the first pair}: \quad \lambda_1 = 1.21, \qquad \mathbf{e}_1 = \begin{pmatrix} 0.42 \\ 0.91 \end{pmatrix},$$

$$\text{the second pair}: \quad \lambda_2 = 0.039, \qquad \mathbf{e}_2 = \begin{pmatrix} -0.91 \\ 0.42 \end{pmatrix}.$$

The two largest variance directions are represented by the two eigenvectors, $\mathbf{e}_1$ and $\mathbf{e}_2$, respectively, as marked in Fig. 9.2a. The eigenvalues are in fact the variances along the respective directions. The total value of eigenvalues, $1.21 + 0.039 = 1.249$, is the total variance of the dataset. The relative variance explained by the first principal component is

$$\frac{\text{variance due to the first principal component}}{\text{total variance}} = \frac{1.21}{1.249} = 96.8\%.$$

If one wants to reduce the original two dimensional space into a single dimension, then using the first principal component is considered a good approximation, as using it accounts for nearly 97% of the variance in the original data, or loosely speaking, using the first principal component represents almost 97% of the information contained in the original data.

In the two-dimensional space, PCA has a clear geometric interpretation. What PCA does is to rotate the coordinates in which the original data resides to a new coordinate system, in which the new horizontal axis is along the first principal direction, $\mathbf{e}_1$, and the new vertical axis is along the second principal direction, $\mathbf{e}_2$. Suppose that in the new coordinate system, the new variables are denoted by z_1 and z_2. Each data point, once projected onto the new coordinate, has their z_1 and z_2 values; these values are known as the *scores* of the data point's corresponding principal components. For instance, the data point, $\boldsymbol{x}^{obs} = (2.6, 2.8)^T$, in the original space, once projected onto the principal component coordinate, its two coordinates become $z_1 = 3.6$ and $z_2 = -1.2$; see Fig. 9.2b. These two coordinate values are obtained by using the following formula, which is to materialize the projection,

$$z_i = \mathbf{e}_i^T(\boldsymbol{x} - \boldsymbol{\mu}_x), \quad i = 1, 2. \tag{9.2}$$

Obviously, the same projection can be performed on every data point in the original dataset, yielding the corresponding principal component scores.

Figure 9.2c presents the two time series of x_1 and x_2, which are the data in the original space. Let us use PCA to reduce the data dimension to one, and then keep the first principal component and discard the second principal component. Figure 9.2d plots the first principal component scores of the data points—this univariate time series is supposed to replace the original two time series in the subsequent analysis, fulfilling the objective of reducing the number of variables to be monitored.

9.2 Detection of Size Changes

We start with the detection task for the size change only case. In this section, the size change model is based on that presented in Sect. 7.3.1. As the model in Sect. 7.3.1 is a retrospective model, the analysis presented in this section is naturally a Phase I analysis.

Recall that the pdf of the normalized particle size is represented in Eq. (7.10) through a B-splines model, rewritten below as

$$\log f_t(x_i) := \eta_{it} = \sum_{j=1}^{n} \alpha_{jt} B_j(x_i),$$

and the estimated NPSD is denoted by $\hat{f}_t(x)$. Once estimated, the NPSD is available as a vector at each t, i.e., $\mathbf{f}_t = [\hat{f}_t(x_1), \ldots, \hat{f}_t(x_m)]^T$, so that detecting a change point in $\hat{f}_t(x)$ amounts to a multivariate detection problem. Reducing the dimension of $\mathbf{f}_t$ helps the task of detection, and for that, Qian et al. (2017) use PCA, much as what we explain in Sect. 9.1.3.

9.2.1 Size Detection Approach

When applying PCA to the NPSD of Video 1 (see Sect. 7.1.2 for data description), Qian et al. (2017) find that only the first principal component (PC) is significant. Figure 9.3 plots the first 10 eigenvalues as well as the scores of the 1st and 2nd PCs. The eigenvalue of the 1st PC is much larger than that of the other PCs. In fact, the 1st PC explains 86.5% of the total variance of the original data. In addition to considering the numerical percentage of the 1st PC, one may also notice that the score of the 1st PC exhibits a clear pattern, whereas that of the 2nd PC appears random, reassuring the soundness of the decision of using only the 1st PC for detection. The scores of the 1st PC is denoted by $\hat{p}_t$.

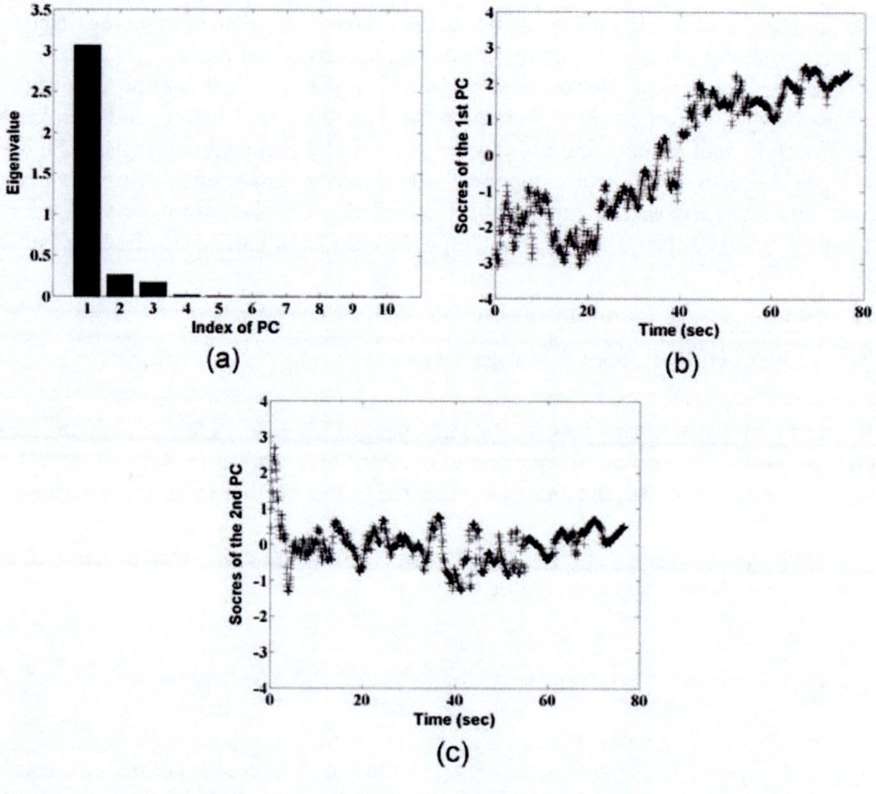

Fig. 9.3 PCA of the NPSD: (**a**) the eigenvalues corresponding to the first ten PCs; (**b**) the scores of the first PC; (**c**) the scores of the second PC. (Reprinted with permission from Qian et al. 2017)

We do want to stress that there can be more than one significant PCs. In that case, one could apply a multivariate change point detection method on the vector formed by significant PCs of a much reduced dimension (Zamba and Hawkins 2006) or conduct multiple univariate detections on individual PC scores simultaneously.

Without knowing the exact number of possible change points in the process, a popular treatment, known as the binary segmentation process (Yao 1988, BSP), is to detect the most significant change point first and then continue applying the same detection method to the two subsequences split at the detected change point. The dominating criterion used in the existing BSP methods for deciding the number of the change points is the Bayesian information criterion (Schwarz 1978, BIC). Qian et al. (2017) find that should they apply the BSP method to the nano video data, with BIC as the stopping criterion, it could find more than 400 change points, obviously over-segmenting the nanocrystal growth trajectory. Qian et al. (2017) also try some state-of-the-art multiple change point detection methods, such as the pruned exact linear time (Killick et al. 2012, PELT) and the wild binary segmentation (Fryzlewicz 2014, WBS) but those methods still return more change points than what the underlying physical principles can explain (eight change points when using PELT and 49 when using WBS).

Apparently, one needs to reduce the number of change points detected to be consistent with the physical understanding. Qian et al. (2017) find that a robust criterion to select the change points for the nano video data is the reduction rate in the sum of squared errors (SSE) of the piecewise constant model before and after a change point is added. Recall that NPSD is supposed to stay stable within each growth stage so that the scores of NPSD's principal component fluctuate around a constant within a growth stage. If all the change points are correctly identified, the piecewise constant model that fit the scores of the NPSD's principal component should produce the lowest SSE.

Once an existing detection method is used to produce a set of candidate change points (Qian et al. (2017) uses PELT, as PELT returns the fewest change points among the methods explored), one can start with a constant model for the entire process and then test each of those change point candidates. Pick the first potential change point to be the place where the largest reduction of SSE can be achieved by using two piecewise constant models. If the reduction of SSE is large enough, Qian et al. (2017) believe this change point is genuine and will continue the selection process. Then, visit all the remaining candidates to find the next change point which gives the largest reduction rate of SSE. Repeat the same steps until the reduction of SSE is no longer significant, suggesting that the difference between the two piecewise constant models is most likely due to random noise rather than to a substantial change in the process. The detailed steps are described as follows.

Suppose that one has already found $c-1$ change points, denoted as $\hat{t}_1, \ldots, \hat{t}_{c-1}$, while there are g remaining candidates, denoted as $\tilde{t}_1, \ldots, \tilde{t}_g$. The next possible change point chosen from $\tilde{t}_1, \ldots, \tilde{t}_g$ is denoted as t_c. They together segment the whole data sequence into $c+1$ subsequences, denoted by S_e, where $e = \{1, \ldots, c+1\}$ and S_e is the set containing the sequence of the data in the eth segment. The overall SSE of the piecewise constant model fitting of $\hat{p}_t$ is computed as:

$$V(\hat{t}_1, \ldots, \hat{t}_{c-1}, t_c) = \sum_{e=1}^{c+1} \sum_{t \in S_e} (\hat{p}_t - b_0^{(e)})^2, \tag{9.3}$$

where $b_0^{(e)}$ is the mean of $\hat{p}_t$'s in S_e. The position of the next potential change point is determined by:

$$\hat{t}_c = \arg \min_{t_c \in (\tilde{t}_1, \ldots, \tilde{t}_g)} V(\hat{t}_1, \ldots, \hat{t}_{c-1}, t_c). \tag{9.4}$$

Once the c-th change point is decided, Qian et al. (2017) suggest deleting $\hat{t}_c$ from the candidate set of $\{\tilde{t}_1, \ldots, \tilde{t}_g\}$ and continuing the selecting process, until there is no remaining change point candidate. Qian et al. (2017) try this on Video 1 data. When using the PELT method, they find eight candidate change points, as shown in Fig. 9.4a. Figure 9.4b presents the profile of the SSE, $V(\hat{t}_1, \ldots, \hat{t}_c)$, in which $\hat{t}_{(1)}, \ldots, \hat{t}_{(8)}$ represents the order of the selection. Qian et al. (2017) deem a potential change point, $\hat{t}_c$, a genuine change point if the reduction rate of SSE is larger than a threshold κ:

$$\frac{V(\hat{t}_1, \ldots, \hat{t}_{c-1}) - V(\hat{t}_1, \ldots, \hat{t}_c)}{V(\hat{t}_1, \ldots, \hat{t}_{c-1})} > \kappa, \tag{9.5}$$

In other words, if including $\hat{t}_c$ reduces the SSE by more than κ, Qian et al. (2017) consider that the change point is due to a true process change rather than random noise. Then Qian et al. (2017) continue the selection for the next potential change point. If the criterion in Eq. (9.5) is not satisfied for all remaining candidate change points, Qian et al. (2017) consider that all the significant change points have been found and thus stop the process.

Borrowing the language used in Sect. 9.1, the ratio in the left hand of the inequality in Eq. (9.5) is the test statistic, and κ is the upper control limit. In this case, one uses only the upper control limit, as the lower control limit is simply zero.

The same strategy can also be applied to a simple statistic, such as the median or mean particle size, and then the detection problem becomes univariate. As the median radius is less sensitive to outliers, Qian et al. (2017) apply the aforementioned method to the median particle size $\tilde{r}_t$. Unlike the NPSD, which remains relatively stable without a change point, $\tilde{r}_t$ exhibits an increasing trend along the growth process, so that one needs to revise the detection process to handle the trend.

Qian et al. (2017) adopt the `strucchange` package in `R` (Kleiber et al. 2002), which detects change points after a regression. Using this package finds 15 candidate change points in the time series of the median particle size. To select the significant change points, Qian et al. (2017) first apply a de-trending operation before performing the change point detection. Following the idea presented in Chen and Gupta (2001) and Qian et al. (2017) use a linear model to de-trend the median

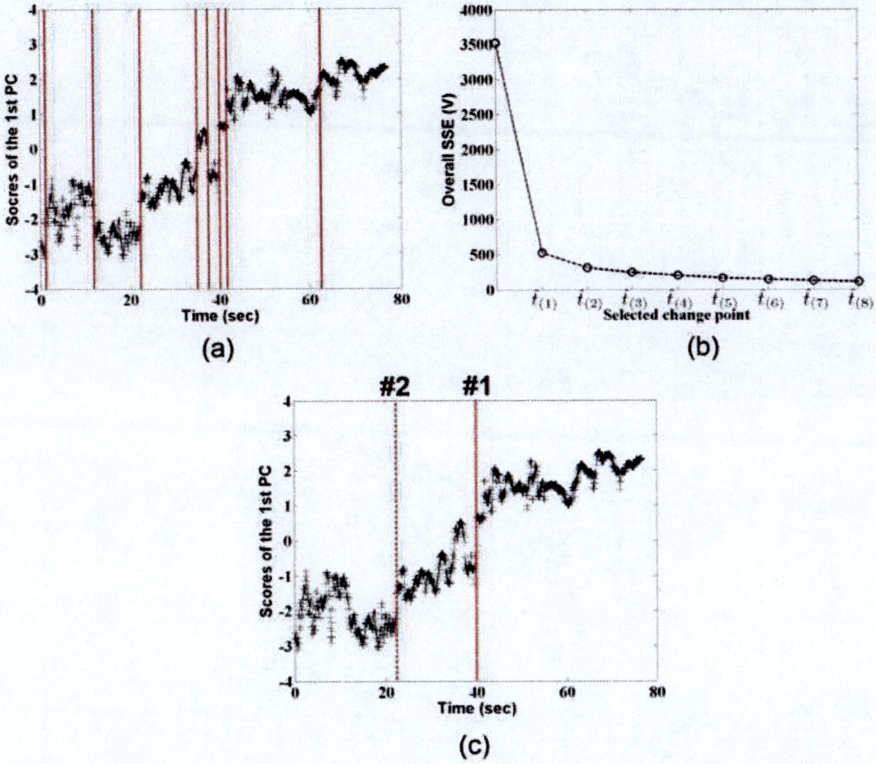

Fig. 9.4 Results of the change point detection of NPSD: **(a)** eight potential change point candidates detected by PELT; **(b)** Change in $V(\cdot)$ when selecting a change point at a time; **(c)** Two significant change points are detected when κ is varied in the range of (0.2, 0.8). Points #1 and #2 mark their positions. (Reprinted with permission from Qian et al. 2017)

particle size. Qian et al. (2017) revise the SSE by using the residuals after fitting a piecewise linear trend model, as follows:

$$V(\hat{t}_1, \ldots, \hat{t}_{c-1}, t_c) = \sum_{e=1}^{c+1} \sum_{t \in S_e} (\tilde{r}_t - b_0^{(e)} - b_1^{(e)} t)^2, \tag{9.6}$$

where $b_0^{(\cdot)}$ and $b_1^{(\cdot)}$ are the coefficients of the respective linear model. After the definition of SSE is revised, the rest change point detection procedure, devised earlier in the context of NPSD, can be adapted for selecting the significant change points in $\tilde{r}_t$.

Figure 9.5 present the intermediate and final detection results in the same case (Video 1) while using the median size.

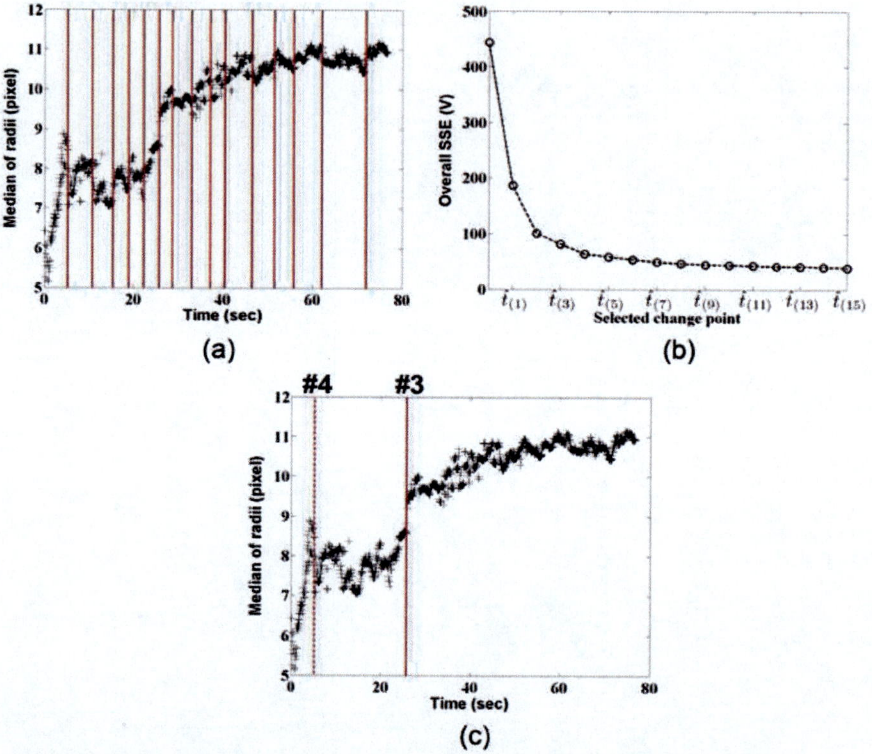

Fig. 9.5 Results of change point detection using median particle size: (**a**) 15 potential change points detected by the `strucchange` package; (**b**) Change in $V(\cdot)$ when selecting a change point at a time; (**c**) Two significant change points are detected when κ is varied in the range of (0.2, 0.8). Points #3 and #4 mark their positions. (Reprinted with permission from Qian et al. 2017)

The key tuning parameter in the above-described change detection procedure is κ. In the application to nano video data, Qian et al. (2017) set $\kappa = 0.5$ for both NPSD and the median particle size. The choice of $\kappa = 0.5$ implies that Qian et al. (2017) deem a candidate change point genuine if its selection reduces the SSE by half or more. By using this value for κ, Qian et al. (2017) detect one change point in NPSD and another one in median size; the two change points are shown as, respectively, "#1" in Fig. 9.4c and "#3" in Fig. 9.5c. Qian et al. (2017) further recommend to use 0.5 as the default value for κ.

Setting $\kappa = 0.5$, the change point detection method produces two change points: at 25.8s in $\tilde{r}_t$ ("#3" in Fig. 9.5c) and at 39.9 s in NPSD ("#1" in Fig. 9.4c). These two change points segment the whole sequence into three stages: (0, 25.8 s), (25.8, 39.9 s) and (39.9, 76.6 s). The delineated stages make it immediately clear how the nanocrystals grow—they go through two major growth stages with a transition stage in between. For this particular process, the two dominating mechanisms have been

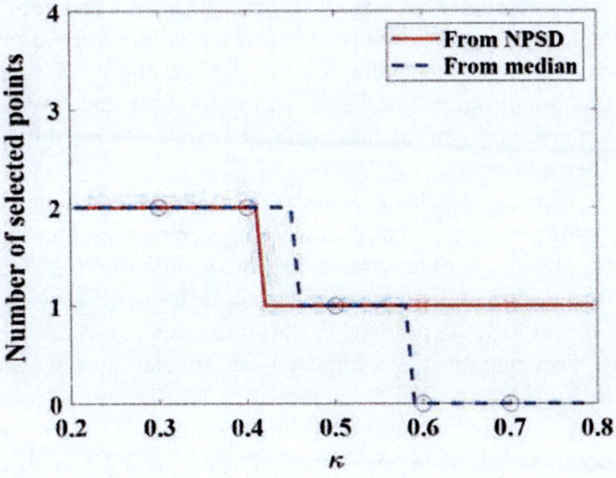

Fig. 9.6 The number of change points detected in NPSD and median particle size when κ varies in (0.2, 0.8). (Reprinted with permission from Qian et al. 2017)

studied and understood (Zheng et al. 2009; Wang et al. 2013; Zhang et al. 2010; Bian et al. 2013). According to these domain science studies, in the period of (0, 25.8 s), the orientated attachment mechanism dominates, whereas in the period of (39.9, 76.6 s), the Ostwald ripening mechanism dominates. It is understandable that the mechanism change does not happen suddenly. As one mechanism gradually takes over from the other, a short transition period naturally exists, which is the period of (25.8, 39.9 s) in this process.

9.2.2 *Sensitivity of Control Limit* κ

Considering the critical role played by κ, Qian et al. (2017) conduct a sensitivity analysis. Figure 9.6 shows the number of change points detected in both NPSD and median size, as κ varies in the range of (0.2, 0.8). The NPSD-based detection produces either one change point or two change points. The first change point detected in the two-point outcome is the same as the change point detected in the single-point case, shown as "#1" in Fig. 9.4c. The second change point is shown as "#2" in the same figure. The median size based detection is more sensitive to the value of κ—it could produce from zero to two change points over the same κ range. The two change points that could have been detected are marked as "#3" and "#4", respectively, in Fig. 9.5c. Aside from the sensitivity issue, another drawback of using the median size statistic is that one would not be able to detect "#1" unless setting κ to some extreme value (like 0.1). Given the analysis done by Zheng et al. (2009), it is known that a stage change indeed occurred around the time of "#1," so that missing this change point is a serious limitation.

When looking closely at the four possible change points, it is apparent that the change points "#2" and "#3" are the outcome of the same change, as their time stamps are only 3.2 s apart. By merging "#2" and "#3," the change point detection outcomes could possibly segment the whole growth into four stages, three stages, or two stages, depending on the specific choice of κ. But an important message is that the difference in the detection outcome does not lead to a drastically different understanding of the basic science behind. To see this point, consider the following alternatives.

When a smaller κ is used, all four change points could have been detected. Having "#4" apparently suggests the existence of an initial nucleation stage, which is generally hard to observe because its duration is short, data variability is high, and the number of nanocrystals is small. Missing this initial stage is understandable and not seriously detrimental to the subsequent analysis.

Had one chosen a large κ (say, greater than 0.6), only one change point (#1 in Fig. 9.4c) would be detected in NPSD and no change point in median size. Consequently, the transition stage could have been missed. Despite that, one would not miss out the big picture of two dominating growths, i.e., OA and OR.

The overall analysis shows that NPSD-based detection outcome is robust, as it captures the important change points consistently in a broad range of the control limit value. To avoid missing potentially important change points in future applications, one should vary κ in a reasonable range and then choose a manageable number of change points.

The fact that NPSD-based detection produces a rather robust detection separating the whole growth trajectory into two major stages speaks to the benefit of having such a detection approach. Had researchers not known *a priori* the mechanism change in the nanocrystal growth process, this detection outcome would hint strongly where to explore so as to expedite the discovery process.

9.2.3 *Hybrid Modeling*

Since the dominating growth mechanisms for the process captured by Video 1 have been studied by the domain experts (Zheng et al. 2009), one can adopt the existing first principle models for each respective growth stage and then use an interpolation to model the transition period. Qian et al. (2017) produce such a unified growth model, a hybrid of the first principle-based model and the empirical model, for the whole nanocrystal growth trajectory.

Taken from the work of Aldous (1999), the models for NPSD, $\hat{f}_t(x)$, and the mean particle size, $\bar{r}_t$, during the OA growth in the first stage of (0, 25.8 s) are, respectively,

$$\begin{aligned} f_{t_{\mathrm{OA}}}(x) &= \frac{2W_{\mathrm{OA}}}{\Gamma(a_{\mathrm{OA}}+1)}(W_{\mathrm{OA}}x)^{2a_{\mathrm{OA}}+1}e^{-(W_{\mathrm{OA}}x)^2}, \\ \bar{r}_{t_{\mathrm{OA}}}^{2(a_{\mathrm{OA}}+1)} &= b_{\mathrm{OA}}(t - t_{\mathrm{OA}}), \end{aligned} \tag{9.7}$$

where $W_{OA} = (a_{OA} + 1)\Gamma(a_{OA} + 3/2)/\Gamma(a_{OA} + 1)$ and $\Gamma(\cdot)$ is the gamma function defined as $\Gamma(z) = \int_0^\infty x^{z-1}e^{-x}dx$. Altogether there are three parameters used in the two models, i.e., a_{OA}, indicating the variance of the process (a larger a means, however, a smaller variance), b_{OA}, indicating the growth rate, and t_{OA}, indicating the initial size of nanocrystals.

The kinetics of OR growth in the third stage of (39.9, 76.6 s) is usually described by the LSW model (Lifshitz and Slyozov 1961). Qian et al. (2017) do choose to use the LSW model to represent the mean particle size ($\bar{r}_t$) growth in the OR stage. For the $\bar{r}_t$ growth, the LSW model is to model the cube of $\bar{r}_t$ with a linear function. The model of $\bar{r}_t$ growth in the OR stage bears a similar appearance as the model of $\bar{r}_t$ in the OA stage but the key difference is the different power term on $\bar{r}_t$.

To model $\hat{f}_t(x)$ in the OR growth stage, however, Qian et al. (2017) find that the LSW model cannot obtain a good fit for $\hat{f}_t(x)$. Figure 9.7 presents a comparison of the empirical NPSDs, estimated at 45 and 70 s, with the NPSD derived from the

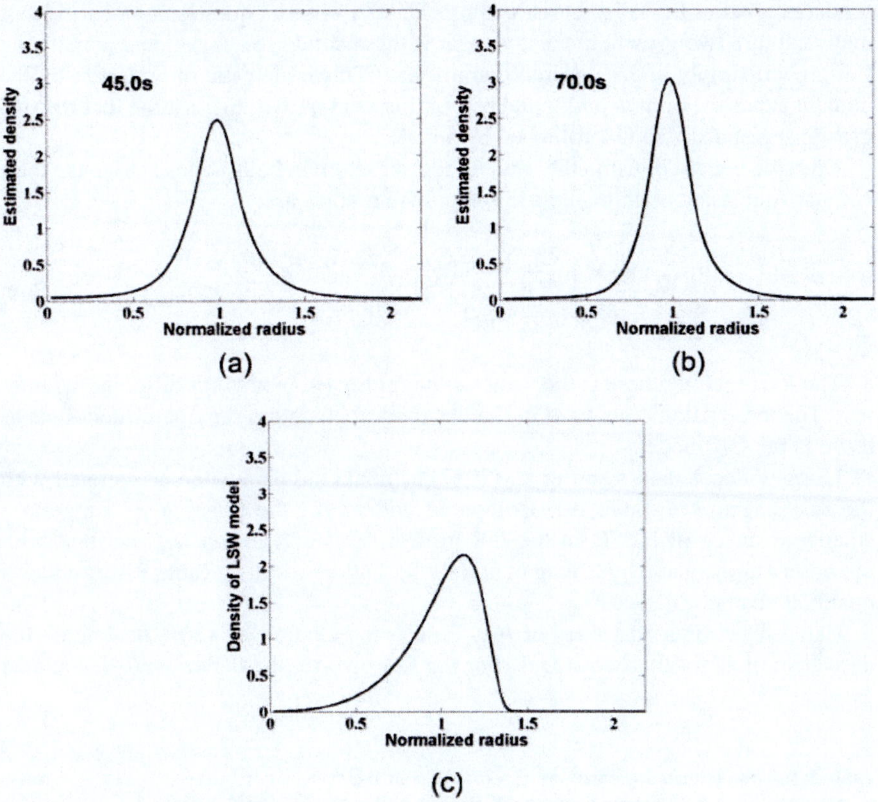

Fig. 9.7 (**a**) The empirical NPSD estimated at 45s; (**b**) The empirical NPSD estimated at 70s; (**c**) The theoretical NPSD derived from the LSW model. (Reprinted with permission from Qian et al. 2017)

LSW model. The two empirical NPSDs are similar, and both of them look rather symmetric. By contrast, the LSW-based NPSD is more skewed with a long lower tail and has larger variance compared with the empirical NPSDs. The long lower tail of the LSW-based NPSD presents a clear contrast with the NPSDs estimated directly from the data.

Qian et al. (2017) hypothesize that there may be two reasons for the mismatch. First, the smaller particles are difficult to track under the current resolution of the in situ TEM, yet the LSW model, with a long left tail, is more sensitive to the missed detection of these particles. Second, the LSW model has been known as inconsistent with some experimental results (Voorhees 1985), even before the study of Qian et al. (2017). For this reason, other researchers proposed modified models to improve the fitting accuracy (Hardy and Voorhees 1988; Lo and Skodje 2000; Baldan 2002), but when these models are tested against the TEM video data at hand, Qian et al. (2017) state that they do not produce more competitive fitting quality, either. Because of this, Qian et al. (2017) decide to use the OA growth model structure (derived by the Smoluchowski equation) to fit for NPSD in the OR growth; doing so indeed produces a better fit. This heuristic approach later raised a question—does Video 1 really capture two growth mechanisms or is the second growth segment actually an OA growth simply under different parameters? This issue calls for a debate by the domain experts. For now and in the rest of this section, we still assume that the two growth segments is an OA followed by an OR.

One side benefit of using the same model structure in both stages is to make their comparison easier. Specifically the OR growth models are:

$$\begin{aligned} f_{t_{\text{OR}}}(x) &= \frac{2W_{\text{OR}}}{\Gamma(a_{\text{OR}}+1)}(W_{\text{OR}}x)^{2a_{\text{OR}}+1}e^{-(W_{\text{OR}}x)^2}, \\ \bar{r}^3_{t_{\text{OR}}} &= b_{\text{OR}}(t - t_{\text{OR}}). \end{aligned} \tag{9.8}$$

The first equation here is the same as that in Eq. (9.7) but with different parameters. The three parameters used in the OR models share the same interpretations as those in the OA model.

Using Video 1 data, Qian et al. (2017) estimate the parameters associated with the two stages; see Table 9.1. Compared with a_{OA}, the larger a_{OR} suggests a smaller variance of NPSD in the OR growth. This conclusion is consistent with the observations made by Zheng et al. (2009), but the result in Table 9.1 provides a quantitative contrast.

Using the estimated values of b_{OA} and b_{OR}, Qian et al. (2017) calculate the derivative of $\bar{r}_t$ for the two stages. For the OA growth, the derivative is calculated as:

Table 9.1 The estimated parameters associated with the two stages in the nanocrystal growth captured by Video 1. (*Source*: Qian et al. 2017)

a_{OA}	b_{OA}	t_{OA}	a_{OR}	b_{OR}	t_{OR}
1.47	42.2	−429.3	7.31	0.55	−1342.7

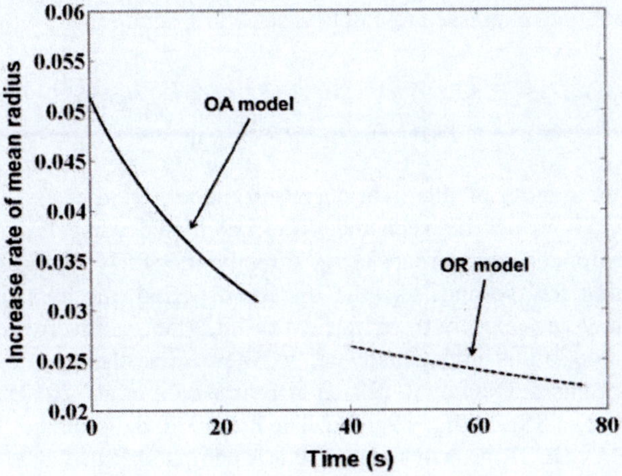

Fig. 9.8 The comparison of the first derivative of $\bar{r}_t$ in the OA and OR growth stages. (Reprinted with permission from Qian et al. 2017)

$$\frac{d\bar{r}_t}{dt} = \frac{1}{2(a_{OA}+1)} b_{OA}[b_{OA}(t - t_{OA})]^{\frac{1}{2(a_{OA}+1)}-1}, \tag{9.9}$$

and for the OR growth, the derivative is calculated as:

$$\frac{d\bar{r}_t}{dt} = \frac{1}{3} b_{OR}[b_{OR}(t - t_{OR})]^{\frac{1}{3}-1}. \tag{9.10}$$

Figure 9.8 compares the derivatives of the OA and OR growth. The gap between the two curves corresponds to the transition period in which no theoretical model is yet available. The two curves make it clear that in the nanocrystal growth, the mean radius growth rate in the OA stage is faster than that in the OR stage, just as the estimated b_{OA} and b_{OR} values suggested. This was again stated by Zheng et al. (2009), and again, the analysis here provides a quantitative picture of the mean radius evolution in the two stages.

The difference in t_{OA} and t_{OR} suggests that the initial nanocrystal sizes are different, and a more negative quantity implies a large initial size. The t_{OA} and t_{OR} values in Table 9.1 make perfect sense, as the OR growth follows the OA growth, so that the initial nanocrystals in OR have a bigger size.

To include the transition period between 25.8 and 39.9 s, Qian et al. (2017) introduce the weighting functions, $\gamma_N(t)$ and $\gamma_R(t)$, for NPSD and mean particle size, respectively, to combine the two stage-wise models of respective statistics (NPSD or mean particle size). The two weighting functions take the value of zero when $t < 25.8$ s, one when $t > 39.9$ s, and increase from zero to one quadratically in between, with their quadratic function coefficients fitted from the corresponding

NPSD or mean particle size in the transition period. The overall growth models of $f_t(x)$ and $\bar{r}_t$ are both expressed in a hybrid structure, such as:

$$\begin{aligned} f_t(x) &= (1 - \gamma_N(t)) f_{t_{\text{OA}}}(x) + \gamma_N(t) f_{t_{\text{OR}}}(x), \\ \bar{r}_t &= (1 - \gamma_R(t)) \bar{r}_{t_{\text{OA}}} + \gamma_R(t) \bar{r}_{t_{\text{OR}}}. \end{aligned} \tag{9.11}$$

To verify the quality of this hybrid growth model, Qian et al. (2017) show in Fig. 9.9a the SSE values between the $f_t(x)$ calculated using Eq. (9.11) and its empirically estimated counterpart using directly the TEM observations. Except in the beginning few seconds and the transition period, the hybrid model-based outcomes follow very closely the empirical results. The relatively worse fit during the transition period is understandable, as there lack theories to describe the transition mechanism. Qian et al. (2017) also fit Woehl et al. (2013)'s single-stage model and show its SSE in Fig. 9.9a, too. The hybrid model produces smaller SSE's for both the OA and OR growth stages and it is comparable to Woehl et al.'s model in the transition period.

The above learning results provide a quantitative model to describe the whole growth trajectory. Using the learned results, one can compute the evolution of particle size distribution (PSD, not normalized), $G_t(r)$, by using the hybrid model of $\bar{r}_t$ and $f_t(x)$, as:

$$G_t(r) = \frac{1}{\bar{r}_t} f_t\left(\frac{r}{\bar{r}_t}\right). \tag{9.12}$$

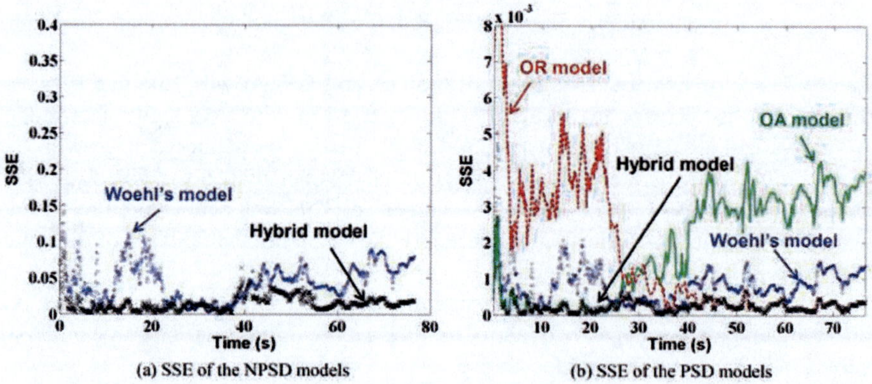

Fig. 9.9 The comparison of the NPSDs based on the first-principal models and the empirical estimation from the data: (**a**) the SSE curve between the hybrid model-based NPSD and the empirical NPSD, and the SSE curve between the Woehl's model-based NPSD and the empirical NPSD; (**b**) the SSE curves between either the hybrid model-based PSD, Whoehl's model-based PSD, OA model-based PSD, or OR model-based PSD, and the empirical PSD. (Reprinted with permission from Qian et al. 2017)

One can apparently also estimate the PSD directly from the TEM observations of the particle radius. The SSE curve between the model-based PSD and the empirically estimated PSD is shown in Fig. 9.9b. In addition to the PSD calculated using the hybrid model, Fig. 9.9b includes the SSE curves between the empirically estimated PSD and the PSDs calculated by using, respectively, Woehl et al. (2013)'s single-stage model, the OA growth model alone, and the OR growth model alone. The hybrid growth model fits the observed data consistently well throughout the entire growth trajectory, while other models appear to have deficiencies in certain periods.

9.3 Phase I Analysis of Shape Changes

In this and the subsequent sections, we discuss change point detection concerning nanoparticle shapes. Many of the descriptions here based on the shape model presented in Chap. 8.

9.3.1 *Recap of the Shape Model and Notations*

Suppose that J_t different individual objects and their outlines in $\mathbb{R}^2$ are taken for each time $t \in T$, where T is the set of the time indices that indicate the imaging times. Following the shape representation used in Sect. 8.5, we represent the jth outline taken at time t as a parametric curve $z_{tj} : \mathbb{S}^1 \mapsto \mathbb{C}$ in the form of Eq. (4.14). The parametric curve $z_{tj}(\theta)$ is then transformed to the centroid distance function,

$$\tilde{r}_{tj}(\theta) = |z_{tj}(\theta) - c_{tj}|, \theta \in \mathbb{S}^1,$$

where $c_{tj} = \int_{\mathbb{S}^1} z_{tj}(\theta) d\theta$ represents the centroid of z_{tj}. In digital images, the outline is only observed at a finite number of locations $\theta \in \Theta_{tj}$. The available data are thus the set of the discrete observations, namely

$$\tilde{\boldsymbol{r}}_{tj} := \{\tilde{r}_{tj}(\theta), \theta \in \Theta_{tj}\}.$$

The centroid distance functions are rotationally aligned to remove any effect of object rotation on the representation, using the general Procrustes analysis described in Sect. 8.5. Let r_{tj} denote the rotationally aligned version of $\tilde{r}_{tj}$, and $\boldsymbol{r}_{tj}$ represent the discrete observations of the aligned function,

$$\boldsymbol{r}_{tj} := \{r_{tj}(\theta), \theta \in \Theta_{tj}\}.$$

The full set of the aligned centroid distance function data is denoted by $\mathcal{X} = \{\boldsymbol{r}_{tj}; t \in T, j \in J_t\}$.

A nanoparticle growth process is characterized by the random expansion of a nanoparticle's outline in time. In Chap. 8, the expansion of an outline is modeled as a monotonity of the centroid distance function of the outline. The centroid distance represents the distance from the centroid of the outline to the points on the outline. As the outline expands, the centroid distances monotonically increase. Let $r(\theta, t)$ represent the random centroid distance function value of an object outline observed at time t at a point $\theta \in \mathbb{S}^1$ on the outline. It can be represented by a B-spline basis expansion representation,

$$r(\theta, t) = \sum_{m=1}^{M} \sum_{i=1}^{n} \alpha_{m,i} \phi_m(t) \gamma_i(\theta), \, t \geq 0 \text{ and } \theta \in \mathbb{S}^1, \tag{9.13}$$

where the product B-spline basis functions $\phi_m(t)\gamma_i(\theta)$ are fixed, and the random coefficients $\alpha_{m,i}$'s determines the random function. This representation is essentially the same as Eq. (8.15). For the change detection task, Park and Shrivastava (2014) use a B-splines of order two, i.e. they use piecewise quadratic polynomials as its basis functions, and choose $M = 4$ and $n = 12$. As $r(\theta, t)$ must be a periodic function in θ, Park and Shrivastava (2014) use a uniform periodic B-spline basis functions for $\gamma_i(\theta)$ (Mortenson 1985). Because $r(\theta, t)$ must also be a monotonically increasing function in t, it implies that the first derivative of $r(\theta, t)$ with respect to t should be positive, i.e., for a fixed ω,

$$\frac{\partial r}{\partial t} = \frac{1}{h} \sum_{m=1}^{M-1} \sum_{i=1}^{n} (\alpha_{m+1,i} - \alpha_{m,i}) \phi_m(t) \gamma_i(\theta) \geq 0, \forall t, \theta, \tag{9.14}$$

where h is the distance between the uniformly spaced knots used for constructing the B-spline basis functions of $\phi_m(t)$'s. Since all B-spline basis functions are positive, i.e. $\gamma_i(\theta) \geq 0$ and $\phi_m(t) \geq 0$, a sufficient condition for monotonic increment of $r(\theta, t)$ is

$$\alpha_{m+1,i} \geq \alpha_{m,i}, \quad m = 1, \ldots, M-1 \text{ and } i = 1, \ldots, n. \tag{9.15}$$

Recall the same condition is expressed in Eq. (8.17) of Chap. 8.

9.3.2 Mixture Priors for Multimode Process Characterization

Let $\boldsymbol{\alpha}$ be an $(M-1)n \times 1$ column vector of random coefficients of $\{\alpha_{m,i}\}$. Considering the sufficient condition for monotonicity, the same as in Sect. 8.5, Park and Shrivastava (2014) restrict the random coefficients to a truncated prior, i.e.,

$$\boldsymbol{\alpha} \sim \mathbb{1}_Q P, \tag{9.16}$$

where $Q := \{\boldsymbol{\alpha} : \alpha_{m+1,i} \geq \alpha_{m,i}, m = 1, \ldots, M-1, i = 1, \ldots, n\}$ and P is a probability measure over the sigma field of $\mathbb{R}^{(M-1)n}$.

The simplest choice for probability measure, P, is a multivariate parametric distribution, e.g., a multivariate normal distribution. However, as argued in Sect. 7.1.1, continuous physical processes such as the nanoparticle self-assembly process often exhibits multiple modes of changes throughput the growth process. It is easy to imagine that if one chooses a parametric distribution which has a single mode in its density for P, the resulting process of $r(\theta, t)$ also has single mode in its density, suggesting that only a single mode of process change is expected. This is rather restrictive and does not meet the practical needs. It is more appropriate to choose a finite mixture of parametric distributions or use a nonparametric form of probability distributions for P, like the treatment undertaken in both Chaps. 7 and 8. But unlike the technique used in Chap. 7, a Bayesian nonparametric form of P (Ferguson 1983) is used here.

In the Bayesian nonparametric approach, the probability measure, P, is random in and by itself. A prior distribution over the probability measure P is modeled as a Dirichlet process with a concentration parameter $\eta_0 > 0$ and a base probability measure, P_0. The base probability measure is chosen to be a Gaussian density, $\mathcal{N}(0, \sigma_0^2 \boldsymbol{I})$.

This prior is denoted by $P \sim \mathrm{DP}(\eta_0 P_0)$. A more intuitive description of the Dirichlet process prior on P is the stick breaking representation of the Dirichlet process prior (Sethuraman 1994), which lets P be represented in an infinite mixture form, so that

$$P = \sum_{s=1}^{\infty} \omega_s \delta_{\boldsymbol{\alpha}_s}, \tag{9.17}$$

where $\boldsymbol{\alpha}_s$ follows an i.i.d. truncated $\mathcal{N}_Q(0, \sigma_0^2 \boldsymbol{I})$, $\omega_s = \zeta_s \prod_{s'<s}(1 - \zeta_{s'})$ with $(\zeta_1, \zeta_2, \ldots)$ being i.i.d. Beta$(1, \eta_0)$, and $\delta_{\boldsymbol{\alpha}_s}$ is a Dirac delta function of $\boldsymbol{\alpha}$—its value is one if $\boldsymbol{\alpha} = \boldsymbol{\alpha}_s$ and zero otherwise. Applying Eqs. (9.17)–(9.16) leads to that $\boldsymbol{\alpha}$ has the infinite mixture of truncated normal distributions as its prior, i.e.,

$$\boldsymbol{\alpha} \sim \sum_{s=1}^{\infty} \omega_s \delta_{\boldsymbol{\alpha}_s}, \tag{9.18}$$

In the mixture model, the random vector, $\boldsymbol{\alpha}_s$, fully describes the s-th mode of nanoparticle growth, and the random weight, ω_s, represents the proportion of the s-th mode in the mixture. Estimating the two random components, namely $\boldsymbol{\alpha}_s$ and ω_s, yields the full characterization of a multimode growth.

9.3.3 Block Gibbs Sampler

Given the training dataset expressed in $\mathcal{X} = \{\boldsymbol{r}_{tj}; t \in T, j \in J_t\}$, the specific objective of this Phase I analysis is to characterize the in-control process condition by obtaining the Bayesian estimates of $\boldsymbol{\alpha}_s$'s and ω_s's. This is to say, one assumes the out-of-control data points have been weeded out and all data are in control in the training set, so that the task left in the Phase I analysis is the parameter estimation for the in-control process.

The observed data are naturally considered as a noisy realization of the stochastically expanding particle profile, i.e.,

$$r_{tjk} = r(\theta_{tjk}, t) + \epsilon_{tj}(\theta_{tjk}), \quad \forall t \in T, j \in J_t, \theta_{tjk} \in \Theta_{tj}, \tag{9.19}$$

where r_{tjk} is the k-th element of $\boldsymbol{r}_{tj}$ and $\epsilon_{tj}(\theta_{tjk})$ is assumed an i.i.d $\mathcal{N}(0, \sigma^2)$ with $\sigma^2 \sim \text{IG}(\varphi_0, \lambda_0)$, where "IG" means an inverse-gamma distribution. The model can be written in a matrix/vector form of

$$\boldsymbol{r}_{tj} = \boldsymbol{\Phi}_{tj}\boldsymbol{\alpha} + \boldsymbol{\epsilon}_{tj}, \quad \forall t \in T, j \in J_t,$$

where $\boldsymbol{\Phi}_{tj}$ is the matrix of the B-spline basis functions with its (k, im)-th element equal to $\phi_m(t)\gamma_i(\theta_{tjk})$, and $\boldsymbol{\epsilon}_{tj}$ is the vector of $\epsilon_{tj}(\theta_{tjk})$'s. If a nanoparticle j observed at time t follows growth mode s then

$$\boldsymbol{r}_{tj} = \boldsymbol{\Phi}_{tj}\boldsymbol{\alpha}_s + \boldsymbol{\epsilon}_{tj}, \quad \forall t \in T, j \in J_t. \tag{9.20}$$

Park and Shrivastava (2014) introduce a hidden mixture component variable, s_{tj}, for indicating the mode of growth that nanoparticle j at time t follows. In other words, $s_{tj} = s$ implies that the growth of nanoparticle j observed at time t is governed by growth mode s. Conditioned on s_{tj}, the conditional likelihood is

$$L(\boldsymbol{r}_{tj}|s_{tj}, \{\boldsymbol{\alpha}_s\}, \sigma^2) = f_G(\boldsymbol{r}_{tj}; \boldsymbol{\Phi}_{tj}\boldsymbol{\alpha}_{s_{tj}}, \sigma^2\boldsymbol{I}), \tag{9.21}$$

where $f_G(\boldsymbol{x}; \boldsymbol{\mu}, \boldsymbol{\Sigma})$ is a multivariate Gaussian density function with mean $\boldsymbol{\mu}$ and covariance $\boldsymbol{\Sigma}$. With the mixing distribution $P(s_{tj} = s) = \omega_s$ from Eq. (9.17), the joint likelihood becomes

$$L(\boldsymbol{r}_{tj}, s_{tj} = s|\{\boldsymbol{\alpha}_s, \omega_s\}, \sigma^2) = \omega_{s_{tj}} f_G(\boldsymbol{r}_{tj}; \boldsymbol{\Phi}_{tj}\boldsymbol{\alpha}_{s_{tj}}, \sigma^2\boldsymbol{I}). \tag{9.22}$$

One needs to obtain the posterior estimate of $\{\boldsymbol{\alpha}_s\}$ for characterizing the s-th growth mode as well as the mixing proportion for that mode, $\{\omega_s\}$.

The posterior estimation can be attained by using a Gibbs sampling approach, developed for a general Dirichlet process mixture model. Park and Shrivastava (2014) follow the exact block Gibbs sampler algorithm (Papaspiliopoulos 1994), which combines the advantages of the Dirichlet process mixture slice sampler

(Walker 2007) and the retrospective MCMC sampler (Papaspiliopoulos and Roberts 2008). The detail of this approach is discussed in Sect. 8.5.2 and the resulting algorithm is in fact Algorithm 8.1.

In the exact Gibbs sampler as outlined in Algorithm 8.1, σ_0^2 is fixed to one. The choice of σ_0^2 does not affect the sampling result much but only affect Step C in the block Gibbs sampler via the ratio of σ^2/σ_0^2. Although σ_0^2 is fixed, the ratio of σ^2/σ_0^2 is adjusted as a result of the changing σ^2 in Step D. The other tuning parameter, η_0, is also fixed to one. Park and Shrivastava (2014) state that using this value leads to a faster convergence of the Gibbs sampler. The hyperparameters for the inverse-gamma prior on σ^2, i.e., φ_0 and λ_0, are both chosen to be 0.1. The choice might need to be tuned when a different nano video data is analyzed. If a larger σ^2 is expected, then smaller φ_0 and λ_0 should be used.

Let $\{\boldsymbol{\alpha}_s^{(q)}, \omega_s^{(q)}; s = 1, \ldots, S^{(q)}\}$ be the posterior samples, taken at the q-th iteration of the exact block Gibbs sampler, of the mixture component parameters ($\boldsymbol{\alpha}_s$ and ω_s) and the number of mixture components (S). Given the posterior samples, one is interested in estimating the posterior means and variances of the mixture component parameters.

In the estimation process of mixture component parameters, the label switching problems can be serious. To address this issue, Park and Shrivastava (2014) first apply the relabeling algorithm of Stephens (1997) to minimize the label switching. The relabeling algorithm does a re-indexing of $\{\boldsymbol{\alpha}_s^{(q)}, \omega_s^{(q)}; s = 1, \ldots, S^{(q)}\}$ for making sure that $\boldsymbol{\alpha}_s^{(q)}$'s and $\omega_s^{(q)}$'s are sampled from the same mixture component. After the relabeling, the posterior mean and variance estimates of the random coefficients, $\boldsymbol{\alpha}_s$, and those of the mixing proportion, ω_s, are simply the sample average and sample variance, respectively, of the corresponding posterior samples.

Let $\hat{\boldsymbol{\alpha}}_s$ denote the posterior mean estimate of $\boldsymbol{\alpha}_s$, and let $\boldsymbol{\phi}_{t,\theta}$ represent the $(M-1)n \times 1$ column vector of $\phi_m(t)\gamma_i(\theta)$'s evaluated at a testing location θ and at time t. Then, the posterior mean of $r_s(t,\theta)$, the centroid distance function of mode s, is $\boldsymbol{\phi}_{t,\theta}^T \hat{\boldsymbol{\alpha}}_s$.

The pointwise confidence interval of $r_s(t,\theta)$ can be estimated by proper quantiles of the samples of $r_s(t,\theta)$. For the q-th iteration, draw B samples from $\mathcal{N}(\boldsymbol{\phi}_{t,\theta}^T \boldsymbol{\alpha}_s^{(q)}, \sigma^2)$. The b-th sample is denoted by $r_s^{(q,b)}(\theta,t)$ for $b = 1, \ldots, B$. Given the type-I error of ν, take the $(1-\nu/2)$-th and $(\nu/2)$-th sample quantiles of $\{r_s^{(q,b)}(\theta,t); \forall q, b\}$ as the upper limit and the lower limit of the $(1-\nu)$ confidence interval for $r_s(\theta,t)$.

9.4 Phase II Analysis of Shape Changes

The Phase II analysis is the test to decide whether a process is in control, given a new set of quality characteristic measurements from the process.

Park and Shrivastava (2014) consider the following statistical hypothesis test based on a new set of nano data observed at time t. The newly arrived data is denoted by $\mathcal{X}^t = \{\dot{\boldsymbol{r}}_{tj}; j \in \dot{J}_t\}$, to differentiate them from the historical training data. Following the notation convention used in Sect. 9.3, $\dot{\boldsymbol{r}}_{tj}$ is the random vector of the aligned centroid distanec function for nanoparticle j observed at time t. Assume that $\dot{\boldsymbol{r}}_{tj}$ follows a probability density of f_{tj}. Then, the hypothesis test is stated as

$$\begin{aligned} H_0: &\quad f_{tj} = f_{tj}^{(0)} \\ H_1: &\quad f_{tj} \neq f_{tj}^{(0)}, \end{aligned} \tag{9.23}$$

where $f_{tj}^{(0)}$ is a probability density function of $\dot{\boldsymbol{r}}_{tj}$ obtained under the in-control process condition, i.e., during the Phase I analysis.

In a Bayesian hypothesis test, one first assumes the prior beliefs for the two hypotheses, i.e., $P(H_0)$ and $P(H_1)$, and then updates the beliefs using the data of $\mathcal{X}^t$ through the Bayes rule, i.e.,

$$P(H_0|\mathcal{X}^t) = \frac{P(\mathcal{X}^t|H_0)P(H_0)}{P(\mathcal{X}^t)}, \quad P(H_1|\mathcal{X}^t) = 1 - P(H_0|\mathcal{X}^t).$$

The updated beliefs, $P(H_0|\mathcal{X}^t)$ and $P(H_1|\mathcal{X}^t)$, are compared with each other and then a decision is made—reject H_0 if $P(H_0|\mathcal{X}^t) < P(H_1|\mathcal{X}^t)$ and cannot reject H_0 otherwise. When there is no strong prior belief favoring either one of the two hypotheses, the test is then equivalent to testing the Bayes factor, i.e., whether

$$B(\mathcal{X}^t) = \frac{P(\mathcal{X}^t|H_0)}{P(\mathcal{X}^t|H_1)} < 1. \tag{9.24}$$

Park and Shrivastava (2014) consider that $P(H_0)$ and $P(H_1)$ unknown and can be possibly unequal. The hypothesis testing is to test if $B(\mathcal{X}^t)P(H_0)/P(H_1) < 1$. If denote $1/b = \frac{P(H_0)}{P(H_1)}$, then the hypothesis testing is done through the test of

$$B(\mathcal{X}^t) < b. \tag{9.25}$$

Here, using the language of Sect. 9.1, $B(\mathcal{X}^t)$ is the test statistic and b is the lower control limit (and here, the upper control limit is infinity).

Park and Shrivastava (2014) split the in-control process data $\mathcal{X}$ into two parts: $\mathcal{X}^{(1)}$ and $\mathcal{X}^{(2)}$, and then use $\mathcal{X}^{(1)}$ and $\mathcal{X}^t$ for estimating $P(\mathcal{X}^t|H_0)$ and $P(\mathcal{X}^t|H_1)$ in the Bayes factor and use $\mathcal{X}^{(2)}$ for estimating the test threshold, b.

The estimation of $P(\mathcal{X}^t|H_0)$ can be done by deriving a conditional likelihood, $P(\mathcal{X}^t|\Theta, H_0)$, conditioned on some parameters of Θ, and then computing $P(\mathcal{X}^t|H_0)$ after taking the expectation of the conditional likelihood with respect to $P(\Theta|H_0)$, as follows,

$$P(\mathcal{X}^t|H_0) = \int P(\mathcal{X}^t|\Theta)dP(\Theta|H_0).$$

Park and Shrivastava (2014) choose the parameter set as $\Theta = \{\{\boldsymbol{\alpha}_s, \omega_s : s = 1, \ldots, S\}, \sigma^2\}$, i.e., the set of parameters that describe the in-control process condition in the growth model in Eqs. (9.19) and (9.20). The conditional likelihood of $\dot{\boldsymbol{r}}_{tj}$ is attained by marginalizing the conditional joint likelihood in Eq. (9.22) with respect to s_{tj}:

$$L(\dot{\boldsymbol{r}}_{tj}|\Theta) = \sum_s \omega_s f_G(\dot{\boldsymbol{r}}_{tj}; \boldsymbol{\Phi}_{tj}\boldsymbol{\alpha}_s, \sigma^2 \boldsymbol{I}).$$

Since all $\dot{\boldsymbol{r}}_{tj}$'s are conditionally independent given the parameters in Θ, the joint conditional likelihood function of $\mathcal{X}^t$ is

$$P(\mathcal{X}^t|\Theta) = \prod_{j \in \dot{J}_t} \sum_s \omega_s f_G(\dot{\boldsymbol{r}}_{tj}; \boldsymbol{\Phi}_{tj}\boldsymbol{\alpha}_s, \sigma^2 \boldsymbol{I}).$$

As such, the marginal likelihood, $P(\mathcal{X}^t|H_0)$, can be obtained by taking the expectation of $P(\mathcal{X}^t|\Theta)$ with respect to $P(\Theta|H_0)$. The $P(\Theta|H_0)$ is estimated by the posterior distribution of Θ given the in-control process data of $\mathcal{X}^{(1)}$, i.e., $P(\Theta|\mathcal{X}^{(1)})$.

The expectation of $P(\mathcal{X}^t|\Theta)$ is numerically approximated by the sample mean of the conditional likelihood $P(\mathcal{X}^t|\Theta)$ over the samples from $P(\Theta|\mathcal{X}^{(1)})$ (Newton and Raftery 1994), as such:

$$\begin{aligned} P(\mathcal{X}^t|H_0) &= \int P(\mathcal{X}^t|\Theta) dP(\Theta|\mathcal{X}^{(1)}) \\ &\approx \frac{1}{N_{q,0}} \sum_q \prod_{j \in \dot{J}_t} \sum_{s=1}^{S^{(q,0)}} \omega_s^{(q,0)} f_G(\dot{\boldsymbol{r}}_{tj}; \boldsymbol{\Phi}_{tj}\boldsymbol{\alpha}_s^{(q,0)}, \sigma^{2(q,0)} \boldsymbol{I}), \end{aligned} \tag{9.26}$$

where $\boldsymbol{\alpha}_s^{(q,0)}$, $\omega_s^{(q,0)}$, $\sigma^{2(q,0)}$ and $S^{(q,0)}$ are the Gibbs outputs of $\boldsymbol{\alpha}_s$, ω_s, σ^2 and S obtained from the q-th iteration of the exact block Gibbs sampler, and $N_{q,0}$ is the number of the Gibbs iterations executed after the burn-in period.

In principle, $P(\mathcal{X}^t|H_1)$ can be likewise computed. But one practical difficulty is that under H_1, unlike under H_0, one does not have much information of the probability distribution of Θ. Hence, for describing $P(\Theta|H_1)$, one should rather consider a rich class of probability distribution functions not equal to $P(\Theta|H_0)$. In the literature, such hypothesis is called a (composite) nonparametric alternative (Berger and Guglielmi 2001; Conigliani et al. 2000). In the Bayesian treatment of the nonparametric alternative, a rich family of nonparametric probability distribution functions are considered with certain prior probabilities assigned to the member distribution functions. The Dirichlet process (Carota and Parmigiani 1996; Basu and Chib 2003), Polya tree process (Berger and Guglielmi 2001), and Dirichlet process mixture (Tokdar and Martin 2021) were used for this purpose. Among them, the Dirichlet process mixture of normal distributions has many advantages over other options—nice convergence properties to a broad class of density functions

(Ghosal and Van Der Vaart 2007; Dass and Lee 2004), and easy and efficient computation (Tokdar and Martin 2021). Hence, Park and Shrivastava (2014) choose to use the Dirichlet mixture model for assigning prior probabilities to nonparametric probability distributions.

In the Dirichlet models, there are lots of uncertainties in $P(\Theta|H_1)$, so the probability distribution of $P(\Theta|H_1)$ is regarded as a random quantity; let us denote it by $\mathcal{P}$. Park and Shrivastava (2014) assign a Dirichlet process prior, $\text{DP}(\rho_1 P_1)$, to the random probability measure, $\mathcal{P}$, where $P_1(\boldsymbol{\alpha}, \sigma^2) = \mathcal{N}_Q(\boldsymbol{\alpha}; 0, \sigma_1^2 \boldsymbol{I})\text{IG}(\sigma^2; \varphi_1, \lambda_1)$ is the base probability distribution of the Dirichlet process, and ρ_1 is a Dirichlet concentration parameter. Park and Shrivastava (2014) set $\rho_1 = 1, \sigma_1^2 = 1, \varphi_1 = 0.1$, and $\lambda_1 = 0.1$ for the video data at hand.

To express the marginal likelihood, $P(\mathcal{X}^t|H_1)$, Park and Shrivastava (2014) invoke the conditional expectation formula, generally expressed as

$$\mathbb{E}_Z\mathbb{E}_Y[P(X|Y, Z, H_1)] = P(X|H_1).$$

Let $X = \mathcal{X}^t$, $Y = \Theta$, and $Z = \mathcal{P}$, the above expression can be written as

$$\begin{aligned} P(\mathcal{X}^t|H_1) &= \mathbb{E}_{\mathcal{P}}\mathbb{E}_{\Theta}[P(\mathcal{X}^t|\Theta, \mathcal{P}, H_1)] \\ &= \mathbb{E}_{\mathcal{P}}\left[\int_{\Theta} P(\mathcal{X}^t|\Theta, H_1)d\mathcal{P}(\Theta|H_1)\right] \\ &= \mathbb{E}_{\mathcal{P}}\left[\int_{\Theta} P(\mathcal{X}^t|\Theta)d\mathcal{P}(\Theta|H_1)\right]. \end{aligned}$$

The H_1 is dropped from $P(\mathcal{X}^t|\Theta, H_1)$ in the last line above because $\mathcal{X}^t$, once conditioned on Θ, is independent of H_1. By the Fubini theorem, which exchanges the order of the two integrations, one has

$$P(\mathcal{X}^t|H_1) = \int_{\Theta} P(\mathcal{X}^t|\Theta)\mathbb{E}_{\mathcal{P}}\left[d\mathcal{P}(\Theta|H_1)\right].$$

Applying the basic property of the Dirichlet process that $\mathbb{E}_{\mathcal{P}}[\mathcal{P}(\Theta|H_1)] = P_1$ (Tokdar and Martin 2021) simplifies the previous expression of $P(\mathcal{X}^t|H_1)$ to

$$P(\mathcal{X}^t|H_1) = \int \prod_{j \in \mathcal{J}_t} f_G(\dot{\boldsymbol{r}}_{tj}; \boldsymbol{\Phi}_{tj}\boldsymbol{\alpha}, \sigma^2\boldsymbol{I})\mathcal{N}_Q(\boldsymbol{\alpha}; 0, \sigma_1^2\boldsymbol{I})\text{IG}(\sigma^2; \varphi_1, \lambda_1)d\boldsymbol{\alpha}d\sigma^2$$

Then, $P(\mathcal{X}^t|H_1)$ is further approximated by the sample mean of the likelihoods over the iid samples of $(\boldsymbol{\alpha}^{(q)}, \sigma^{2(q)}) \sim \mathcal{N}_Q(\boldsymbol{\alpha}; 0, \sigma_1\boldsymbol{I})\text{IG}(\sigma^2; \varphi_1, \lambda_1)$, i.e.,

$$P(\mathcal{X}^t|H_1) = \frac{1}{N_{q,1}}\sum_q \prod_{j \in \mathcal{J}_t} f_G(\dot{\boldsymbol{r}}_{tj}; \boldsymbol{\Phi}_{tj}\boldsymbol{\alpha}^{(q)}, \sigma^{2(q)}\boldsymbol{I}), \tag{9.27}$$

where $N_{q,1}$ is the number of the i.i.d. samples of $(\boldsymbol{\alpha}^{(q)}, \sigma^{2(q)})$.

Lastly, the control limit in Eq. (9.25), b, is chosen such that $P(B(\mathcal{X}^t) < b|H_0) < \nu$, i.e., the probability of the type-I error of the test is less than ν. In order to find such a b, Park and Shrivastava (2014) make use of the second part of the training data collected for the Phase-I analysis, i.e., $\mathcal{X}^{(2)}$. Park and Shrivastava (2014) first take R bootstrap samples from $\mathcal{X}^{(2)}$, where the r-th bootstrap sample is denoted by $\mathcal{X}^{(2,r)}$. Given $\mathcal{X}^{(2,r)}$, compute $b_r = B(\mathcal{X}^{2,r})$ by taking the ratio of Eq. (9.26) and Eq. (9.27) for $\mathcal{X}^t = \mathcal{X}^{2,r}$. The estimate for b is the $(1-\nu)$-th sample quantile of $\{b_r : r = 1, \ldots, R\}$, which is denoted by $\hat{b}_\nu$.

In summary, when a new image frame arrives, $\mathcal{X}^t$ is obtained at time t. To test whether the process is out of control, first compute the Bayes factor score of the new data point, $B(\mathcal{X}^t)$, and then check if $B(\mathcal{X}^t) < \hat{b}_\nu$. Conclude the process is out of control if the inequality is true.

9.5 Case Study

Park and Shrivastava (2014) first simulate a multimode process of nanoparticle growth by using a generative procedure. They randomly generate a nanoparticle outline at time t, as follows. Given the inputs of $t, \omega_1, \omega_2, \boldsymbol{\alpha}_1, \boldsymbol{\alpha}_2, \sigma_0^2, \sigma^2$, generate samples of $\{r(t,\theta) : \theta \in \{0, 0.01, 0.02, \ldots, 2\pi\}\}$ through the following two steps:

1. Sample the coefficients in the shape model

$$\boldsymbol{\alpha} \sim \omega_1 \mathcal{N}_Q(\boldsymbol{\alpha}_1, \sigma_0^2 \boldsymbol{I}) + \omega_2 \mathcal{N}_Q(\boldsymbol{\alpha}_2, \sigma_0^2 \boldsymbol{I}). \tag{9.28}$$

2. Sample the centroid distance functions

$$r(t,\theta) \sim \mathcal{N}(\boldsymbol{\phi}_{t,\theta}^T \boldsymbol{\alpha}, \sigma^2) \text{ for } \theta \in \{0, 0.01, 0.02, \ldots, 2\pi\}, \tag{9.29}$$

To simulate an in-control process, Park and Shrivastava (2014) choose $\boldsymbol{\alpha}_1$ and $\boldsymbol{\alpha}_2$ randomly from $\mathcal{N}_Q(\boldsymbol{0}, \boldsymbol{I})$, and set $\sigma_0^2 = 1$, $\sigma^2 = 1$, $\omega_1 = 0.7$, and $\omega_2 = 0.3$ in the generative procedure. For each $t \in \{0, 1, \ldots, 50\}$, Park and Shrivastava (2014) generate a random sample of 200 nanoparticle outlines. Seventy percent of the nanoparticle outlines are reserved as the inputs to the exact block Gibbs sampler for obtaining the posterior samples of $\{\{\boldsymbol{\alpha}_s^{(q,0)}, \omega_s^{(q,0)} : s = 1, \ldots, S^{(q,0)}\}, \sigma^{(q,0)} : q = 1, \ldots, 20,000\}$. The burn-in period is set as the first 10,000 samples. The results of the block Gibbs sampler are summarized in Sect. 9.5.1. The remaining 30% of samples are used for estimating $\hat{b}_\nu$ under a specified type-I error of ν. For estimating b, Park and Shrivastava (2014) set $R = 2,500$.

To simulate the out-of-control processes, Park and Shrivastava (2014) consider four different out-of-control cases and set $\boldsymbol{\alpha}_s$'s, ω_s's, σ_0^2, and σ^2 accordingly. The specific values of these parameters are described in Sects. 9.5.2–9.5.5. Park and Shrivastava (2014) perform Phase-II analysis of these test cases and measure the average delay in detection under three different levels of the type-I error, i.e., $\nu = 0.05, 0.01$, and 0.0027.

9.5.1 Phase I Result

In the Gibbs sampling, the hyperparameters, φ_1 and λ_1, are set to 0.1 and the Dirichlet concentration parameter, α_0, is set to one. After discarding the initial 10,000 samples in the burn-in period, the number of process modes, $S^{(q,0)}$, is constantly two, the same as the number of the process modes used to simulate the data in the first place. The estimate of the posterior mean of ω_s is 0.7095 for $s = 1$ and 0.2905 for $s = 2$. The posterior mean estimates are very close to the true values used, i.e., $\omega_1 = 0.7$ and $\omega_2 = 0.3$.

The posterior mean estimate of $\boldsymbol{\alpha}_s$, denoted by $\hat{\boldsymbol{\alpha}}_s$, is also compared with the true value of $\boldsymbol{\alpha}_s$. Since the variable $\boldsymbol{\alpha}_s$ is high dimensional, to help appreciate the quality of the estimate, Park and Shrivastava (2014) present a visual comparison using the particle outlines derived from $\boldsymbol{\alpha}_s$ and those from $\hat{\boldsymbol{\alpha}}_s$. Given $\boldsymbol{\alpha}_s$ or $\hat{\boldsymbol{\alpha}}_s$, the centroid distance function of the nanoparticle outline observed at t can be reconstructed as $r(t,\theta) = \boldsymbol{\phi}_{t,\theta}^T \boldsymbol{\alpha}_s$ or $\hat{r}(t,\theta) = \boldsymbol{\phi}_{t,\theta}^T \hat{\boldsymbol{\alpha}}_s$. Figure 9.10 presents both $r(t,\theta)$ and $\hat{r}(t,\theta)$ for the two process modes, where one may notice that the two centroid distance functions are rather similar to each other.

9.5.2 Case I: $\boldsymbol{\alpha}_s$ Changed

The first out-of-control case is when the average particle profiles see a change. Park and Shrivastava (2014) simulate this change by changing the mean parameter $\boldsymbol{\alpha}_1$ to $\boldsymbol{\alpha}_1 + \delta\sigma_0 \mathbf{1}$ for $\delta \in \{0, 0.5, \ldots, 3.0\}$, where $\mathbf{1}$ is a column vector of all ones. The different choices of δ yield different magnitudes for the change. Other parameters in the generative procedure of Eqs. (9.28) and (9.29) are kept unchanged from the in-control case.

Given a δ, Park and Shrivastava (2014) generate 500 random samples at different times, each of which contains 200 nanoparticle outlines. The Bayes factor score for each random sample, $B(X^t)$, is computed, and the inequality, $B(X^t) < \hat{b}_\nu$, is checked to see whether the process is out of control.

Figure 9.11 presents the negative Bayes factor scores, namely $-B(X^t)$, under different magnitudes of mean shift. In each plot, the initial 500 points are generated using $\delta = 0$, representing the in-control process. The change point takes place at the sample number of 500. The next 500 points are from an out-of-control process of its respective δ value. Apparently, this change is the sustained mean shift type, as explained in Sect. 9.1.1.

The average delay in detection, as mentioned in Sect. 9.1.1, is usually characterized by a measure known as the average run length (ARL). A run length is

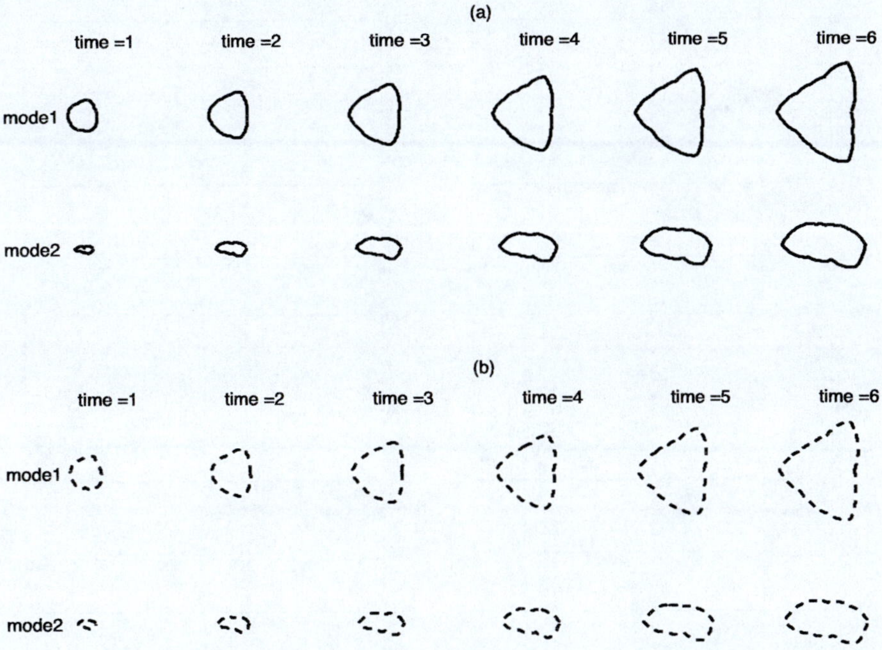

Fig. 9.10 Comparison of the nanoparticle outlines. The solid outlines are derived by using the true value of $\boldsymbol{\alpha}_s$ and the dotted outlines are derived by using the posterior mean estimates of $\boldsymbol{\alpha}_s$. (**a**) True mean outlines of particles. (**b**) Estimated mean outlines of particles. (Reprinted with permission from Park and Shrivastava 2014)

the number of samples until the control chart sees a test sample exceeding its control limits since the process change happens. The average run length is simply the expectation of the run length under a given circumstance. When the process is out-of-control under a $\delta > 0$, ARL implies how soon, on average, one sees a detection since the process has shifted. This ARL is referred to as the "out-of-control" ARL. Under the circumstance of $\delta = 0$, the ARL means that how frequently, on average, one sees a wrong detection or a false alarm. This ARL is referred to as the "in-control" ARL. Understandably, decision makers would like to have the out-of-control ARL as short as possible and the in-control ARL as long as possible.

Table 9.2 presents the average run length results of the multimode profile detection method under the different values of δ, including the in-control ARL. For all three values of ν, the average run lengths under $\delta = 0$ are similar to or longer than $1/\nu$, suggesting that the ARL performance under the in-control process condition is maintained as desired. On the other hand, the multimode profile detection method is sensitive to mean shifts. For a mean shift as small as $\delta = 0.5$, the control chart flags such a change almost immediately after it has taken place. When ν is set to 0.0027, the average delay in detecting a mean shift of $0.5\sigma_0$ is 1.35 samples. If one takes a

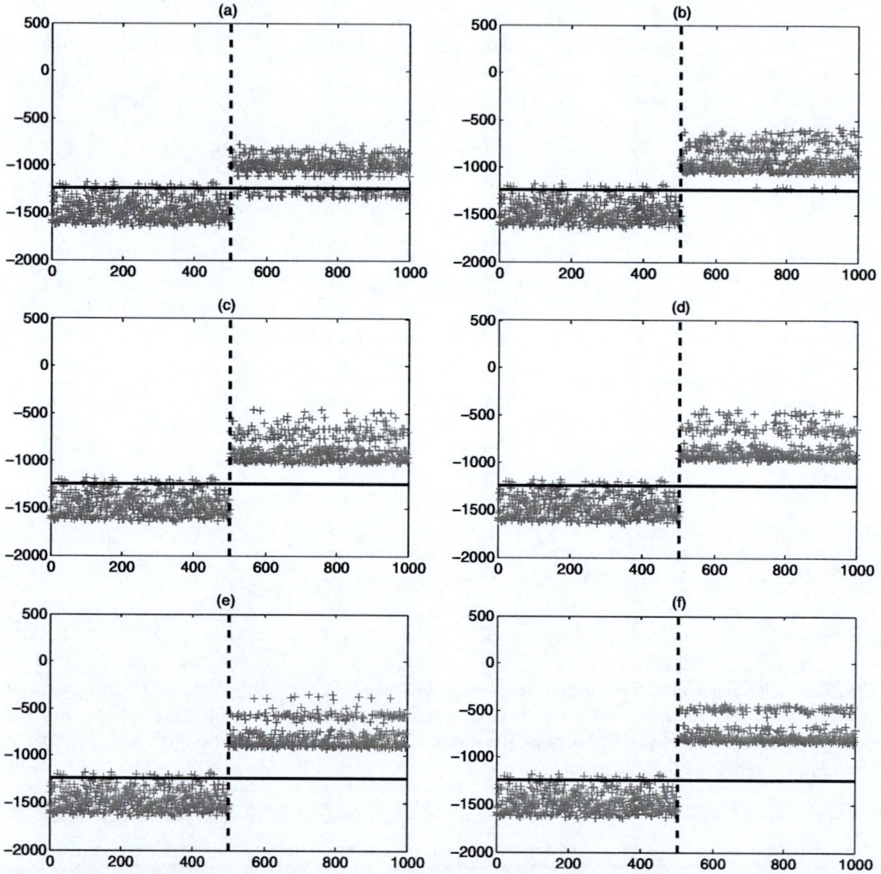

Fig. 9.11 Examples of the multimode profile detection method for detecting a mean shift in the simulated process. The horizontal axis is the sample number and the vertical axis is the negative Bayes factor score. In each plot, the initial 500 samples are generated with $\delta = 0$, whereas the next 500 samples are generated with the δ specified in the caption of that plot. The solid horizontal lines are at $-\hat{b}_\nu$ for $\nu = 0.05$ and the dotted vertical lines are at sample number of 500. **(a)** delta = 0.5. **(b)** delta = 1. **(c)** delta = 1.5. **(d)** delta =2. **(e)** delta =2.5. **(f)** delta =3. (Reprinted with permission from Park and Shrivastava 2014)

sample from a process of interest every 50 ms, which is the practical sampling rate of the in situ TEM, the average time to signal is 67.5 ms. Considering that the total time length of nanoparticle synthesis is in the order of minutes for a fast synthesis process (Ojea-Jiménez et al. 2010) and in the order of hours for slow ones (Song et al. 2005; Rioux et al. 2006), the average time to signal appears rapid enough for detecting process mean shifts before the process deviates too far from its in-control status.

Table 9.2 ARL of the multimode profile detection method for detecting a process mean shift. (*Source*: Park and Shrivastava 2014)

	Shift magnitude (δ)						
Type-I error (ν)	0.0	0.5	1.0	1.5	2.0	2.5	3.0
0.05	21.74	1.25	1.02	1.00	1.00	1.00	1.00
0.01	166.7	1.31	1.04	1.00	1.00	1.00	1.00
0.0027	> 500	1.35	1.06	1.00	1.00	1.00	1.00

Table 9.3 ARL of the multimode profile detection method for detecting a partial mean shift in the process. (*Source*: Park and Shrivastava 2014)

	Shift magnitude (δ)						
Type-I error (ν)	0.0	0.5	1.0	1.5	2.0	2.5	3.0
0.05	21.74	13.51	3.03	1.55	1.09	1.01	1.00
0.01	166.7	50.00	3.88	1.88	1.45	1.16	1.10
0.0027	> 500	71.43	4.03	1.89	1.45	1.17	1.10

9.5.3 Case II: Only Part of $\boldsymbol{\alpha}_s$ Changed

The second out-of-control case that Park and Shrivastava (2014) simulate is also a mean shift case. Instead of having all the elements in $\boldsymbol{\alpha}_s$ shifted the same amount, they randomly select 10% of the elements to be shifted while keeping the other 90% elements unchanged. This study is to test how robust the change detection method is when different varieties of mean shifts take place.

The partial mean shift is created by changing the mean parameter $\boldsymbol{\alpha}_1$ to $\boldsymbol{\alpha}_1+\delta\sigma_0\boldsymbol{e}$, where $\boldsymbol{e}$ is an $(M-1)n \times 1$ column vector with 10% randomly chosen elements valued at one while the rest is set to zero. The δ is still in the range of (0.0, 3.0) with an increment of 0.5. The ARL is computed in the same way as in Case I.

Table 9.3 presents the ARL results. The multimode profiling detection method is less sensitive in detecting a partial mean shift, but it still has reasonably short ARLs.

9.5.4 Case III: σ^2 Changed

The third out-of-control case is to change the process variance, σ^2. Park and Shrivastava (2014) simulate the changes in process variance by changing the variance parameter σ^2 to $\sigma^2 + \delta\sigma^2$ in the generative procedure of Eqs. (9.28) and (9.29). The change magnitude coefficient, δ, is again in the range of (0.5, 3.0) with an increment of 0.5. The computation of ARL is done the same way as in Cases I and II.

Table 9.4 shows the average run length results. The multimode profile detection method is less sensitive in detecting a change in process variance than in detecting a process mean shift. When σ^2 is increased to $2\sigma^2$, the ARL is 15.63 for $\nu = 0.0027$.

Table 9.4 ARL of the multimode profile detection method for detecting a change in process variance. (*Source*: Park and Shrivastava 2014)

	Shift magnitude (δ)						
Type-I error (ν)	0.0	0.5	1.0	1.5	2.0	2.5	3.0
0.05	21.74	13.89	7.35	6.41	3.68	1.79	1.00
0.01	166.7	29.41	12.50	11.63	6.76	3.01	1.00
0.0027	> 500	38.46	15.63	14.29	9.43	3.07	1.00

Table 9.5 ARL of the multimode profile detection method for detecting a change in the mixture proportion of the process modes. (*Source*: Park and Shrivastava 2014)

	Shift magnitude (δ)						
Type-I error (ν)	0.0	0.1	0.2	0.3	0.4	0.5	0.6
0.05	21.74	8.06	4.00	3.05	2.27	1.07	1.00
0.01	166.7	13.51	6.49	4.72	3.11	1.44	1.00
0.0027	> 500	16.67	7.46	5.05	3.21	1.59	1.00

When the sampling interval is 50 ms, the average time to signal is 0.78 s, much increased as compared to the mean shift case, albeit still short as compared with the total time of a growth process.

9.5.5 Case IV: ω_s Changed

The fourth and last out-of-control case is a change in the mixing proportion of the process modes. Park and Shrivastava (2014) simulate this change by changing the proportion parameter ω_1 to $\omega_1 - \delta$ in the generative procedure, where δ is now in the range of (0, 0.6) with an increment of 0.1.

Table 9.5 presents the ARL results. The numerical results show that the change detection method has done a fair job in detecting a change in the mixing proportions of the two process modes. The original mixing proportions are $\omega_1 = 0.7$ and $\omega_2 = 0.3$. When ω_1 is shifted from 0.7 to 0.6, the ARL is 16.67 under type-I error of $\nu = 0.0027$. When the sampling interval of the in situ TEM is 50 ms, the average time to signal is around 0.83 s.

9.5.6 Application to Nanoparticle Self-Assembly Processes

Park and Shrivastava (2014) apply the multimode profile detection method to the nano video data described in Sect. 7.1.2. The data used here are from Video 1 and Video 2 of the three video clips. Park and Shrivastava (2014) extract a subset of the process for the change detection purpose. Specifically, Park and Shrivastava (2014) extract the video data from Video 1 every 0.5 s from 2 to 52 s, and from Video 2

Table 9.6 ARL of the multimode profile detection method when it is applied to a real nanoparticle synthesis process. The "> 50" implies that no case is flagged among 50 test samples from the in-control process. (*Source*: Park and Shrivastava 2014)

	Type-I error (ν)		
	0.05	0.01	0.0027
In-control ARL	50	> 50	> 50
Out-of-control ARL	1.22	1.95	4.33

every 0.5 s from 2 to 21 s. The Video 1 subset is referred to as the first dataset in the sequel, and the Video 2 subset as the second dataset.

The two processes are different when it comes to the strength of the electron beam applied to nanoparticles for initiating their growth, and the difference in the beam strength led to different nanoparticle growths. The geometrical changes in the first dataset are mainly driven by the nucleation and the subsequent growth of nucleus and a second round nucleation of nanoparticles is often observed, whereas the second round of nucleation is rare in the second dataset. Park and Shrivastava (2014) regard the first dataset as the in-control process data and treat the second dataset as the potentially out-of-control process data.

Fifty percent of the first dataset (51 image frames) is used to conduct the Phase I analysis, i.e., learn the in-control model parameters for the multimode profile detection method. The remaining 50% (50 image frames) is kept for calculating the in-control ARL. The whole second dataset (39 image frames) is used for calculating the out-of-control ARL.

Through the Phase I analysis, the multimode profile detection method identifies eighteen different growth modes. Among them, seven are most frequently observed, and they altogether account for 73% of the nanoparticle growth. Figure 8.7 demonstrates the seven major growth modes of nanoparticles.

In the Phase II analysis, Park and Shrivastava (2014) use the 50% of the first dataset (50 samples) that is not used in the Phase-I analysis to assess the in-control ARL, and use the whole set of the second dataset (39 samples) to assess the performance of average run length.

Table 9.6 summarizes the average run lengths under different ν's. Among the 50 test samples from the in-control process, only one case is flagged as 'abnormal' when $\nu = 0.05$ and no case is flagged when a smaller ν is used. The change detection method flags more frequently among the 39 samples from the out-of-control process.

References

Aldous D (1999) Deterministic and stochastic models for coalescence (aggregation and coagulation): A review of the mean-field theory for probabilists. Bernoulli 5(1):3–48

Baldan A (2002) Review progress in Ostwald ripening theories and their applications to nickel-base super alloys, Part I: Ostwald ripening theories. Journal of Materials Science 37(11):2171–2202

Basu S, Chib S (2003) Marginal likelihood and Bayes factors for Dirichlet process mixture models. Journal of the American Statistical Association 98(461):224–235

Berger J, Guglielmi A (2001) Bayesian and conditional frequentist testing of a parametric model versus nonparametric alternatives. Journal of the American Statistical Association 96(453):174–184

Bian B, Xia W, Du J, Zhang J, Liu JP, Guo Z, Yan A (2013) Growth mechanisms and size control of FePt nanoparticles synthesized using $Fe(Co)_x(x<5)$-oleylamine and platinum(II) acetylacetonate. Nanoscale 5(6):2454–2459

Carota C, Parmigiani G (1996) On Bayes factors for nonparametic alternatives, Oxford University Press, London, chap Bayesian Statistics 5, pp 507–511

Chen J, Gupta AK (2001) On change point detection and estimation. Communications in Statistics—Simulation and Computation 30(3):665–697

Conigliani C, Castro J, O'Hagan A (2000) Bayesian assessment of goodness of fit against nonparametric alternatives. Canadian Journal of Statistics 28(2):327–342

Dass S, Lee J (2004) A note on the consistency of Bayes factors for testing point null versus non-parametric alternatives. Journal of Statistical Planning and Inference 119(1):143–152

Ferguson T (1983) Bayesian density estimation by mixtures of normal distributions, Academic Press, New York, chap Recent Advances in Statistics, pp 287–303

Fryzlewicz P (2014) Wild binary segmentation for multiple change-point detection. The Annals of Statistics 42(6):2243–2281

Ghosal S, Van Der Vaart A (2007) Posterior convergence rates of Dirichlet mixtures at smooth densities. The Annals of Statistics 35(2):697–723

Hardy S, Voorhees P (1988) Ostwald ripening in a system with a high volume fraction of coarsening phase. Metallurgical Transactions A 19(11):2713–2721

Jolliffe IT (2002) Principal Component Analysis. Springer-Verlag, New York

Killick R, Fearnhead P, Eckley I (2012) Optimal detection of change-points with a linear computational cost. Journal of the American Statistical Association 107(500):1590–1598

Kleiber C, Hornik K, Leisch F, Zeileis A (2002) `strucchange`: An `R` package for testing for structural change in linear regression models. Journal of Statistical Software 7(2):1–38

Lifshitz I, Slyozov V (1961) The kinetics of precipitation from supersaturated solid solutions. Journal of Physics and Chemistry of Solids 19(1):35–50

Lo A, Skodje RT (2000) Kinetic and Monte Carlo models of thin film coarsening: Cross over from diffusion-coalescence to Ostwald growth modes. The Journal of Chemical Physics 112(4):1966–1974

Montgomery DC (2009) Introduction to Statistical Quality Control, 6th edn. John Wiley and Sons, New York

Mortenson M (1985) Geometrie Modeling. John Wiley and Sons, New York

Newton M, Raftery A (1994) Approximate Bayesian inference with the weighted likelihood bootstrap. Journal of the Royal Statistical Society Series B 56(1):3–48

Ojea-Jiménez I, Romero F, Bastús N, Puntes V (2010) Small gold nanoparticles synthesized with sodium citrate and heavy water: Insights into the reaction mechanism. The Journal of Physical Chemistry C 114(4):1800–1804

Papaspiliopoulos O (1994) A note on posterior sampling from Dirichlet mixture models. Unpublished Technical Report, Centre for Research in Statistical Methodology, University of Warwick

Papaspiliopoulos O, Roberts G (2008) Retrospective Markov chain Monte Carlo methods for Dirichlet process hierarchical models. Biometrika 95(1):169–186

Park C, Shrivastava AK (2014) Multimode geometric-profile monitoring with correlated image data and its application to nanoparticle self-assembly processes. Journal of Quality Technology 46(3):216–233

Qian Y, Huang JZ, Ding Y (2017) Identifying multi-stage nanocrystal growth using *in situ* TEM video data. IISE Transactions 49(5):532–543

Rioux R, Song H, Grass M, Habas S, Niesz K, Hoefelmeyer J, Yang P, Somorjai G (2006) Monodisperse platinum nanoparticles of well-defined shape: Synthesis, characterization, catalytic properties and future prospects. Topics in Catalysis 39(3):167–174

Schwarz G (1978) Estimating the dimension of a model. The Annals of Statistics 6(2):461–464

Sethuraman J (1994) A constructive definition of Dirichlet priors. Statistics Sinica 4(2):639–650

Song H, Kim F, Connor S, Somorjai G, Yang P (2005) Pt nanocrystals: Shape control and langmuir-blodgett monolayer formation. The Journal of Physical Chemistry B 109(1):188–193

Stephens M (1997) Bayesian methods for mixtures of normal distributions. PhD Thesis, University of Oxford

Tokdar S, Martin R (2021) Bayesian test of normality versus a Dirichlet process mixture alternative. Sankhyā B : The Indian Journal of Statistic in press

Voorhees PW (1985) The theory of Ostwald ripening. Journal of Statistical Physics 38(1-2):231–252

Walker S (2007) Sampling the Dirichlet mixture model with slices. Communications in Statistics—Simulation and Computation 36(1):45–54

Wang F, Richards V, Shields S, Buhro W (2013) Kinetics and mechanisms of aggregative nanocrystal growth. Chemistry of Materials 26(1):5–21

Woehl TJ, Park C, Evans JE, Arslan I, Ristenpart WD, Browning ND (2013) Direct observation of aggregative nanoparticle growth: Kinetic modeling of the size distribution and growth rate. Nano Letters 14(1):373–378

Yao YC (1988) Estimating the number of change-points via Schwarz's criterion. Statistics & Probability Letters 6(3):181–189

Zamba KD, Hawkins DM (2006) A multivariate change-point model for statistical process control. Technometrics 48(4):539–549

Zhang J, Huang F, Lin Z (2010) Progress of nanocrystalline growth kinetics based on oriented attachment. Nanoscale 2(1):18–34

Zheng H, Smith RK, Jun YW, Kisielowski C, Dahmen U, Alivisatos AP (2009) Observation of single colloidal platinum nanocrystal growth trajectories. Science 324(5932):1309–1312

Chapter 10
Multi-Object Tracking Analysis

10.1 Basics of Multi-Object Tracking Analysis

Object tracking analysis is a general term used in computer vision that refers to a machine learning task of automatically tracking individual objects in an image sequence (Smeulders et al. 2013). As illustrated in Fig. 10.1, object tracking analysis can be classified into single-object tracking and multi-object tracking. Multi-object tracking analysis is further categorized into parallel tracking (Smeulders et al. 2013) and interacting multi-object tracking (Park et al. 2015), depending on whether the objects to be tracked interact or not. By 'object interaction,' we mean that objects mutually affect their movement and visual appearance, e.g., one object interferes with the movement of another object or two objects merge into one object. Naturally, parallel tracking analysis is to track independent and non-interacting objects in an image sequence, whereas interacting multi-object tracking analysis is to track interacting objects.

The multi-object tracking analysis (MOTA) in material image analysis falls predominantly in the category of interacting multi-object tracking analysis, because material objects often evolve through mutual interactions. In other words, tracking the interactions is important to understand the material evolutionary process. Consider, for an example, the bottom-up synthesis of nanomaterials, which is based on the self-organization of atomic scale elements into nanoscale elements in a liquid or gas environment (Arole and Munde 2014). Imaging the bottom-up synthesis could create a sequence of image observations, showing that smaller scale elements are merged into larger scale objects, such as illustrated in Fig. 10.2. Tracking the merging events is essential for understanding the bottom-up synthesis process.

Park and Ding (2019) formalized the MOTA in material image analysis as a three-step procedure. As depicted in Fig. 10.3, the first step is image segmentation that identifies individual material objects from noisy images, the second step is data association that associates the objects identified in different image frames for

C. Park, Y. Ding, *Data Science for Nano Image Analysis*, International Series in Operations Research & Management Science 308,
https://doi.org/10.1007/978-3-030-72822-9_10

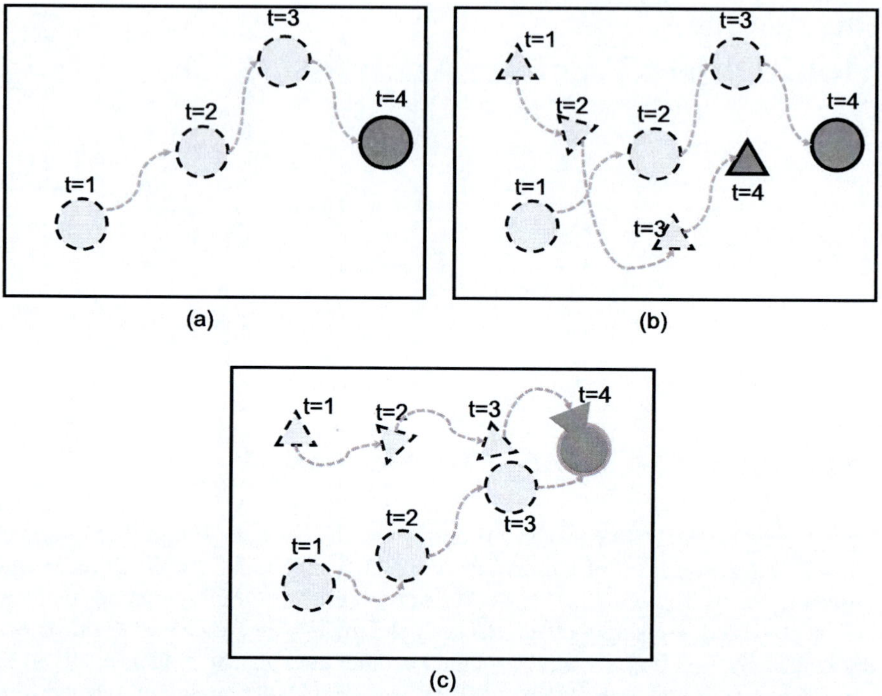

Fig. 10.1 Classification of object tracking analysis. The main criteria for categorization are the number of objects being tracked in an image sequence and whether the objects interact or not. (**a**) Single-object tracking analysis. (**b**) Parallel tracking analysis. (**c**) Interacting multi-object tracking analysis

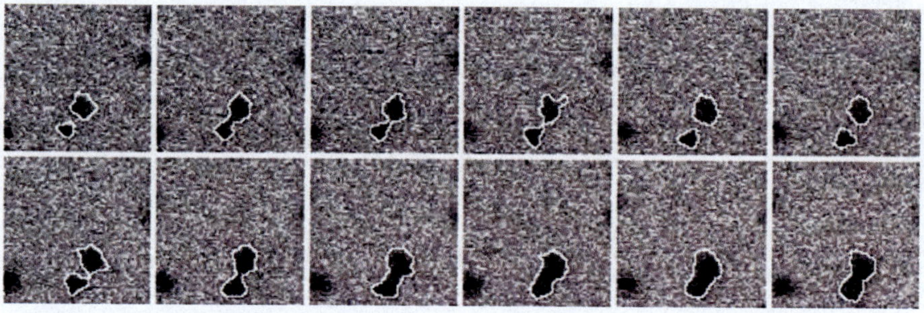

Fig. 10.2 Material image sequence of interacting silver nanoparticles. The image is a portion of the figure presented in Park et al. (2015)

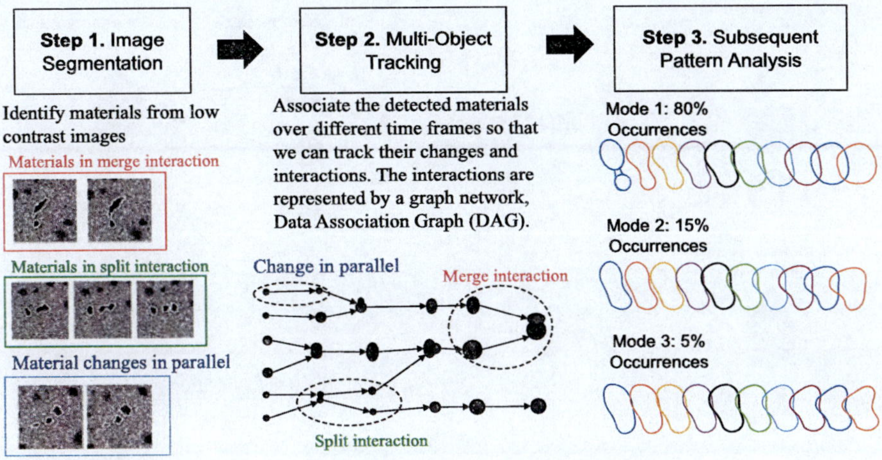

Fig. 10.3 Three steps in multi-object tracking analysis of material evolutions and interactions. Step 1 detects individual material objects or structures in noisy images. Step 2 associates the detections in different image frames with each other for estimating the trajectories of material evolutions and interactions. Step 3 finds frequent and rare patterns in the trajectories (reprint with permission from Park and Ding (2019))

forming the trajectories of material evolution, and the last step is a pattern analysis of the resulting trajectories over the course of material evolution.

Among the three steps, the first step is in fact discussed in details in Chap. 3. As for the third step, various possible analysis can be performed. One example is to identify the patterns of the material evolution followed by either the majority of material objects (principal modes) or the minority of material objects (outliers). Statistical modeling necessary for this type of analysis is similar to the approach introduced in Sect. 8.5. Another analysis in the third step is about mutual object interactions, addressing both issues of classifying the mutual interactions and modeling their occurrences. We will present an interaction analysis in Sect. 10.7.

Between the first and third steps is the data association step, which plays a crucial role in the object tracking analysis. It is not hard to see that once a quality data association is done, many existing methods can be used to unearth the pattern in the associated object sequences, not limited to the use of what is described in Sect. 10.7. But the data association step is unique from the rest of the contents in this book, as it is only so keenly needed in MOTA. For this reason, we devote much of this chapter to the discussion of the modeling and solution approaches related to the data association problem.

The data association problem is a problem of establishing a correspondence among the visual measurements of foregrounds in a sequence of images. A complete sequence of all the associated visual measurements is referred to as a *track*, while a subset of the track is referred to as a *tracklet*. The outcome of association can be

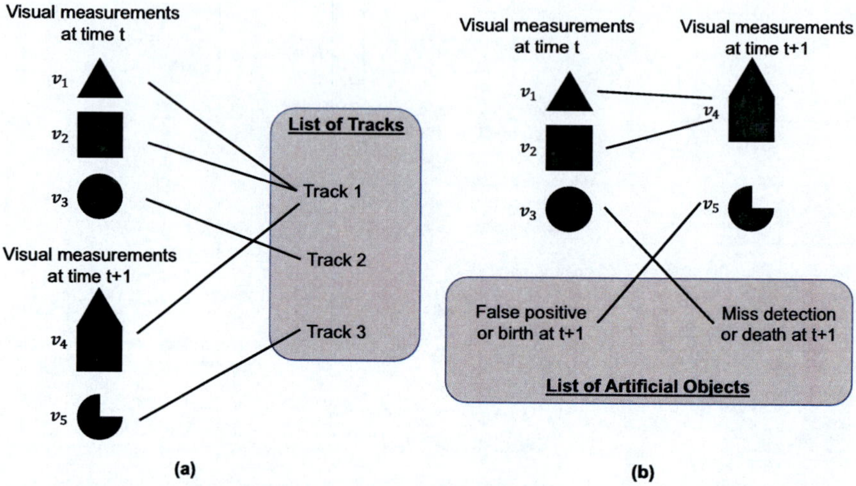

Fig. 10.4 Representation of data association graphs. (**a**) Measurement-to-object association. (**b**) Inter-measurement association

represented by a graph. By and large, two graph representations are popularly used, as illustrated in Fig. 10.4, namely, a graph associating visual measurements (from images) to tracks (Fig. 10.4a) or a graph associating visual measurements among each other (Fig. 10.4b). The first representation is a bipartite graph connecting visual measurements with tracks, and the second representation is a multipartite graph, in which each vertex represents a visual measurement and then all vertices are grouped by the time instances when they appear in the image sequence. Between the two representations, one can be converted to the other.

Data association in material image analysis can be complicated by a multitude of factors, such as

- Image objects may be falsely detected (when there is no object) or not successfully detected (when there are objects), due to the low contrast or poor quality of material images;
- The foreground objects are too numerous and may crowd the image;
- The foreground objects share similar appearances, as they are all grayscales and have similar sizes and shapes;
- The appearance of new material objects, which were not present previously;
- The disappearance of previously present objects;
- Material objects aggregate, or the foreground objects merge, and
- Material objects degenerate into small pieces, or the foreground objects split.

It would be ideal to incorporate all these complexities into a problem formulation leading to an optimal data association solution. Such modeling is understandably complicated, and its efficient solution can be challenging.

The earliest work in data association includes the multiple hypothesis tracking method (Reid 1979, MHT) and joint probabilistic data association filters (Rasmussen and Hager 2001, JPDAF). MHT solves the data association problem sequentially—based on the visual measurements archived up to time t, MHT makes a prediction of the tracks in the next time frame, so that the prediction becomes the base to shortlist the promising data associations to the tracks from new measurements arriving at time $t+1$. The promising associations are referred to as the hypotheses. When the hypotheses are conflicting, i.e., the same visual measurement is involved in multiple associations, they are resolved by solving an optimization problem that chooses among the conflicting hypotheses optimally.

JPDAF is the probabilistic version of MHT, assigning a confidence score to each hypothesis, and then, averaging the prediction over the promising hypotheses with their confidence scores as the weights. The two approaches work efficiently when objects spread widely and sparsely over images, so that the number of conflicting hypotheses is rather few. If objects densely locate, however, the number of conflicting hypotheses could escalate quickly. The number conflicting hypotheses increases further in the presence of object merging and splitting. Under such circumstances, solving MHT and JPDAF becomes computationally prohibitive. For this reason, the application of MHT and JPDAF is practically limited to tracking a small number of objects, like fewer than ten objects.

Another data association approach is sampling-based, which is a modified version of MHT or JPDAF. The sampling-based approach samples possible data associations from a posterior probability. This line of methods differs primarily in terms of what type of posterior probabilities is used and how to sample from the posterior probability (Genovesio and Olivo-Marin 2004; Khan et al. 2006; Storlie et al. 2009). Once the posterior probability is available, it can be naturally incorporated into some filtering algorithms, such as particle filters and the Kalman filter, for estimating the final state of the tracks.

The major limitation of the sampling-based approach is its computational inefficiency when the number of objects is more than ten, as the number of possible data associations to be sampled increases exponentially in the number of objects. It is not surprising that all aforementioned methods were applied to fewer than 10 objects—in fact, fewer than five objects in most cases. The computational complexity of the sampling-based approach might be alleviated by using a dynamic programming approach for certain special cases (Kumar et al. 2006), or by limiting the sampling space of possible data associations based on the spatial and temporal proximity of data (Yu and Medioni 2009). Because the sampling-based approach sequentially solves the data association problem for each time frame, another limitation of the approach is that the resulting solution is often entrapped in local optima within a few neighboring time frames. Although a globally optimal solution is theoretically achievable with an infinite number of samples (by the theoretical property of MCMC sampling), the solution is almost always locally optimal in finite samples. In light of all these, the sampling-based approach has limitations in handling long-term data associations, such as in the case of long-term object occlusions (Perera et al. 2006).

When the number of objects to be tracked is large, like in the material image analysis, the optimization-based data association is more appropriate. The optimization approach formulates the data association problem as a minimum-cost network flow problem and pursues a global optimal solution, or at least some (globally) sub-optimal solutions. The approach recasts the problem of assigning the measurements in each video frame to an object into a problem of associating the measurements in different video frames with each other. In the recast problem formulation, a group of the associated measurements are considered from the same object. The main advantage of the optimization approach over the sampling-based approach is that the former casts data associations for all time frames in a video into a single problem formulation, so that the optimization approach generally provides a better solution for identifying data associations spanning over a long time frame, e.g., the case of long-term object occlusion.

The minimum-cost network flow problem is typically formulated as a binary linear integer programming problem, solved by either a linear programming relaxation (Jiang et al. 2007; Zhang et al. 2008) or by some heuristic algorithms such as the successive shortest path algorithms (Pirsiavash et al. 2011; Berclaz et al. 2011). These methods impose the constraint that one measurement is associated with only one object. For tracking one-to-two merges or two-to-one splits, some methods adopted postprocessing steps (Jaqaman et al. 2008; Sergé et al. 2008). The more general methods of handling one-to-many or many-to-one associations are based on a binary quadratic programming formulation (Choi and Savarese 2012; Henriques et al. 2011), which can be solved by an integer programming solver such as the branch-and-bound algorithm (Choi and Savarese 2012). However, the branch-and-bound algorithm still suffers from a combinatorial computation complexity, so using it is not appropriate for data association problems of big sizes. A linear relaxation reformulation of the binary quadratic programming was studied (Henriques et al. 2011). Previously, the binary quadratic programming formulation can only handle one-to-two or two-to-one data associations. Park et al. (2015) extended the approach to data association problems involving more complex interactions. The resulting approach is referred to as the multi-way minimum-cost data association. To solve the multi-way minimum-cost data association, Park et al. (2015) also devised an efficient Lagrangian dual relaxation solution.

Among those briefly reviewed above, we will describe in the sequel three MOTA solutions for material image analysis—all from the category of optimization-based approaches. Section 10.3 describes the linear assignment formulation (Henriques et al. 2011), Sect. 10.4 describes the two-stage assignment approach (Jaqaman et al. 2008), and Sect. 10.5 describes the multi-way minimum-cost data association (Park et al. 2015).

10.2 Linear Assignment Problem for Data Association

We first introduce a common set of notations, to be used hereinafter. Denote by V_r the set of all visual measurements retrieved from an image sequence, and by $v \in V_r$ an element that represents a visual measurement. The visual measurements are grouped by the times of their appearances. Let $T(v)$ represent the time of appearance for $v \in V_r$. Association of visual measurements to tracks is represented by a graph, $G = (V, E)$, where V is the set of vectices and E is the set of edges connecting two vectices in V. The vertex set, V, includes all the measurements in V_r, plus a source node s and a sink node t, i.e., $V = V_r \bigcup \{s, t\}$. Each element e of E is an ordered 2-tuple of V, i.e.,

$$e = (v1, v2),$$

where $v1, v2 \in V$ are the two ends of the directed edge. Being a directed edge means that by the notation of $(v1, v2)$, we postulate that $v1$ appears before $v2$ does. Depending on the membership of the two ends, the edge $e = (v1, v2)$ has different meanings. Specifically,

- When $v1, v2 \in V_r$, the edge represents an association between the two corresponding visual measurements, and $T(v1) < T(V2)$, Note that the two ends on an edge are not necessarily from two adjacent time frames in the image sequence.
- When $v1 = s$, $v2 \in V_r$, the edge represents the *birth* of $v2$, i.e., $v2$ did not exist earlier and just came in the video scene.
- When $v1 \in V_r$, $v2 = t$, the edge represents the *death* of $v1$, i.e., $v1$ disappears from the video scene.
- The case of $v1 = s$ and $v2 = t$ is not allowed. At least one end of an edge must belong to V_r, because linking the source directly to the sink without anything else in between does not have any practical implication.

Image artifacts or clutters can be falsely identified as visual measurements, so they may be included in V_r. It is generally assumed that those faulty measurements only appear in a very short time period of the image sequence. Therefore, a measurement $v \in V$ associated with only s and t (i.e. birth and immediate death) is regarded as a faulty measurement. The definition of faulty measurements can be broadened by extending the time period of appearances from a single period to multiple periods. The faulty measurements can be removed in a post-processing step, performed after data association.

As the vertex set V of the data association graph G is fixed, the edge set E itself naturally distinguishes different graph structures. Let $\mathbb{E}$ denote the set of all possible edges, i.e.,

$$\mathbb{E} := \left\{(v1, v2) \in (V_r \times V) \bigcup (V \times V_r);\ T(v1) < T(v2)\right\},$$

and let $\mathbb{B} = \{0, 1\}$. For $e \in \mathbb{E}$, we define a binary indicator $z_e \in \mathbb{B}$, such that

$$z_e = 1 \text{ only if } e \in \mathbb{E}.$$

The values of z_e can be used to specify the presence of an edge in the edge set $\mathbb{E}$ and thus represent the final structure of graph G.

In a parallel tracking analysis, there is no object interaction, so that each $v \in V$ should have one in-degree and one out-degree. Let $O(v) \subset \mathbb{E}$ represent a set of the outgoing edges from $v \in V$, and $I(v) \subset \mathbb{E}$ represent a set of the incoming edges to $v \in V$. The in-degree and out-degree constraints are described by

$$\text{In-degree constraint: } \sum_{e \in I(v)} z_e = 1 \text{ for } v \in V,$$

$$\text{Out-degree constraint: } \sum_{e \in O(v)} z_e = 1 \text{ for } v \in V.$$

The graph G satisfying these degree constraints is referred to as a degree-one data association graph. We denote by $\mathcal{G}_1$ the space of all such degree-one data association graphs.

Data association problem can be formulated as a problem of finding a graph $G \in \mathcal{G}_1$ that minimizes the cost function,

$$\sum_{e \in \mathbb{E}} c_e z_e$$

where c_e is the cost of edge e. When the start and end nodes of e are all visual measurements in V_r, c_e stands for the cost of associating two visual measurements with each other. This cost depends on the visual appearances, the appearance times of the two visual measurements, as well as the spatial locations of them. When either one of the start and end nodes is s or t, c_e stands for the cost of linking a visual measurement to s or t (i.e. birth or death), rather than with another visual measurement. The cost of birth or death sets the upper limit of the association cost of two visual measurements. Considering that the objective is to minimize the overall association cost, the pairs of two visual measurements, of which the association cost is above the upper limit, will not be associated.

When the values of $\{z_e; e \in \mathbb{E}\}$ is constrained by $\mathcal{G}_1$, the data association problem is formulated as an optimization problem as follows:

$$
\begin{aligned}
\text{Minimize} \quad & \sum_{e \in \mathbb{E}} c_e z_e \\
\text{subject to} \quad & \sum_{e \in I(v)} z_e = 1 \text{ for } v \in V \\
& \sum_{e \in O(v)} z_e = 1 \text{ for } v \in V \\
& z_e \in \mathbb{B} \text{ for } e \in \mathbb{E}.
\end{aligned}
$$

This optimization problem is referred to as the linear assignment problem (Martello and Toth 1987). It is a combinatorial problem, but it has been proven that its constraint matrix is totally modular, so that the optimal solution of the problem is equivalent to that of its linear relaxation version, namely

$$
\begin{aligned}
\text{Minimize} \quad & \sum_{e \in \mathbb{E}} c_e z_e && (10.1a) \\
\text{subject to} \quad & \sum_{e \in I(v)} z_e = 1 \text{ for } v \in V && (10.1b) \\
& \sum_{e \in O(v)} z_e = 1 \text{ for } v \in V && (10.1c) \\
& 0 \le z_e \le 1 \text{ for } e \in \mathbb{E}.
\end{aligned}
$$

The relaxed problem is a special case of the minimum-cost network flow problem (Ahujia et al. 1993). It can be solved effectively through a linear programming (LP) solver. The major limitation of the standard linear assignment approach is that it can track only non-interaction objects (i.e., objects with degree-one interactions).

10.3 Linear Assignment Approach for Tracking Objects with Degree-Two Interactions

To alleviate the limitation of the linear assignment approach described in Sect. 10.2, i.e., it can only solve the parallel tracking problem, we describe in this section an extension that can track object interactions of mild complexities. Here, the complexity of object interactions is quantified by the number of objects involved in an interaction, or equivalently, the degree of object interaction.

Recall that the linear assignment formulation in Eq. (10.1) constrains $G \in \mathcal{G}_1$, that is, a visual measurement has only one in-degree and one out-degree. The degree-one constraint needs to be relaxed for modeling object interactions, such as object merges or object splits. In a data association graph, a k-to-one merge can be represented by a k-to-one association, in which k visual measurements are associated with another visual measurement. Likewise, a one-to-k split is represented by a one-to-k association, in which a single visual measurement is associated with k visual measurements. To represent such interactions, the in-degree and out-degree constraints are allowed to be greater than one, that is,

$$\text{In-degree constraint:} \sum_{e \in I(v)} z_e \geq 1 \text{ for } v \in V,$$

$$\text{Out-degree constraint:} \sum_{e \in O(v)} z_e \geq 1 \text{ for } v \in V.$$

A k-to-one association ending at $v \in V$ can be represented by a set of k edges having the same end v, i.e., a k-subset of $I(v)$. Let $I_k(v)$ denote a collection of all k-subsets of $I(v)$, whose element

$$\{e\} \in I_k(v) \text{ is a } k\text{-to-one association ending at } v \in V.$$

Similarly, a one-to-k association starting at $v \in V$ can be represented by a set of k edges having the same start node v, i.e., a k-subset of $O(v)$. Let $O_k(v)$ denote a collection of all k-subsets of $O(v)$, whose element

$$\{e\} \in O_k(v) \text{ is a one-to-}k \text{ association starting at } v \in V.$$

The objective function also needs to be revised, so that it includes the costs for the one-to-k and k-to-one associations. The cost of a one-to-k or k-to-one association is not simply an addition of the costs of the edges involved, meaning that

$$c_{\{e\}} \neq \sum_{e \in \{e\}} c_e \text{ for } \{e\} \in O_k(v) \text{ or } I_k(v).$$

For this reason, extra items need to be introduced in the objective function in Eq. (10.1a) to account for the costs of one-to-k and k-to-one associations.

Henriques et al. (2011) formulated a general linear assignment problem to track objects with degree-two object interactions, which correspond to two-to-one merges or one-to-two splits. He augmented the decision variables, $\{z_e \in \mathbb{B} \text{ for } e \in \mathbb{E}\}$, by introducing a set of additional binary decision variables related to degree-two object interactions. Let $z_{e1,e2} = z_{e1}z_{e2}$ represent the decision variables representing a two-to-one association for $e1, e2 \in I(v)$ and $v \in V_r$. All such two-to-one associations are given by

$$\{z_{e1,e2} \in \mathbb{B};\, e1, e2 \in I(v), v \in V_r\}. \tag{10.2}$$

Similarly, the decision variables representing one-to-two associations are

$$\{z_{e1,e2} \in \mathbb{B}; e1, e2 \in O(v), v \in V_r\}. \tag{10.3}$$

The association cost, corresponding to a two-to-one or a one-to-two association, is denoted by $c_{e1,e2}$. With these notations, the object function is augmented with the new cost term as

$$\sum_{e \in \mathbb{E}} c_e z_e + \sum_{v \in V_r} \sum_{e1,e2 \in I(v) \cup O(v)} c_{e1,e2},$$

The in-degree and out-degree constraints are also revised to ensure that each visual measurement $v \in V_r$ is associated with either a one-to-one, a one-to-two, or a two-to-one association, that is,

$$\sum_{e \in I(v)} z_e + \sum_{e1,e2 \in I(v)} z_{e1,e2} = 1 \quad \text{for } v \in V_r, \tag{10.4}$$

and

$$\sum_{e \in O(v)} z_e + \sum_{e1,e2 \in O(v)} z_{e1,e2} = 1 \quad \text{for } v \in V_r. \tag{10.5}$$

The problem of linear assignment for degree-two interactions is formally expressed by a new binary integer linear programming formulation as

$$\begin{aligned}
\text{Minimize} \quad & \sum_{e \in \mathbb{E}} c_e z_e + \sum_{v \in V_r} \sum_{e1,e2 \in I(v) \cup O(v)} c_{e1,e2}, \\
\text{subject to} \quad & \sum_{e \in I(v)} z_e + \sum_{e1,e2 \in I(v)} z_{e1,e2} = 1 \quad \text{for } v \in V_r, && \text{(10.6a)} \\
& \sum_{e \in O(v)} z_e + \sum_{e1,e2 \in O(v)} z_{e1,e2} = 1 \quad \text{for } v \in V_r, && \text{(10.6b)} \\
& z_e \in \mathbb{B} \text{ for } e \in \mathbb{E}, \\
& z_{e1,e2} \in \mathbb{B} \text{ for } e1, e2 \in I(v), v \in V_r, \\
& z_{e1,e2} \in \mathbb{B} \text{ for } e1, e2 \in O(v), v \in V_r.
\end{aligned}$$

This formulation has a computational advantage because its constraint coefficient matrix is totally unimodular like in the problem described in Sect. 10.2. As such, solving its linear relaxation with an LP solver could attain the optimal solution. On the other hand, the formulation still has limitations in its modeling power, in terms of expressing complex object interactions.

First, the degree of object interactions considered in the aforementioned linear assignment approach is limited to two, i.e., accounting only for either two-to-

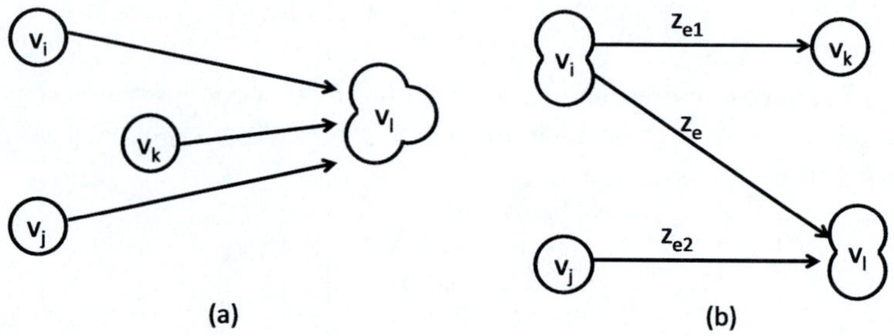

Fig. 10.5 Limited modeling power of the linear assignment approach

one merges or one-to-two splits. The approach cannot model higher degrees of interactions, such as the three-to-one merges as illustrated in Fig. 10.5a.

Second, in the linear assignment approach, a visual measurement cannot be associated with both a one-to-two and a two-to-one association, due to the constraints in Eqs. (10.6a) and (10.6b). A typical example for a visual measurement being involved with both types of the associations is a scenario in which one object splits into two, while one of which immediately merges with another object; see the illustration in Fig. 10.5b.

10.4 Two-Stage Assignment Approach for Tracking Objects with Degree-Two Interactions

Still modeling degree-two object interactions in a data association graph, Jaqaman et al. (2008) chose to use a two-stage approach. In the first stage, the visual measurements between successive frames are initially associated using the simple linear assignment approach, as formulated in Eq. (10.1). Since the linear assignment approach is capable of handling only one-to-one associations, some visual measurements involved in the degree-two interactions may not be associated with any other visual measurements, e.g., v_2 in Fig. 10.6a. As a result, the first stage produces many tracks that are not fully associated. Those partially associated tracks are referred to as *track segments*. In Fig. 10.6a, each track segment is depicted as a dotted-line box.

The second stage of the two-stage approach, as shown in Fig. 10.6b, is to connect the track segments, so that they form complete tracks. Suppose that K track segments are generated in the first stage. Recall that each of the resulting track segments is a simple sequence of visual measurements, since the first stage only allows one-to-one associations. Let $S_i = \{\tilde{v}_{ij}; j = 1, \ldots, J_i\}$ denote a sequence of visual measurements for the ith track segment, where $\tilde{v}_{ij}$ denotes the jth visual element in the sequence. The first and last visual measurements of a track segment can be associated with s, t or a visual measurement from another track segment. When the first measurement is associated with s, or the last measurement

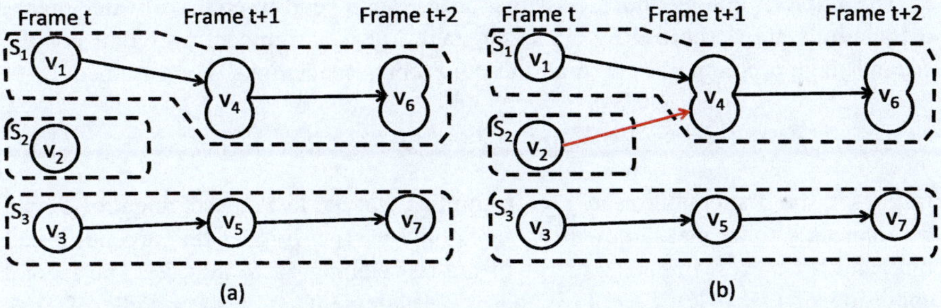

Fig. 10.6 Two-stage approach for tracking degree-two object interactions. **(a)** Stage 1. **(b)** Stage 2

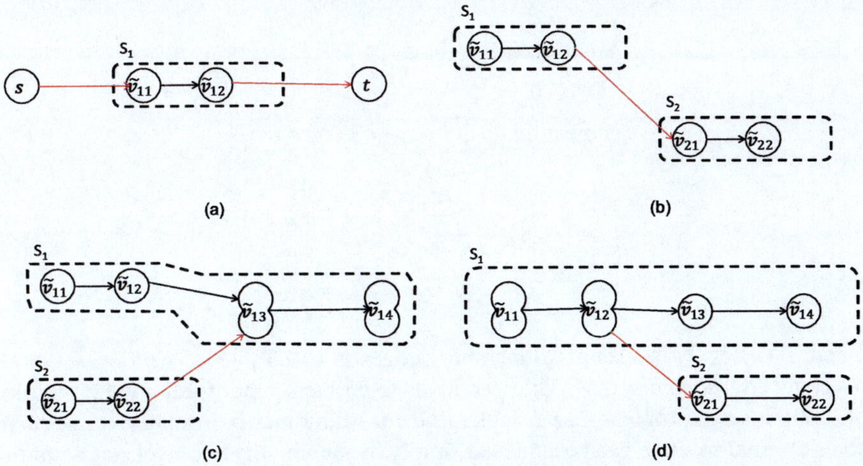

Fig. 10.7 Possible associations in the second stage of Jaqaman et al. (2008). **(a)** Initiation or termination of a track segment. **(b)** End-to-end link: the last node of a track segment is linked to the start node of another track segment. **(c)** Merge: the last node of a track segment is linked to a node from another track segment. **(d)** Split: a node in a track segment is linked to the start node of another track segment

is associated with t, it implies that the track segment is either initiated (a birth) or terminated (a death), as illustrated in Fig. 10.7a. When the first visual measurement of a track segment is connected to the last measurement of another track segment, it is an end-to-end link of the two track segments, as shown in Fig. 10.7b. When the last measurement of a track segment is linked to a visual measurement in the middle of another track segment, it implies a merge, as shown in Fig. 10.7c. Similarly, when a visual measurement in the middle of a track segment is linked to the first measurement of another track segment, it implies a split, as shown in Fig. 10.7d. Visual measurements in the middle of track segments cannot be connected to visual measurements in other track segments, s, or t.

The choices among those possible birth, death, end-to-end, split and merge associations are optimized by a second-stage linear assignment problem. For the second stage problem, the set of all these possible associations is constrained to

$$\mathbb{E}_2 = \{(\tilde{v}_{ij}, \tilde{v}_{kl}) \in \mathbb{E} : (j = J_i) \vee (j > 1 \wedge l = 1)\}. \tag{10.7}$$

In the set, the first condition of $j = J_i$ implies that the last measurement of S_i can be connected to an measurement $\tilde{v}_{kl}$ in S_k. If $l = 1$, it implies that an end-to-end link, and if $l > 1$, it implies a merge of the last element of S_i into $\tilde{v}_{kl}$. The second condition of $(j > 1 \wedge l = 1)$ is that a measurement $\tilde{v}_{ij}$ in the middle of S_i is connected to the first element of S_k, implying a split of $\tilde{v}_{ij}$ into $\tilde{v}_{ij+1}$ and the first element of S_k. The second-stage linear assignment problem under the constraints is formulated as

$$\begin{aligned}
\text{Minimize} \quad & \sum_{e \in \mathbb{E}_2} c_e z_e \\
\text{subject to} \quad & \sum_{e \in I_2(v)} z_e = 1 \text{ for } v \in V \\
& \sum_{e \in O_2(v)} z_e = 1 \text{ for } v \in V \\
& z_e \in \mathbb{B} \text{ for } e \in \mathbb{E}_2,
\end{aligned} \tag{10.8}$$

where $I_2(v) \subset \mathbb{E}_2$ is a set of all incoming edges to $v \in V$, and $O_2(v)$ is a set of all outgoing edges from $v \in V$. The second stage problem inherits all good properties of the first stage problem. In particular, the constraint matrix is totally unimodular, so its optimal solution can be achieved simply by solving the linear relaxation of the binary integer programming problem in Eq. (10.8).

A major issue with this two-stage approach is that the result of the first stage affects that of the second stage. Errors in the first stage cannot be fixed in the second stage, which possibly induces secondary errors down the road, and consequently, these cascading errors are hard to predict and analyze.

10.5 Multi-Way Minimum Cost Data Association

Park et al. (2015) proposed a generic data association approach, addressing the limitations of the approaches described in the preceding sections. They introduced a new optimization formulation for a data association, which allows object interactions of a general degree of M.

To extend the simple linear assignment approach in Sect. 10.2, the new approach relaxes the in-degree and out-degree constraints in Eqs. (10.1c) and (10.1b) to allow a general degree of M

$$\text{In-degree constraint: } 1 \leq \sum_{e \in I(v)} z_e \leq M \text{ for } v \in V,$$
$$\text{Out-degree constraint: } 1 \leq \sum_{e \in O(v)} z_e \leq M \text{ for } v \in V. \tag{10.9}$$

With the relaxed constraints, a merge of m ($\leq M$) objects to $v \in V$ can be represented by a set of m edges having the same end node v, namely

$$\{e\} \in I_m(v). \tag{10.10}$$

Here we use a set notation $\{e\}$ to denote an element in $I_m(v)$, because the element $\{e\}$ is a set of m edges. Similarly, a split to l ($\leq M$) objects can be represented by a set of l edges having the same start node $v \in V$, namely

$$\{e\} \in O_l(v). \tag{10.11}$$

In addition, one-to-l and m-to-one associations can share common edges; one such example is illustrated in Fig. 10.5b. Park et al. (2015) proved that every pair of one-to-l and m-to-one associations can share at most one edge. Furthermore, there exists $\{e1\} \in I_m(v_2)$ and $\{e2\} \in O_l(v_1)$ such that they share a common edge,

$$\{e1\} \cap \{e2\} = (v1, v2).$$

For an edge $(v1, v2) \in \mathbb{E}$, the collection of the one-to-l and m-to-one associations that share the edge is denoted by

$$C_{m,l}(v1, v2) = \{\{e1\} \cup \{e2\}; \{e1\} \cap \{e2\} = (v1, v2), \{e1\} \in I_m(v_2), \{e2\} \in O_l(v_1)\}.$$

The binary decision variable that indicates whether $\{e\} \in I_m(v)$ or $\{e\} \in O_l(v)$ activates in a data association graph is defined as a product of the binary variables,

$$z_{\{e\}} = \prod_{e1 \in \{e\}} z_{e1},$$

which is one only if all of the element edges in the set are activated, and thus, the corresponding z_{e1}'s are all one. It is not difficult to see the following inequality between $z_{\{e\}}$ and z_{e1},

$$z_{\{e\}} \leq z_{e1}, \quad \forall e1 \in \{e\}. \tag{10.12}$$

In addition, $z_{\{e\}}$ is also bounded by

$$\sum_{e1\in\{e\}} (z_{e1} - 1) + 1 \le z_{\{e\}} \text{ and } 0 \le z_{\{e\}}. \tag{10.13}$$

These inequalities as well as the degree constraints in Eq. (10.9) constitute the constraints for the new data association formulation.

To define the cost of the data associations in the new formulation, we first introduce a few notations. Denote by c_{e1} the associated cost for an edge $e1 \in \mathbb{E}$ and by $c_{\{e\}}$ the associated cost for $\{e\}$. The total association cost includes the following four terms.

C1. The cost terms for one-to-one associations:

$$\sum_{e1\in\mathbb{E}} c_{e1} z_{e1}.$$

Please note that when the start node of $e1$ is s (i.e., a birth) or the end node is t (i.e., a death), $c_{e1} = c_{\max}$.

C2. The Cost terms for one-to-l associations, for $1 < l <= M$:

$$\sum_{v\in V_r} \sum_{\{e\}\in O_l(v)} \tilde{c}_{\{e\}} z_{\{e\}}$$

C3. The cost terms for m-to-one associations, for $1 < m <= M$:

$$\sum_{v\in V_r} \sum_{\{e\}\in I_m(v)} \tilde{c}_{\{e\}} z_{\{e\}}.$$

C4. The cost terms for the cases that an m-to-one association and an one-to-l association share a common edge:

$$\sum_{(v1,v2)\in\mathbb{E}} \sum_{\{e\}\in C_{m,l}(v1,v2)} \tilde{c}_{\{e\}} z_{\{e\}}.$$

In the above description, we used the adjusted cost term $\tilde{c}_{\{e\}}$ for a merge, a split or their combination because when an association $\{e\}$ is activated, all of its sub-associations $\{e2\} \subset \{e\}$ are also activated. The cost terms corresponding to the sub-associations are consequently added to the total cost. But these sub-association costs need to be taken out. The adjusted cost term is the remaining cost of $c_{\{e\}}$, after excluding the cost terms corresponding to sub-associations, i.e.,

$$\tilde{c}_{\{e\}} = c_{\{e\}} - \sum_{j=1}^{m-1} \sum_{\{e2\}\in C_j(\{e\})} \tilde{c}_{\{e2\}}. \tag{10.14}$$

In summary, the objective function of an M-way data association problem is the summation of the cost terms in C1 through C4.

To put all constraints and cost terms into a vectorial form, we introduce the following vectorial notations, $\boldsymbol{z}_1 = (z_{e1}; e1 \in \mathbb{E})$, $\boldsymbol{z}_{1l} = (z_{\{e\}}; \{e\} \in O_l(v), v \in V_r)$, $\boldsymbol{z}_{m1} = (z_{\{e\}}; \{e\} \in I_m(v), v \in V_r)$, and $\boldsymbol{y}_{ml} = (\{e\} \in C_{m,l}(v1, v2); (v1, v2) \in \mathbb{E})$, with the corresponding cost coefficient vectors of $\boldsymbol{c}_1$, $\boldsymbol{c}_{1l}$, $\boldsymbol{c}_{m1}$ and $\boldsymbol{d}_{ml}$, respectively. Then, the objective function is

$$\boldsymbol{c}_1^T \boldsymbol{z}_1 + \sum_{l=2}^{M} \boldsymbol{c}_{1l}^T \boldsymbol{z}_{1l} + \sum_{m=2}^{M} \boldsymbol{c}_{m1}^T \boldsymbol{z}_{m1} + \sum_{l=2}^{M} \sum_{m=2}^{M} \boldsymbol{d}_{ml}^T \boldsymbol{y}_{ml}. \tag{10.15}$$

The M-way data association problem includes the following in-degree and out-degree constraints:

D1. $\boldsymbol{A}_1 \boldsymbol{z}_1 \geq \boldsymbol{b}_1$ includes a set of all in-degree and out-degree constraints in Eq. (10.9).

In addition, we have the following constraints connecting the lower order variables to the higher order variables.

D2. $\boldsymbol{A}_{1l} \boldsymbol{z}_1 + \boldsymbol{B}_{1l} \boldsymbol{z}_{1l} \geq \boldsymbol{b}_{1l}$ includes the constraints in Eqs. (10.12) and (10.13) for $\{e\} \in O_l(v)$ and $v \in V_r$.
D3. $\boldsymbol{A}_{m1} \boldsymbol{z}_1 + \boldsymbol{C}_{m1} \boldsymbol{z}_{m1} \geq \boldsymbol{b}_{m1}$ includes the constraints in Eqs. (10.12) and (10.13) for $\{e\} \in I_m(v)$ and $v \in V_r$.
D4. $\boldsymbol{P}_{1l} \boldsymbol{z}_{1l} + \boldsymbol{Q}_{m1} \boldsymbol{z}_{m1} + \boldsymbol{R}_{ml} \boldsymbol{y}_{ml} \geq \boldsymbol{b}_{ml}$ includes the constraints (10.12) and (10.13) for $\{e\} \in C_{m,l}(v1, v2)$ and $(v1, v2) \in \mathbb{E}$.

In the summary, the general data association problem for tracking object interactions of degree M is formulated as

$$\text{Minimize} \quad \boldsymbol{c}_1^T \boldsymbol{z}_1 + \sum_{l=1}^{M} \boldsymbol{c}_{1l}^T \boldsymbol{z}_{1l} + \sum_{m=1}^{M} \boldsymbol{c}_{m1}^T \boldsymbol{z}_{m1} + \sum_{m,n=1}^{M} \boldsymbol{d}_{ml}^T \boldsymbol{y}_{ml},$$

$$\text{subject to} \quad \boldsymbol{A}_1 \boldsymbol{z}_1 \geq \boldsymbol{b}_1, \tag{10.16a}$$

$$\boldsymbol{A}_{1l} \boldsymbol{z}_1 + \boldsymbol{B}_{1l} \boldsymbol{z}_{1l} \geq \boldsymbol{b}_{1l}, \tag{10.16b}$$

$$\boldsymbol{A}_{m1} \boldsymbol{z}_1 + \boldsymbol{C}_{m1} \boldsymbol{z}_{m1} \geq \boldsymbol{b}_{m1}, \tag{10.16c}$$

$$\boldsymbol{P}_{ml} \boldsymbol{z}_{1l} + \boldsymbol{Q}_{ml} \boldsymbol{z}_{m1} + \boldsymbol{R}_{ml} \boldsymbol{y}_{ml} \geq \boldsymbol{b}_{ml}, \tag{10.16d}$$

$$\boldsymbol{z}_1 \in B^{p_1}, \boldsymbol{z}_{1l} \in B^{p_{1l}}, \boldsymbol{z}_{m1} \in B^{p_{m1}}, \boldsymbol{y}_{ml} \in B^{q_{ml}}.$$

10.5.1 Special Properties of the Constraint Coefficient Matrix

Consider the constraint coefficient matrix of the general data association problem in Eq. (10.16),

$$\boldsymbol{R} = \begin{bmatrix} \boldsymbol{A}_1 & \mathbf{0} & \mathbf{0} & \mathbf{0} \\ \boldsymbol{A}_{1l} & \boldsymbol{B}_{1l} & \mathbf{0} & \mathbf{0} \\ \boldsymbol{A}_{m1} & \mathbf{0} & \boldsymbol{C}_{m1} & \mathbf{0} \\ \mathbf{0} & \boldsymbol{P}_{ml} & \boldsymbol{Q}_{ml} & \boldsymbol{R}_{ml} \end{bmatrix}.$$

Theorem 10.1 below states that $\boldsymbol{R}$ matrix is not totally unimodular; its proof can be found in the appendix of Park et al. (2015). Therefore, the linear programming (LP) relaxation of problem in Eq. (10.16) does not produce an integer optimal solution.

Theorem 10.1 *$\boldsymbol{R}$ is not totally unimodular.*

However, the constraint coefficient matrix $\boldsymbol{A}_1$ in Eq. (10.16a) is totally unimodular, since the matrix is the constraint coefficient matrix for a network flow constraint (Schrijver 2003, Theorem 13.9). Moreover, the constraint coefficient matrix for the constraint in Eq. (10.16d) is also totally unimodular, as stated formally in Theorem 10.2 below.

Theorem 10.2 *$[\boldsymbol{P}_{mn} \quad \boldsymbol{Q}_{mn} \quad \boldsymbol{R}_{ml}]$ is totally unimodular.*

Its proof can also be found in the appendix of Park et al. (2015).

10.5.2 Lagrange Dual Solution

The constraint coefficient matrix for Eq. (10.16a) and that for Eq. (10.16d) are totally unimodular. While the two 'bad' constraints in Eqs. (10.16b) and (10.16c) are not totally unimodular, we can relax them through Lagrange relaxation, so that the constraint coefficient matrices of the relaxed problem are totally unimodular, and thus, they form a nice block diagonal matrix as follows:

$$\text{(LR) Minimize} \quad \boldsymbol{c}_1^T \boldsymbol{z}_1 + \sum_{l=1}^{M} \boldsymbol{c}_{1l}^T \boldsymbol{z}_{1l} + \sum_{m=1}^{M} \boldsymbol{c}_{m1}^T \boldsymbol{z}_{m1} + \sum_{l,m=1}^{M} \boldsymbol{d}_{ml}^T \boldsymbol{y}_{ml},$$

$$+ \sum_{l=1}^{M} \boldsymbol{\lambda}_{1l}^T (\boldsymbol{b}_{1l} - \boldsymbol{A}_{1l}\boldsymbol{z}_1 - \boldsymbol{B}_{1l}\boldsymbol{z}_{1l}),$$

$$+ \sum_{m=1}^{M} \boldsymbol{\lambda}_{m1}^T (\boldsymbol{b}_{m1} - \boldsymbol{A}_{m1}\boldsymbol{z}_1 - \boldsymbol{B}_{m1}\boldsymbol{z}_{m1}),$$

$$
\begin{aligned}
&\boldsymbol{A}_1 \boldsymbol{z}_1 && \geq \boldsymbol{b}_1,\\
&\qquad \boldsymbol{P}_{ml}\boldsymbol{z}_{1l} + \boldsymbol{Q}_{ml}\boldsymbol{z}_{m1} + \boldsymbol{R}_{ml}\boldsymbol{y}_{ml} && \geq \boldsymbol{b}_{ml},\\
&\boldsymbol{z}_1 \in B^{p_1}, \boldsymbol{z}_{1l} \in B^{p_{1l}}, \boldsymbol{z}_{m1} \in B^{p_{m1}}, \boldsymbol{y}_{ml} \in B^{q_{ml}},\\
&\boldsymbol{\lambda}_{1l} \geq \mathbf{0}, \boldsymbol{\lambda}_{m1} \geq 0,
\end{aligned}
$$

where $\boldsymbol{\lambda}_{1l}$'s and $\boldsymbol{\lambda}_{m1}$'s are the Lagrange multipliers. Problem (LR) has exactly the same optimal solution as the original problem in Eq. (10.16). Unfortunately, solving problem (LR) is still NP-hard since it is a non-convex quadratic programming problem (Pardalos and Vavasis 1991).

For solving Problem (LR), Park et al. (2015) pursued a Lagrangian dual solution of the problem, which repeats the iterative runs of two steps. The first step is to optimize Problem (LR) for all binary variables, while keeping the Lagrange multipliers unchanged; this first step is referred to as the *sub problem (SP)*. The second step is to optimize Problem (LR) for the Lagrange multipliers, while keeping the binary variables unchanged; this second step is referred to as the *master problem (MP)*.

The sub problem solves

$$
\begin{aligned}
&\text{(SP) } z_{LR}(\boldsymbol{\lambda}_{1l}, \boldsymbol{\lambda}_{m1}) =\\
&\text{Minimize} \quad \boldsymbol{c}_{1,\lambda}^T \boldsymbol{z}_1 + \sum_{l=1}^{M} \boldsymbol{c}_{1l,\lambda}^T \boldsymbol{z}_{1l} + \sum_{m=1}^{M} \boldsymbol{c}_{m1,\lambda}^T \boldsymbol{z}_{m1} + \sum_{l,m=1}^{M} \boldsymbol{d}_{ml}^T \boldsymbol{y}_{ml},\\
&\qquad \boldsymbol{A}_1 \boldsymbol{z}_1 \qquad\qquad\qquad\qquad \geq \boldsymbol{b}_1,\\
&\qquad\qquad \boldsymbol{P}_{ml}\boldsymbol{z}_{1l} + \boldsymbol{Q}_{ml}\boldsymbol{z}_{m1} + \boldsymbol{R}_{ml}\boldsymbol{y}_{ml} \geq \boldsymbol{b}_{ml},\\
&\qquad \boldsymbol{z}_1 \in B^{p_1}, \boldsymbol{z}_{1l} \in B^{p_{1l}}, \boldsymbol{z}_{m1} \in B^{p_{m1}},\\
&\qquad \boldsymbol{y}_{ml} \in B^{q_{ml}},
\end{aligned}
$$

where $\boldsymbol{c}_{1,\lambda} = (\boldsymbol{c}_1 - \sum_{l=1}^{M} \boldsymbol{A}_{1l}^T \boldsymbol{\lambda}_{1l} - \sum_{m=1}^{M} \boldsymbol{A}_{m1}^T \boldsymbol{\lambda}_{m1})^T$, $\boldsymbol{c}_{1l,\lambda} = (\boldsymbol{B}_{1l}^T \boldsymbol{\lambda}_{1l} - \boldsymbol{c}_{1l})^T$, and $\boldsymbol{c}_{m1,\lambda} = (\boldsymbol{B}_{m1}^T \boldsymbol{\lambda}_{m1} - \boldsymbol{c}_{m1})$. Due to the nice block diagonal structure in the constraint coefficient matrix of Problem (SP), the sub problem can be decomposed further into Problem (SP1) and Problem (SP2), in which the binary constraints of (SP1) and (SP2) are relaxed to be general linear boundary constraints, thanks to the total unimodularities of (SP1)'s and (SP2)'s constraint coefficient matrices. This is to say

$$
\text{(SP1)} \quad \boldsymbol{z}_1^* = \text{argmin}\{\boldsymbol{c}_{1,\lambda}^T \boldsymbol{z}_1;\ \boldsymbol{A}_1 \boldsymbol{z}_1 \geq \boldsymbol{b}_1, \mathbf{0} \leq \boldsymbol{z}_1 \leq \mathbf{1}\},
$$

and

$$(\text{SP2}) \quad \{z^*_{1l}, z^*_{m1}, y^*_{ml}\}$$
$$= \operatorname{argmin}\Big\{\sum_{l=1}^{M} c^T_{1l,\lambda} z_{1l} + \sum_{m=1^M} c^T_{m1,\lambda} z_{m1} + \sum_{l,m=1}^{M} d^T_{ml} y_{ml},$$
$$P_{ml} z_{1l} + Q_{ml} z_{m1} + R_{ml} y_{ml} \geq b_{ml},$$
$$0 \leq z_{1l}, z_{m1}, y_{ml} \leq 1\Big\}.$$

Problem (SP) above is a general linear programming problem, which is solvable in polynomial time.

Problem (SP) is a relaxation of the original problem in Eq. (10.16), meaning that its optimal value $z_{LR}(\boldsymbol{\lambda}_{1l}, \boldsymbol{\lambda}_{m1})$ is always smaller than or equal to the optimal value of the original problem. In the master problem, the values of the Lagrange multipliers $(\boldsymbol{\lambda}_{1l}, \boldsymbol{\lambda}_{m1})$ are updated so as to minimize the gap between the optimal values of the relaxation problem and the original problem; this is accomplished by maximizing $z_{LR}(\boldsymbol{\lambda}_{1l}, \boldsymbol{\lambda}_{m1})$.

Let $z_1^{(k)}$ be the optimal solution of (SP1) at iteration k and let $z_{1l}^{(k)}$'s and $z_{m1}^{(k)}$'s be the optimal solution of (SP2) at iteration k. For iteration k, the master problem solves

$$(\text{MP}) \quad \{\boldsymbol{\lambda}_{1l}, \boldsymbol{\lambda}_{m1}\} = \operatorname{argmax}\Big\{\sum_{l=1}^{M} \boldsymbol{\lambda}^T_{1l}(\boldsymbol{b}_{1l} - \boldsymbol{A}_{1l} z_1^{(k)} - \boldsymbol{B}_{1l} z_{1l}^{(k)})$$
$$+ \sum_{m=1}^{M} \boldsymbol{\lambda}^T_{m1}(\boldsymbol{b}_{m1} - \boldsymbol{A}_{m1} z_1^{(k)} - \boldsymbol{B}_{m1} z_{m1}^{(k)})\Big\} \tag{10.17}$$

for updating $\boldsymbol{\lambda}_{1l}$ and $\boldsymbol{\lambda}_{m1}$. A common approach for solving Problem (MP) is to use a subgradient algorithm. The subgradient algorithm updates $(\boldsymbol{\lambda}_{1l}^{(k-1)}, \boldsymbol{\lambda}_{m1}^{(k-1)})$ in the previous iteration to $(\boldsymbol{\lambda}_{1l}^{(k)}, \boldsymbol{\lambda}_{m1}^{(k)})$ through

$$(\boldsymbol{\lambda}_{1l}^{(k)}, \boldsymbol{\lambda}_{m1}^{(k)}) = (\boldsymbol{\lambda}_{1l}^{(k-1)}, \boldsymbol{\lambda}_{m1}^{(k)}) + \alpha_k \nabla\boldsymbol{\lambda}^{(k-1)},$$

where $\nabla\boldsymbol{\lambda}^{(k-1)} = (\boldsymbol{b}_{1l} - \boldsymbol{A}_{1l} z_1^{(k)} - \boldsymbol{B}_{1l} z_{1l}^{(k)}, \boldsymbol{b}_{m1} - \boldsymbol{A}_{m1} z_1^{(k)} - \boldsymbol{B}_{m1} z_{m1}^{(k)})$. Since the objective function of problem (MP) is piecewise linear and concave (Nemhauser and Wolsey 1988, Section II.3.6), the subgradient algorithm should converge with a proper choice of α_k. In practice, α_k was chosen as

$$\alpha_k = s \frac{|z_{MP}^{(k-1)} - z_{SP}^{(k-1)}|}{||\nabla\boldsymbol{\lambda}^{(k-1)}||_2},$$

where $z_{MP}^{(k-1)}$ is the best objective value of Problem (MP) found by the $(k-1)$th iteration, $z_{SP}^{(k-1)}$ is the objective value of Problem (SP) found at the $(k-1)$th iteration, and s is a constant in between 0 and 2 (Polyak 1987); our recommendation is to use

$$s = \frac{1}{|z_{MP}^{(1)} - z_{SP}^{(1)}|}.$$

The difference between $z_{MP}^{(k-1)}$ and $z_{SP}^{(k-1)}$ is called the duality gap. The iterations of solving Problems (MP) and (SP) are considered 'converged' when the change in the duality gap is less than a certain threshold. The number of iterations to convergence is different for different problems. For our numerical study in Sect. 10.6, the convergence is fast. We have reached the convergence with less than 500 iterations for most of our simulation cases. The solution of Problem (SP) at the convergence provides the Lagrange dual relaxation solution of the original problem in Eq. (10.16). The Lagrangian dual approach always provides a solution closer to the true optimal solution than the LP relaxation (Nemhauser and Wolsey 1988, Section II.3.6).

We would like to add a remark on the computation complexity of the Lagrange dual solution approach. The worst time complexity of solving a linear programming problem is proportional to $n^3/\log(n)$ where n is the number of variables (Anstreicher 1999). For N objects and F time frames in a video, a data association problem involves the association of $N \cdot F$ data features, and the number of simple one-to-one associations among them is less than N^2F^2. Therefore, Problem (SP1) becomes a linear programming problem with at most N^2F^2 decision variables, and its worst time complexity is $O(N^6F^6/\log N^2 \cdot F^2)$. Since a visual measurement can be associated with $(N \cdot F - 1)$ other visual measurements, the possible number of one-to-l and m-to-one associations is at most $\binom{NF-1}{M}$, and the total number of such associations is at most $NF\binom{NF-1}{M}$, which is less than $N^{M+1}F^{M+1}$. In reality, the number should be much smaller than $N^{M+1}F^{M+1}$ if one counts only the plausible associations. As such, Problem (SP2) has at most $N^{M+1}F^{M+1}$ decision variables, and the worst time complexity of the problem becomes $O(N^{3M+3}F^{3M+3}/log(N^{3M+3}F^{3M+3}))$. The complexity of Problem (MP) is comparable to that of Problem (SP2), which is rather expensive, but it is still far cheaper than the complexity of the original binary integer problem, which is $O(2^{NF})$. We also like to note that the complexity of this Lagrange dual approach is comparable to Henriques' and Serge's with $M = 2$. Even when N is a few hundreds, the Lagrange dual solution approach can be solved for $M \leq 3$ within a practically reasonable time.

10.6 Case Study: Data Association for Tracking Particle Interactions

We tested different data association approaches to the problem of particle tracking in a video. The set of videos include some simulated particle videos as well as a real video of nanoparticles obtained by an electron microscope. For each dataset, we tried four state-of-the-art data association methods, i.e., the Henriques' method described in Sect. 10.3, the Jaqaman's method described in Sect. 10.4, the MCMC Data Association method (Yu and Medioni 2009), and the multi-way minimum cost data association approach described in Sect. 10.5. We feed to different methods the same visual measurements so that the comparison can be focused on evaluating the performance of the data association step. To do that, we first ran an existing object detection method to obtain the visual measurements of objects and then used the resulting measurements as inputs for all of the aforementioned methods for obtaining the data association results. The performance metrics used are the false positive rates and false negative rates for the outcome of the data associations, including the 1-to-2 split, 2-to-1 merge, 1-to-3 split, 3-to-1 merge, 1-to-m split ($m > 3$), n-to-1 merge ($n > 3$), faulty measurements, birth events, and death events.

10.6.1 Simulation Study

First we generated several sets of simulated videos data that visualize a number of moving and interacting objects. We recorded the locations and visual measurements of the objects for each image frame and stored the associations among the visual measurements, which serve as the ground truth for the data associations. The ground truth is needed to compute the aforementioned performance metrics. The computation times were also recorded.

Simulation Details We simulated a system of N particles in a bounded two-dimensional space. All particles have circular outlines with a random size in radius. The radius of particle i, denoted by r_i, is sampled from

$$\log r_i \sim \mathcal{N}(0.5, 0.1).$$

The particles are allowed to move randomly and then interact when encountering. At time $t = 1$, all particles are equally spaced over the two-dimensional bounded space, with the distance to their closest neighbors equalling to g. After that, particles start moving independently in a Brownian motion, with its centroid position randomly changing to $(x_{i,t}, y_{i,t})$ at time t, via the sampling of

$$x_{i,t+1}|x_{i,t} \sim \mathcal{N}(x_{i,t}, \sigma^2),\ y_{i,t+1}|y_{i,t} \sim \mathcal{N}(y_{i,t}, \sigma^2).$$

We fixed $\sigma = 0.3$, but please note that the movement speeds of the particles do still vary even with a fixed σ.

We indirectly change the degree of inter-particle interactions by alerting g. With a smaller g, the particles locate close together at the beginning of the simulation process, while with a greater g, the particles have more space in between. The subsequent Brownian motion of the particles will cause some particles to merge or some merged particles to split again. We tested different values of g, from 1.5 to 4.0 with an increment of 0.5. The smallest g is chosen to be 1.5. This choice is made because we would like for the particles not to touch each other at the beginning of the simulation. Considering that the mean radius used at time 1 to simulate the particles is 0.5, a choice of g smaller than 1.5 will result too many particles touch each other at time 1. On the other end, the value of 4.0 is chosen as the largest value for g because too few particles are interacting when g is greater than that value.

To simulate faulty measurements, we randomly generated small spherical particles, with its radius equal to one, at random time frames and locations. To qualify them as faulty measurements, we make these small particles to immediately disappear one frame after their appearance.

To simulate the births of new objects, we randomly generated new particles at random locations and random times. The time between two successive births is set to be an exponential distribution of mean 2. If a particle moves out of the predefined two-dimensional bounded box, we treat the particle as 'dead'.

The number of time frames in the simulated videos is fixed to 20, and the number of particles per frame varies, depending on the number of faulty measurements and the numbers of birth and death events, but the average is 81. The problem of particle tracking in the simulation requires the solution of a data association problem of 1620 ($= 81 \times 20$) measurements, on average. The number of simple associations among all measurements is $19 \times 81 \times 81 = 124,659$.

Figure 10.8 presents an example of the simulated random process from time 1 to time 19 with an increment of 2.

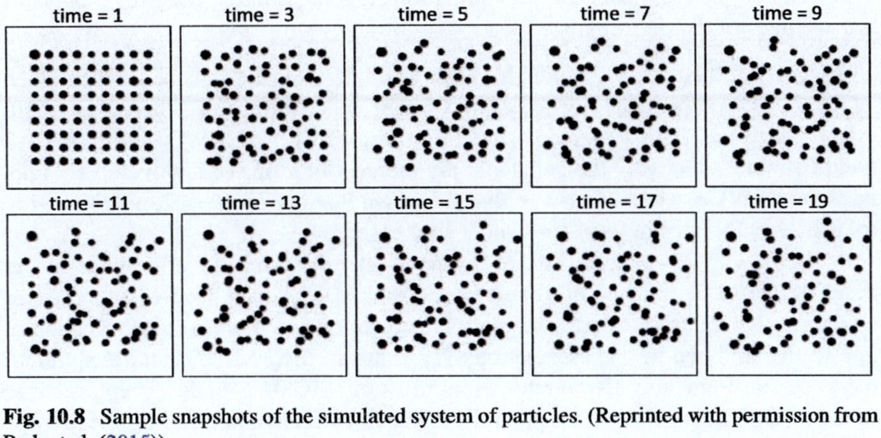

Fig. 10.8 Sample snapshots of the simulated system of particles. (Reprinted with permission from Park et al. (2015))

The cost of associating two particle measurements is computed based on the Hausdorff distance between the interior point sets of the two respective measurements, each set of which is a set of point coordinates inside the outline of an object. Let $\boldsymbol{x}_i$ denote the interior point set for particle measurement i. The cost of associating measurement i with measurement j, i.e., c_e for $e = (i, j)$, is denoted by $d_H(\boldsymbol{x}_i, \boldsymbol{x}_j)$, where

$$d_H(X, Y) = \max\left\{ \sup_{x\in X} \inf_{y\in Y} ||x - y||,\ \sup_{y\in Y} \inf_{x\in X} ||x - y|| \right\}$$

and $||x - y||$ is the Euclidean distance in $\mathbb{R}^2$. The cost of an one-to-l association, or m-to-one, or their combination, is

$$c_{\{e\}} = d_H\left(\bigcup_{i:(i,j)\in\{e\}} \boldsymbol{x}_i,\ \bigcup_{j:(i,j)\in\{e\}} \boldsymbol{x}_j \right).$$

which is the Hausdorff distance between a union of all starting nodes of the edges in $\{e\}$ and a union of all ending nodes of the edges in $\{e\}$. The cost coefficients, f_1 and f_2, for the birth and death events, respectively, are uniformly set to one, close to the 3σ value.

Results For each $g \in \{1.5, 2.0, 2.5, 3.0, 3.5, 4.0\}$, we generated 40 different simulated datasets. For each dataset, we ran the multi-way minimum-cost data association (MWDA) method with both $M = 2$ and $M = 3$, the Henriques' method, the Jaqaman's method, and the MCMC data association (MCMC-DA) method. We computed the false positive (FP) rates and false negative (FN) rates for the following scenarios: overall, 1-to-2 split, 2-to-1 merge, 1-to-3 split, 3-to-1 merge, 1-to-m split ($m > 3$), n-to-1 merge ($n > 3$), faulty measurements, birth events, and death events. We averaged the rates over the 40 simulation runs for each g. Table 10.1 presents the averages of the performance metrics.

In terms of the overall FN rate, MWDA when $M = 3$ performs better than when $M = 2$. The performance gap becomes larger as g decreases. When g is small, 1-to-m splits and n-to-1 merges occur more frequently, and MWDA with $M = 3$ handled more effectively the splits and the merges. In terms of the overall FP rate, however, MWDA when $M = 3$ is slightly worse than when $M = 2$, mainly owing to its higher FP rates in the 1-to-3 and 3-to-1 cases.

MWDA with $M = 2$ was more accurate than the other three methods in the cases of one-to-two and two-to-one associations. MWDA can handle general cases of one-to-two and two-to-one associations, attributing to its better performance in one-to-two and two-to-one associations. By contrast, MCMC-DA handles split and merged measurements based on the reversible jump MCMC, but the acceptance rate of the MCMC step is low. Consequently, it was not effective in handling the merged and split measurements. Jaqaman's method has the FP rates comparable to MWDA, but significantly higher FN rates.

Table 10.1 Data association performance metrics using the simulated videos. The false positive rates (FP) and false negative rates (FN) are averaged over the 40 simulation runs for a given g. *Source*: Park et al. (2015) with permission

		MWDA ($M = 3$)		*MWDA* ($M = 2$)		*Henriques*		*Jaqaman*		*MCMC-DA*	
$g = 1.5$	*Total*	*FN*	*FP*	*FN*	*FP*	*FN*	*FP*	*FN*	*FP*	*FN*	*FP*
Overall	2090.9000	0.0935	0.0628	0.1165	0.0519	0.1357	0.0493	0.1707	0.0916	0.2849	0.0268
1-to-2	196.6500	0.3956	0.2567	0.3692	0.3564	0.3860	0.3503	0.5230	0.4766	1.0000	1.0000
2-to-1	235.3500	0.3701	0.2736	0.3537	0.4036	0.3644	0.3965	0.4353	0.4816	1.0000	1.0000
1-to-3	53.7750	0.4644	0.4887	1.0000	0.0000	1.0000	0.0000	1.0000	0.0000	1.0000	1.0000
3-to-1	66.2250	0.4915	0.5877	1.0000	0.0000	1.0000	0.0000	1.0000	0.0000	1.0000	0.0000
1-to-m	14.3000	1.0000	0.0000	1.0000	0.0000	1.0000	0.0000	1.0000	0.0000	1.0000	0.0000
n-to-1	27.6000	1.0000	0.0000	1.0000	0.0000	1.0000	0.0000	1.0000	0.0000	1.0000	0.0000
Faulty measurements	14.8750	0.5681	0.1569	0.3899	0.1906	0.3782	0.3309	0.7664	0.6559	0.0773	0.6839
Birth	9.6000	0.1380	0.4626	0.1016	0.5556	0.0964	0.7209	0.4167	0.6393	0.0469	0.8742
Death	10.4000	0.3269	0.6353	0.2452	0.7315	0.2356	0.8236	0.5601	0.6914	0.0553	0.9034
		MWDA ($M = 3$)		*MWDA* ($M = 2$)		*Henriques*		*Jaqaman*		*MCMC-DA*	
$g = 2.0$	*Total*	*FN*	*FP*	*FN*	*FP*	*FN*	*FP*	*FN*	*FP*	*FN*	*FP*
Overall	2622.3500	0.0235	0.0167	0.0318	0.0140	0.0421	0.0133	0.0785	0.0311	0.1536	0.0112
1-to-2	152.1500	0.2152	0.1165	0.2106	0.1751	0.2244	0.1699	0.3076	0.2804	1.0000	1.0000
2-to-1	189.3500	0.1933	0.1178	0.1893	0.1906	0.2010	0.1872	0.2509	0.2705	1.0000	1.0000
1-to-3	17.2500	0.2522	0.3175	1.0000	0.0000	1.0000	0.0000	1.0000	0.0000	1.0000	1.0000
3-to-1	22.5750	0.2492	0.3705	1.0000	0.0000	1.0000	0.0000	1.0000	0.0000	1.0000	0.0000
1-to-m	1.8500	1.0000	0.0000	1.0000	0.0000	1.0000	0.0000	1.0000	0.0000	1.0000	0.0000
n-to-1	3.4750	1.0000	0.0000	1.0000	0.0000	1.0000	0.0000	1.0000	0.0000	1.0000	0.0000
Faulty measurements	15.4500	0.3285	0.0326	0.2298	0.0442	0.2233	0.1257	0.5712	0.0804	0.0421	0.6044
Birth	9.3000	0.0968	0.2538	0.0726	0.3868	0.0726	0.6235	0.2285	0.2439	0.0188	0.8475
Death	9.5750	0.1775	0.3879	0.1332	0.5475	0.1305	0.7170	0.3786	0.3568	0.0313	0.8925

(continued)

Table 10.1 (continued)

$g = 2.5$	*Total*	*MWDA* ($M = 3$)		*MWDA* ($M = 2$)		*Henriques*		*Jaqaman*		*MCMC-DA*	
		FN	*FP*	*FN*	*FP*	*FN*	*FP*	*FN*	*FP*	*FN*	*FP*
Overall	2832.5500	0.0067	0.0051	0.0095	0.0039	0.0127	0.0038	0.0254	0.0127	0.0883	0.0065
1-to-2	89.8500	0.1219	0.0505	0.1185	0.0817	0.1247	0.0796	0.1714	0.1832	1.0000	1.0000
2-to-1	115.1000	0.0969	0.0618	0.0956	0.1106	0.1012	0.1097	0.1295	0.1971	1.0000	1.0000
1-to-3	5.4000	0.1111	0.2809	1.0000	0.0000	1.0000	0.0000	1.0000	0.0000	1.0000	1.0000
3-to-1	8.0250	0.1028	0.2558	1.0000	0.0000	1.0000	0.0000	1.0000	0.0000	1.0000	0.0000
1-to-m	0.4000	1.0000	0.0000	1.0000	0.0000	1.0000	0.0000	1.0000	0.0000	1.0000	0.0000
n-to-1	0.6000	1.0000	0.0000	1.0000	0.0000	1.0000	0.0000	1.0000	0.0000	1.0000	0.0000
Faulty measurements	15.3000	0.1601	0.0096	0.0882	0.0089	0.0882	0.0296	0.2892	0.0109	0.0310	0.5407
Birth	9.2500	0.0243	0.1502	0.0135	0.2224	0.0135	0.3976	0.1108	0.1320	0.0135	0.7809
Death	10.7500	0.0837	0.1564	0.0651	0.2852	0.0651	0.4469	0.2558	0.1950	0.0209	0.8364
$g = 3.0$	*Total*	*MWDA* ($M = 3$)		*MWDA* ($M = 2$)		*Henriques*		*Jaqaman*		*MCMC-DA*	
		FN	*FP*	*FN*	*FP*	*FN*	*FP*	*FN*	*FP*	*FN*	*FP*
Overall	2987.9000	0.0017	0.0017	0.0026	0.0013	0.0045	0.0012	0.0114	0.0055	0.0529	0.0054
1-to-2	49.5000	0.0556	0.0209	0.0545	0.0410	0.0616	0.0383	0.0838	0.1337	1.0000	1.0000
2-to-1	67.0500	0.0477	0.0420	0.0470	0.0644	0.0507	0.0640	0.0641	0.1532	1.0000	1.0000
1-to-3	1.6500	0.0000	0.3125	1.0000	0.0000	1.0000	0.0000	1.0000	0.0000	1.0000	1.0000
3-to-1	2.3250	0.0968	0.1515	1.0000	0.0000	1.0000	0.0000	1.0000	0.0000	1.0000	0.0000
1-to-m	0.0000	0.0000	0.0000	0.0000	0.0000	0.0000	0.0000	0.0000	0.0000	0.0000	0.0000
n-to-1	0.1000	1.0000	0.0000	1.0000	0.0000	1.0000	0.0000	1.0000	0.0000	1.0000	0.0000
Faulty measurements	14.8000	0.0777	0.0055	0.0507	0.0071	0.0507	0.0277	0.1622	0.0045	0.0304	0.4693
Birth	9.5750	0.0209	0.0406	0.0000	0.0726	0.0000	0.2426	0.0470	0.0605	0.0078	0.6917
Death	9.3000	0.0403	0.0746	0.0376	0.1322	0.0376	0.3017	0.1532	0.0992	0.0296	0.8041

Overall	3023.1000	0.0005	0.0006	0.0007	0.0005	0.0014	0.0005	0.0033	0.0017	0.0329	0.0052
1-to-2	25.1500	0.0219	0.0180	0.0219	0.0219	0.0258	0.0220	0.0457	0.0419	1.0000	1.0000
2-to-1	35.2000	0.0270	0.0284	0.0270	0.0379	0.0270	0.0379	0.0355	0.1101	1.0000	1.0000
1-to-3	0.2250	0.0000	0.2500	1.0000	0.0000	1.0000	0.0000	1.0000	0.0000	1.0000	1.0000
3-to-1	0.5250	0.0000	0.2222	1.0000	0.0000	1.0000	0.0000	1.0000	0.0000	1.0000	0.0000
1-to-*m*	0.0000	0.0000	0.0000	0.0000	0.0000	0.0000	0.0000	0.0000	0.0000	0.0000	0.0000
n-to-1	0.0000	0.0000	0.0000	0.0000	0.0000	0.0000	0.0000	0.0000	0.0000	0.0000	0.0000
Faulty measurements	15.0500	0.0216	0.0000	0.0150	0.0017	0.0150	0.0017	0.0282	0.0000	0.0465	0.3902
Birth	9.6250	0.0000	0.0100	0.0000	0.0175	0.0000	0.0989	0.0026	0.0071	0.0234	0.6078
Death	10.1000	0.0124	0.0290	0.0124	0.0453	0.0124	0.1361	0.0668	0.0179	0.0371	0.7194
		MWDA ($M = 3$)		*MWDA* ($M = 2$)		*Henriques*		*Jaqaman*		*MCMC-DA*	
$g = 4.0$	*Total*	*FN*	*FP*	*FN*	*FP*	*FN*	*FP*	*FN*	*FP*	*FN*	*FP*
Overall	3073.6500	0.0001	0.0002	0.0001	0.0001	0.0005	0.0001	0.0011	0.0004	0.0222	0.0047
1-to-2	12.4500	0.0161	0.0041	0.0161	0.0041	0.0161	0.0041	0.0241	0.0000	1.0000	1.0000
2-to-1	19.5500	0.0077	0.0152	0.0077	0.0152	0.0077	0.0152	0.0153	0.0587	1.0000	1.0000
1-to-3	0.0000	0.0000	0.0000	0.0000	0.0000	0.0000	0.0000	0.0000	0.0000	0.0000	1.0000
3-to-1	0.0000	0.0000	1.0000	0.0000	0.0000	0.0000	0.0000	0.0000	0.0000	0.0000	1.0000
1-to-*m*	0.0000	0.0000	0.0000	0.0000	0.0000	0.0000	0.0000	0.0000	0.0000	0.0000	0.0000
n-to-1	0.0000	0.0000	0.0000	0.0000	0.0000	0.0000	0.0000	0.0000	0.0000	0.0000	0.0000
Faulty measurements	15.2500	0.0066	0.0000	0.0000	0.0000	0.0000	0.0033	0.0000	0.0000	0.0328	0.3608
Birth	9.6000	0.0000	0.0026	0.0000	0.0026	0.0000	0.0471	0.0000	0.0025	0.0156	0.5013
Death	9.4750	0.0132	0.0027	0.0132	0.0027	0.0132	0.0579	0.0396	0.0082	0.0290	0.6530

The computation time for MWDA method $M = 2$ per simulation case was 51.6 s, slower than Henriques' method (38.6 s) and Jaqaman's method (25.1 s) but faster than MCMC-DA (79.6 s). The computation time for MWDA method with $M = 3$ was 1734 s, considerably slower than all other methods. All computations were run on a personal computer with `Intel-i7` CPU and 8 GB memory.

10.6.2 Tracking Nanoparticles in In Situ Microscope Images

In this section, we apply the same set of data association methods to a real video. This video is a 89-second solution-phase silver nanoparticle growth, captured by *in situ* transmission electron microscopy, in a rate of one frame per second; please refer to Woehl et al. (2012) for details on the imaging technique. For the data association analysis, we took the subset of the video frames from 40 s to 80 s with one frame per two seconds, resulting in a total of 20 image frames. We chose the latter half of the video because the nanoparticles were not actively changing or interacting for the first 40 s. On average, 280 silver nanoparticles were present per image frame. Individual nanoparticles grow with time—some merge into larger aggregates, while some others split into smaller particles. Understanding how the particles interact and grow is crucial to the understanding of the nanoparticle growth mechanism.

In the video frames, nanoparticles are not moving fast but they are actively interacting and growing. We first applied a simple image thresholding method on the video frames to extract the outlines and interiors of nanoparticles. Then, we applied the MWDA method with $M = 3$, Henriques' method, Jaqaman's method, and MCMC-DA to the extracted nanoparticle outlines for associating the individual extractions with their counterparts over different time frames.

We randomly picked 18 nanoparticles and manually annotated their movements and interactions in order to create the ground truth; see Fig. 10.9 for three examples of growth trajectory. We compared the outcomes from the data association methods with the ground truth and computed the false positive rates and false negative rates. The same set of performance metrics, as used for the simulated videos, is presented in Table 10.2. Overall, MWDA with $M = 3$ demonstrates a clear advantage over the other three methods.

10.7 Case Study: Pattern Analysis of Nanoparticle Oriented Attachments

Material scientists have long conjectured that during a chemical synthesis, nanoparticles aggregate with one another in certain preferential directions, a process referred to as the *oriented attachment* (Welch et al. 2016; Zhang et al. 2012). We are

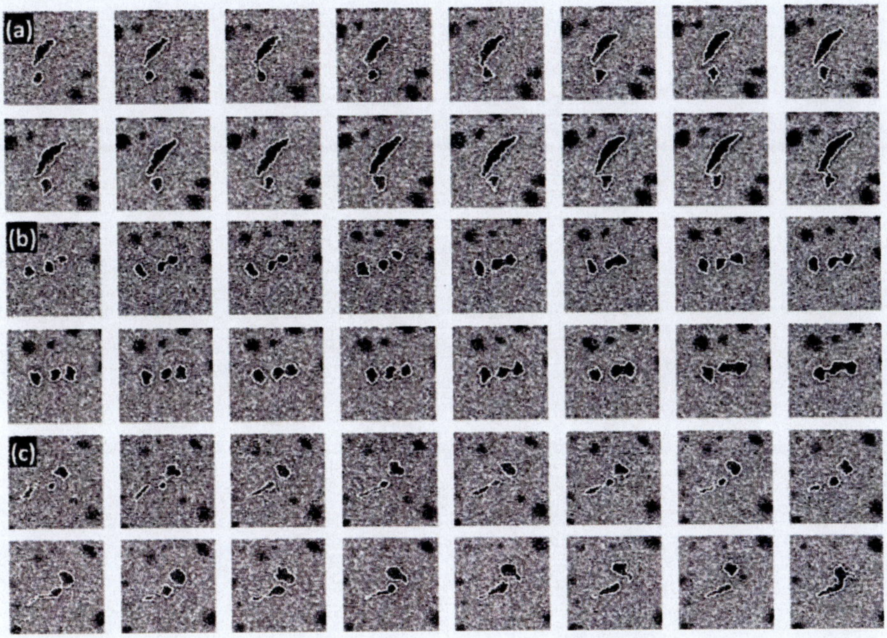

Fig. 10.9 Three samples of nanoparticle trajectories. Each sample goes from top-left to bottom-right, with a time increment of 1 s between two consecutive image frames. (Reprinted with permission from Park et al. (2015))

interested in studying the orientations of two particles involved in these two-to-one aggregation events.

The data association analysis described in Sect. 10.6.2 identified 184 two-to-one aggregation events, of which Fig. 10.10 displays an example. For each aggregation event, we capture the images at two moments: the first is the image of two primary nanoparticles taken immediately before the aggregation, e.g., the image at $t = 2$, and the second is the image of the secondary nanoparticle taken immediately after the aggregation, e.g., the image at $t = 4$. After the aggregation, the orientations of the two primary nanoparticles do not change anymore, due to strong physical forces bonding them together. For this reason, the aggregated image could be taken any time after the aggregation, but our choice is the time immediate after the aggregation just in case that the aggregate might later undergo a significant restructuring.

The time resolution of the imaging process is faster than a normal aggregation speed, so that the moments ‘immediate before the aggregation’ and ‘immediate after the aggregation’ are well defined from the observed image sequences. Each of the before images and after images is two-dimensional, depicting the projection of the three-dimensional geometries of nanoparticles onto a two-dimensional plane. Since the nanoparticles imaged are constrained to a very thin layer of a sample chamber, we assume that the geometrical information along the z-direction is relatively

Table 10.2 Data association performance metrics using the real electron microscope video

Types of data association	Total	MWDA ($M = 3$)		Henriques		Jaqaman		MCMC-DA	
		FN	FP	FN	FP	FN	FP	FN	FP
1-to-1	1208.0000	0.0331	0.0379	0.0861	0.0612	0.4912	0.2614	0.5066	0.2857
1-to-2	200.0000	0.0200	0.1091	0.1000	0.1667	0.9600	0.8000	1.0000	1.0000
2-to-1	228.0000	0.0351	0.0984	0.1140	0.1368	0.8947	0.5862	0.9912	0.9091
1-to-3	0.0000	0.0000	0.0000	0.0000	0.0000	0.0000	0.0000	0.0000	1.0000
3-to-1	1.0000	1.0000	0.0000	1.0000	0.0000	1.0000	0.0000	1.0000	1.0000
1-to-m	0.0000	0.0000	0.0000	0.0000	0.0000	0.0000	0.0000	0.0000	0.0000
n-to-1	0.0000	0.0000	0.0000	0.0000	0.0000	0.0000	0.0000	0.0000	0.0000
Faulty measurements	0.0000	0.0000	0.0004	0.0000	0.0013	0.0000	0.0004	0.0000	0.0134
Birth	3.0000	0.0000	0.0000	0.0000	0.7500	0.3333	0.8333	0.3333	0.9524
Death	0.0000	0.0000	1.0000	0.0000	1.0000	0.0000	1.0000	0.0000	1.0000

Source: Park et al. (2015) with permission

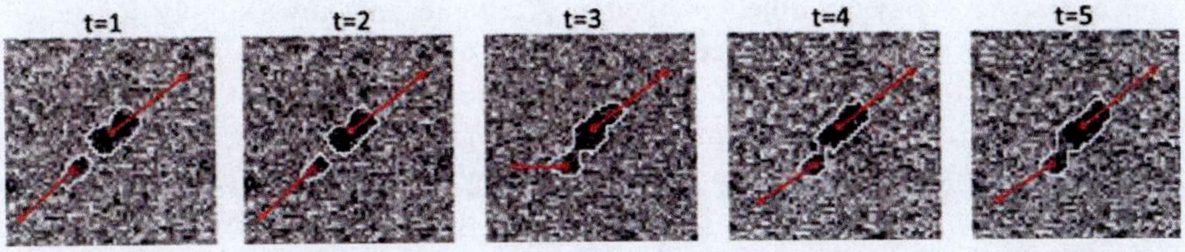

Fig. 10.10 Examples of dynamic microscopy data of particle aggregation. This example consists of a sequence of five microscope images that show the movement and aggregation of two nanoparticles. The second image, labeled 't=2,' is the image of the two nanoparticles taken immediately before the aggregation, whereas the fourth image, labeled 't=4,' is the image right after the aggregation. (Reprinted with permission from Sikaroudi et al. (2018))

Fig. 10.11 A picture illustrating the aggregation of particles. A particle aggregation is a collision of two primary particles followed by a restructuring into a secondary particle. (Reprinted with permission from Sikaroudi et al. (2018))

insignificant. A set of the image pairs for the 184 aggregation events is analyzed for studying how primary nanoparticles are oriented in these events.

10.7.1 Modeling Nanoparticle Oriented Attachments

As illustrated in Fig. 10.11, a particle aggregation is essentially a two-step process—a collision of two primary particles followed by their restructuring into a larger secondary particle. Some collisions effectively lead to a subsequent restructuring (or coalescence), while other collisions may be ineffective. The degree of effectiveness depends on how the primary nanoparticles are spatially oriented in a collision. When the primary particles are oriented ineffectively, they could be separated again or rotate to a preferred orientation (Li et al. 2012).

An aggregation can be mathematically abstracted as a merge of two geometric objects (Sikaroudi et al. 2018). Let us first describe how to model the geometric objects. Let $\mathbb{X}$ denote the set of all image pixel coordinates in an $H \times W$ two-dimensional digital image, such that

$$\mathbb{X} := \{(h, w) : h = 0, 1, 2, \ldots, H, w = 0, 1, 2, \ldots, W\}.$$

This is to say, a geometric object imaged on $\mathbb{X}$ is represented by a simply connected subset of $\mathbb{X}$ that represents the set of all image pixels locating inside the geometric object. This set-based representation has been popularly used in shape analysis (Mémoli and Sapiro 2005; Memoli 2007). Compared with other popular shape representation models such as the representation by landmark points (Kendall 1984; Dryden and Mardia 2016) or the representation by a parametric curve (Younes 1998; Srivastava et al. 2010), the set-based representation meets our analysis needs better in this nanoparticle pattern analysis because an aggregation of two objects can be naturally and readily represented by the union of two pixel sets representing the two objects.

Geometric objects translate and rotate before they aggregate. The translation and rotation of $\mathbb{X}$ are represented by a Euclidean rigid body transformation. Let $\mathbb{SE}(\mathbb{X})$ denote the collection of all Euclidean rigid body transformations defined on $\mathbb{X}$. An element ϕ in $\mathbb{SE}(\mathbb{X})$ is a rigid body transformation that shifts $\boldsymbol{x} \in \mathbb{X}$ by $\boldsymbol{c}_\phi \in \mathbb{X}$ in the negative direction and rotates the shifted object about the origin by $\theta_\phi \in [0, 2\pi]$, that is,

$$\phi(\boldsymbol{x}) = \begin{bmatrix} \cos(\theta_\phi) & -\sin(\theta_\phi) \\ \sin(\theta_\phi) & \cos(\theta_\phi) \end{bmatrix} (\boldsymbol{x} - \boldsymbol{c}_\phi). \tag{10.18}$$

The $\boldsymbol{c}_\phi$ is referred to as the translation vector of ϕ, and the θ_ϕ is referred to as the rotation angle of ϕ. For a set $X \subset \mathbb{X}$ and a transformation $\phi \in \mathbb{SE}(X)$, we use $\phi(X)$ to denote the image of X, transformed by ϕ, i.e.,

$$\phi(X) = \{\phi(\boldsymbol{x}); \boldsymbol{x} \in X\}.$$

When X represents the image of a geometric object, $\phi(X)$ represents the image of the geometric object transformed by the translation and rotation defined by ϕ. The rigid body transformations do not deform the geometric object. To summarize the notations and their meanings, we have the following:

- $X \subset \mathbb{X}$ and $Y \subset \mathbb{X}$ denote two simply connected subsets of $\mathbb{X}$ that represent the two primary objects;
- $Z \subset \mathbb{X}$ denote the simply connected subset of $\mathbb{X}$ that represents the aggregate of the two primary objects;
- $\phi_X \in \mathbb{SE}(\mathbb{X})$ and $\phi_Y \in \mathbb{SE}(\mathbb{X})$ denote the Euclidean rigid body transformations that represent the locations and orientations of X and Y before they aggregate;
- $\boldsymbol{c}_{\phi_X}$ and $\boldsymbol{c}_{\phi_Y}$ denote the translation vectors of the two transformations;
- θ_{ϕ_X} and θ_{ϕ_Y} denote the rotation angles of the transformations.

These notations are illustrated in Fig. 10.12.

Before the aggregate, Z, is fully restructured to a different shape, Z is approximately the overlapping union of $\phi_X(X)$ and $\phi_Y(Y)$, namely

$$Z = \phi_X(X) \cup \phi_Y(Y).$$

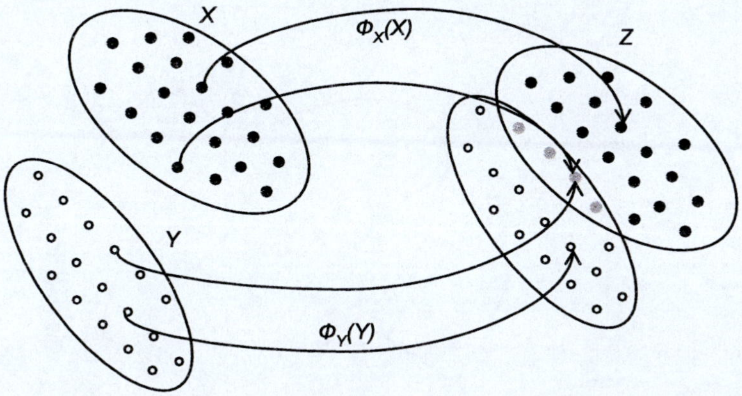

Fig. 10.12 Set-based representations of two particles and the particle aggregate, in which X and Y are the two simply connected sets that represent the two particles, and Z is the aggregation of the two particle, represented by the union of X and Y. In the process of aggregation, there could be a rigid body transformation involved for each of the primary particles. The rigid body transformations are represented by ϕ_X and ϕ_Y, respectively. (Reprinted with permission from Sikaroudi et al. (2018))

In practice, because $\mathbb{X}$ is a digital image, the equality does not hold exactly, due to digitization errors.

The aggregate Z can be partitioned into three pieces: $Z_1 = \phi_X(X)\backslash\phi_Y(Y)$, $Z_2 = \phi_Y(Y)\backslash\phi_X(X)$ and $Z_3 = \phi_X(X)\cap\phi_Y(Y)$, where \ is the set difference operator. We designate the center of Z_3 as the aggregation center and denote it by $\boldsymbol{c}_{X,Y}$. We define the orientation of X in Z as the orientation of $\boldsymbol{c}_{X,Y}$ in the standard coordinate system of $\phi_X(X)$ as described in Fig. 10.13.

Let us represent the standard coordinate system for X by a one-to-one map, $T_X : \mathbb{X} \to \mathbb{R}^2$, that assigns to a point $x \in X$ a pair of numerical coordinates. The map, $T_X \circ \phi_X^{-1}$, defines the standard coordinate system for $\phi_X(X)$ induced by T_X, as for $y \in \phi_X(X)$, $\phi_X^{-1}(y) \in X$ is uniquely determined. This is true because ϕ_X is a bijection and T_X can assign to the point $\phi_X^{-1}(y) \in X$ a pair of the unique coordinate numbers, $T_X \circ \phi_X^{-1}(y)$. As such, the orientation of X in Z is

$$\boldsymbol{v}_X = \frac{T_X \circ \phi_X^{-1}(\boldsymbol{c}_{X,Y})}{||T_X \circ \phi_X^{-1}(\boldsymbol{c}_{X,Y})||}, \text{ or } \theta_X = \text{angle}(\boldsymbol{v}_X),$$

where the angle, $(\boldsymbol{v}_X)$, is the angular part of the polar coordinate of $\boldsymbol{v}_X$. Similarly, the orientation of Y in Z is defined by

$$\boldsymbol{v}_Y = \frac{T_Y \circ \phi_Y^{-1}(\boldsymbol{c}_{X,Y})}{||T_Y \circ \phi_Y^{-1}(\boldsymbol{c}_{X,Y})||}, \text{ or } \theta_Y = \text{angle}(\boldsymbol{v}_Y).$$

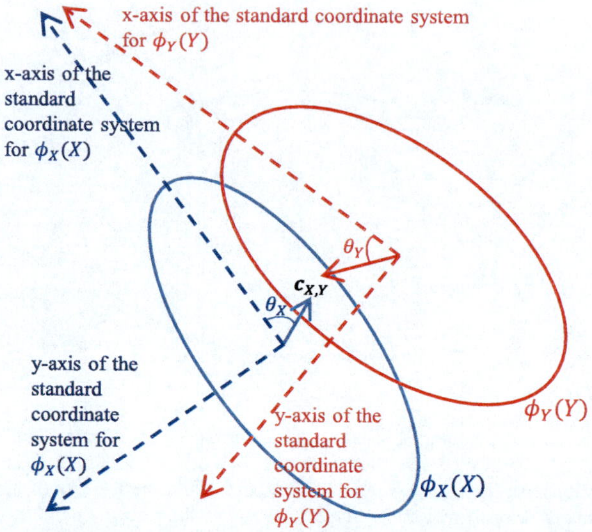

Fig. 10.13 Definition of θ_X and θ_Y. The symbol $\boldsymbol{c}_{X,Y}$ represents the center of the intersection of two aggregating components. It belongs to a part of $\phi_X(X)$ as well as to a part of $\phi_Y(Y)$. The θ_X defines which part of $\phi_X(X)$ intersects with $\phi_Y(Y)$, and likewise, the θ_Y defines which part of $\phi_Y(Y)$ intersects with $\phi_X(X)$. (Reprinted with permission from Sikaroudi et al. (2018))

Our primary interest is to study oriented attachment, i.e., to investigate what angles of θ_X and θ_Y are more frequently observed in multiple aggregation events.

10.7.2 Statistical Analysis of Nanoparticle Orientations

In this section, we present a statistical analysis to answer the following two questions: (1) whether there are preferential orientations of the primary objects when they aggregate, and (2) if so, what the orientations are.

Suppose that we have N aggregation observations,

$$\{(X_n, Y_n, Z_n); n = 1, \ldots .N\},$$

where X_n and Y_n represent the two primary geometric objects for the nth observation and Z_n represents the corresponding aggregate. The $2N$ primary objects are grouped into K shape categories based on their geometric similarities.

Some shape categories have geometrical symmetries around their major axes and minor axes, e.g., a rod and an ellipse. The major axis of a geometric object X_n is defined by the first principal loading vector of the coordinates in X_n, and the minor axis is perpendicular to the major axis. When an object is symmetric around the major and minor axes, the following orientation angles of the object are indistinguishable, that is to say,

$$\theta \equiv -\theta \equiv \pi - \theta \equiv -\pi + \theta \text{ for } \theta \in [0, \pi/2]. \tag{10.19}$$

Because of this, for a symmetric shape category, we normalize the orientation θ to be

$$\tilde{\theta} = \begin{cases} |\theta| & \text{if } |\theta| \leq \pi/2, \\ \pi - |\theta| & \text{otherwise,} \end{cases} \tag{10.20}$$

which is θ's equivalent form in the first quadrant $[0, \pi/2]$.

For a symmetric shape category, we perform our statistical inference on an normalized angle $\tilde{\theta}$, whereas for a non-symmetric shape category, we do so still on the original, unnormalized angle θ. The probability distribution of θ for a non-symmetric case can be modeled as a von Mises distribution, which is popular in describing a unimodal probability density of angular data (Mardia et al. 2012). The statistical inferences on the distribution model have been well studied in circular statistics (Fisher 1995). We will hence spend more effort discussing the symmetric cases in the sequel.

For a symmetric shape category, the equivalence in Eq. (10.19) holds in θ, and the probability density function of θ should have the following symmetries,

$$f(\theta) = f(-\theta) = f(-\pi + \theta) = f(\pi - \theta). \tag{10.21}$$

This means that if f has a mode at $\gamma \in [0, \pi/2]$, it also has the modes at $-\gamma$, $-\pi + \gamma$ and $\pi - \gamma$, respectively. For a unimodel probability density of angular data, we mention above that the von-Mises distribution can be employed. For the probability density function of a symmetric θ, which has four modes, we then take the mixture of four von Mises distributions with equal weights to represent the four modes caused by the four-way symmetry, i.e.,

$$\begin{aligned} f(\theta; \gamma, \kappa) =& \frac{1}{8\pi I_0(\kappa)} \exp\{\kappa \cos(\theta - \gamma)\} + \frac{1}{8\pi I_0(\kappa)} \exp\{\kappa \cos(\theta + \pi - \gamma)\} \\ &+ \frac{1}{8\pi I_0(\kappa)} \exp\{\kappa \cos(\theta + \gamma)\} + \frac{1}{8\pi I_0(\kappa)} \exp\{\kappa \cos(\theta - \pi + \gamma)\} \\ =& \frac{1}{8\pi I_0(\kappa)} \exp\{\kappa \cos(\theta - \gamma)\} + \frac{1}{8\pi I_0(\kappa)} \exp\{-\kappa \cos(\theta - \gamma)\} \\ &+ \frac{1}{8\pi I_0(\kappa)} \exp\{\kappa \cos(\theta + \gamma)\} + \frac{1}{8\pi I_0(\kappa)} \exp\{-\kappa \cos(\theta + \gamma)\} \\ =& \frac{1}{4\pi I_0(\kappa)} \cosh(\kappa \cos(\theta - \gamma)) + \frac{1}{4\pi I_0(\kappa)} \cosh(\kappa \cos(\theta + \gamma)) \\ =& \frac{1}{2\pi I_0(\kappa)} \cosh(\kappa \cos(\gamma) \cos(\theta)) \cosh(\kappa \sin(\gamma) \sin(\theta)), \end{aligned}$$

where $\cosh(\cdot)$ is a hyperbolic cosine function and $\gamma \in [0, \pi/2]$, κ is the concentration parameter of the von Mises distribution, and I_0 is the modified Bessel function of order 0. One can easily check that the density function satisfies the symmetry in Eq. (10.21). Because the normalization in Eq. (10.20) maps θ into the first quadrant $[0, \pi/2]$ and f has the same density for all quadrants, the density function of the normalized angle $\tilde{\theta}$ is simply four times of f, that is,

$$g(\tilde{\theta}; \gamma, \kappa) = \frac{2}{\pi I_0(\kappa)} \cosh(\kappa \cos(\gamma) \cos(\tilde{\theta})) \cosh(\kappa \sin(\gamma) \sin(\tilde{\theta})), \tag{10.22}$$

where $\gamma, \tilde{\theta} \in [0, \pi/2]$. One can show $\int_0^{\pi/2} g(\tilde{\theta}; \gamma, \kappa) = 1$, so that it is indeed a valid probability density function. In the next few subsections, we lay out the rest of the details in our analysis—Sect. 10.7.2.1 describes the maximum likelihood estimation of the two parameters γ and κ; Sect. 10.7.2.2 conducts the goodness-of-fit test for the estimated parameters; and Sects. 10.7.2.3 and 10.7.2.4 describes the statistical procedure for testing the two hypotheses mentioned at the beginning of this section.

10.7.2.1 Maximum Likelihood Estimation

We present a numerical procedure to compute the maximum likelihood estimates of γ and κ in $g(\tilde{\theta}; \gamma, \kappa)$, given the random samples, $\{\tilde{\theta}_1, \ldots, \tilde{\theta}_N\}$, drawn from the density. The log likelihood function is

$$\begin{aligned} L_N(\gamma, \kappa) = \sum_{n=1}^{N} & \log(\cosh(\kappa \cos(\gamma) \cos(\tilde{\theta}_n))) \\ & + \log(\cosh(\kappa \sin(\gamma) \sin(\tilde{\theta}_n))) - N \log(I_0(\kappa)). \end{aligned} \tag{10.23}$$

The first order necessary conditions, $\frac{\partial L_N}{\partial \gamma} = 0$ and $\frac{\partial L_N}{\partial \kappa} = 0$, do not return a closed-form expression for γ and κ. The two parameters γ and κ will have to be numerically optimized by using, for example, the Newton-Raphson algorithm. The optimization algorithm starts with initial guesses on the parameter values and iteratively changes the values toward the increasing direction of the likelihood in Eq. (10.23).

A possible initial guess for κ can be given as the unbiased estimator of $\frac{I_1(\kappa)}{I_0(\kappa)}$, where I_1 is the modified Bessel function of order 1. Let us denote the initial guess of κ by s_κ. A possible initial guess for γ can be the sample angular mean s_γ,

$$s_\gamma = \arctan\left(\frac{\bar{s}}{\bar{c}}\right) \text{ and } \frac{I_1(s_\kappa)}{I_0(s_\kappa)} = \frac{N}{N-1}\bar{c}^2 + \bar{s}^2 - \frac{1}{N-1},$$

where $\bar{s} = \frac{1}{N}\sum_{n=1}^{N} \sin(\tilde{\theta}_n)$ and $\bar{c} = \frac{1}{N}\sum_{n=1}^{N} \cos(\tilde{\theta}_n)$. To avoid the situation that the Newton-Raphson algorithm may be entrapped at a local optimum, we ran the algorithm with different initial guesses, in the range of $\gamma \in \{s_\gamma - 0.1, s_\gamma, s_\gamma + 0.1\}$

Table 10.3 Biases and variances of the maximum likelihood estimates of $\hat{\gamma}$ and $\hat{\kappa}$. A value in each cell is the average of the 100 replications of the random sampling followed by the maximum likelihood estimation

	$\gamma = \pi/6, \kappa = 10$		$\gamma = \pi/4, \kappa = 10$		$\gamma = \pi/6, \kappa = 5$	
Simulation inputs	$\hat{\gamma}$	$\hat{\kappa}$	$\hat{\gamma}$	$\hat{\kappa}$	$\hat{\gamma}$	$\hat{\kappa}$
Bias	0.0017	0.0488	0.0012	0.0274	0.0032	0.0503
Variance	0.0001	0.0036	0.0001	0.0032	0.0005	0.0057

Source: Sikaroudi et al. (2018) with permission

and $\kappa \in \{s_\kappa - 1, s_\kappa, s_\kappa + 1\}$, and then chose the solution that gave the highest likelihood value.

We used three simulation cases to evaluate the bias and variance of the aforementioned maximum likelihood estimates. First, we first drew 1000 random samples from $g(\tilde{\theta}; \gamma, \kappa)$, with γ and κ specified in Table 10.3. Then, we used the random samples to estimate γ and κ using the maximum likelihood estimation. The estimates, $\hat{\gamma}$ and $\hat{\kappa}$, were compared with the values of γ and κ used to simulate the random samples in the first place. Their differences are the biases of the estimates. The procedure of random sampling, followed by the maximum likelihood estimation, was repeated 100 times. The biases of the 100 replications were averaged, and the variances of the 100 replications were also calculated. Table 10.3 summarizes the outcomes. The biases and variances of $\hat{\gamma}$ were very small, and the biases of $\hat{\kappa}$ are higher but still close to zero.

10.7.2.2 Goodness-of-Fit Test

We use the Kolmogorov-Smirnov test (Arnold and Emerson 2011) to test the goodness-of-fit of $g(\tilde{\theta}; \hat{\gamma}, \hat{\kappa})$ to a random sample $\{\tilde{\theta}_1, \ldots, \tilde{\theta}_N\}$. Let $G(\tilde{\theta})$ denote the cumulative distribution function that corresponds to $g(\tilde{\theta}; \hat{\mu}, \hat{\kappa})$, and let $G_n(\tilde{\theta})$ be the empirical cumulative distribution function,

$$G_n(\tilde{\theta}) = \frac{1}{N} \sum_{n=1}^{N} I_{[-\infty, \tilde{\theta}]}(\tilde{\theta}_n).$$

The test statistic for the Kolmogorov-Smirnov test is the difference in between the two cumulative distribution functions, defined as,

$$T_N = \sqrt{n} \sup_{\tilde{\theta}} |G(\tilde{\theta}) - G_n(\tilde{\theta})|.$$

If the test statistic is below a critical value, the fit of G to G_n is good. The critical value is determined, so that the type-I error is set at a prescribed value of α. Let us denote the critical value by $t_{\alpha,N}$. The critical value can be decided through the following Monte Carlo simulation:

Step 1. Take a random sample of size N from $g(\tilde{\theta}; \hat{\gamma}, \hat{\kappa})$ and construct the empirical cumulative distribution function G_n for the random sample.

Step 2. Compute T_N.

Step 3. Repeat Step 1 and Step 2 many times, which results in a number of T_N values. The critical value of the test statistic with type-I error at α is the $1 - \alpha$ quantile of the resulting T_N values.

10.7.2.3 Testing the Uniformity of Distribution

The first hypothesis to test is whether there is a preferential orientation of the primary objects in its aggregate. It is related to testing whether $g(\tilde{\theta}; \gamma, \kappa)$ is uniform, since uniformity implies the lack of preferential orientation. The uniformity of the density function $g(\tilde{\theta}; \gamma, \kappa)$ is determined by its parameter κ. As κ decreases, the density function $g(\tilde{\theta}; \gamma, \kappa)$ becomes closer to an angular uniform distribution—perfectly uniform with $\kappa = 0$ or nearly uniform with $\kappa \leq 0.5$. Therefore, we formulate the uniformity testing as follows,

$$\begin{aligned} \mathrm{H}_0: &\quad \kappa \leq 0.5, \\ \mathrm{H}_1: &\quad \kappa > 0.5. \end{aligned}$$

We can test the hypothesis based on a general likelihood ratio test with a random sample from $g(\tilde{\theta}; \gamma, \kappa)$. Suppose that $\{\tilde{\theta}_1, \ldots, \tilde{\theta}_N\}$ is the random sample. Using the likelihood function in Eq. (10.23), we can define the likelihood ratio test statistic for testing H_0 versus H_1 as

$$R_\kappa = \max_{\kappa > 0.5} L_N(\gamma, \kappa) - \max_{\kappa \leq 0.5} L_N(\gamma, \kappa).$$

Evaluating the test statistic involves evaluating two maximum likelihoods under different linear constraints on κ, which can be solved easily using the Newton-Raphson algorithm, as described in Sect. 10.7.2.1. When the test statistic is above a prescribed critical value, we reject H_0. The critical value for the test at the type-I error of α can be determined, again, through a Monte Carlo simulation, such as

Step 1. Sample $\kappa \sim \text{Uniform}([0, 0.5])$ and $\gamma \sim \text{Uniform}([0, \pi/2])$.

Step 2. Take a random sample of size N from $g(\tilde{\theta}; \gamma, \kappa)$, and evaluate R_κ for the random sample.

Step 3. Repeat Steps 1 and 2 many times, which results in a number of R_κ values. The critical value of the test statistic with type-I error at α is the $1 - \alpha$ quantile of the resulting R_κ values.

10.7.2.4 Testing the Mean Orientation

The second hypothesis to test is whether the mean orientation of a primary object in its aggregate is γ_0. When the orientation follows the probability density $g(\tilde{\theta}; \gamma, \kappa)$, this test can be formulated as testing whether $\gamma = \gamma_0$. Same as in Sect. 10.7.2.3, we can test the hypothesis, again, based on a general likelihood ratio test. The specific likelihood ratio test statistic is

$$R_\gamma = \max_{\gamma,\kappa} L_N(\gamma, \kappa) - \max_{\gamma=\gamma_0,\kappa} L_N(\gamma, \kappa).$$

When the test statistic is below a critical value, it implies that there is no significant evidence to reject the null hypothesis, $\gamma = \gamma_0$. The critical value of the test statistic with type-I error at α can be determined also through a Monte Carlo simulation, i.e.,

Step 1. Sample $\kappa \sim \text{Uniform}([0, 30])$ and $\gamma = \gamma_0$.
Step 2. Take a random sample of size N from $g(\tilde{\theta}; \gamma, \kappa)$, and evaluate R_γ for the random sample.
Step 3. Repeat Steps 1 and 2 many times, which results in a number of R_γ values. The critical value of the test statistic with type-I error at α is the $1 - \alpha$ quantile of the resulting R_γ values.

10.7.3 Results

The nanoparticles involved in the 184 nanoparticle aggregation events are grouped into three shape categories. Figure 10.14 illustrates the images of the three shape categories and some selected nanoparticle images belonging to each of the categories. The three shape categories are distinct in terms of their aspect ratio, which is defined as the ratio of the major axis length of a shape over the minor axis length of the same shape. The mean aspect ratios are 1.99 for the first category, 1.40 for the second category, and 1.22 for the last category. Based on the appearances of the nanoparticles in each category, we name the first category ($k = 1$) as 'Rod' (82 objects), the second category ($k = 2$) as 'Ellipse' (146 objects), and the third category ($k = 3$) as 'NearSphere' (140 objects).

Depending on the shape categories of the primary nanoparticles involved in an aggregation, the aggregation events can be classified into six groups: Rod-Rod (12 cases), Rod-Ellipse (26 cases), Rod-NearSphere (32 cases), Ellipse-Ellipse (33 cases), Ellipse-NearSphere (54 cases), and NearSphere-NearSphere (27 cases). We retrieved the orientation angles of the primary nanoparticles as described in Sect. 10.7.2, that is,

$$\{(\tilde{\theta}_X^{(n)}, \tilde{\theta}_Y^{(n)}); n = 1, \ldots .N\},$$

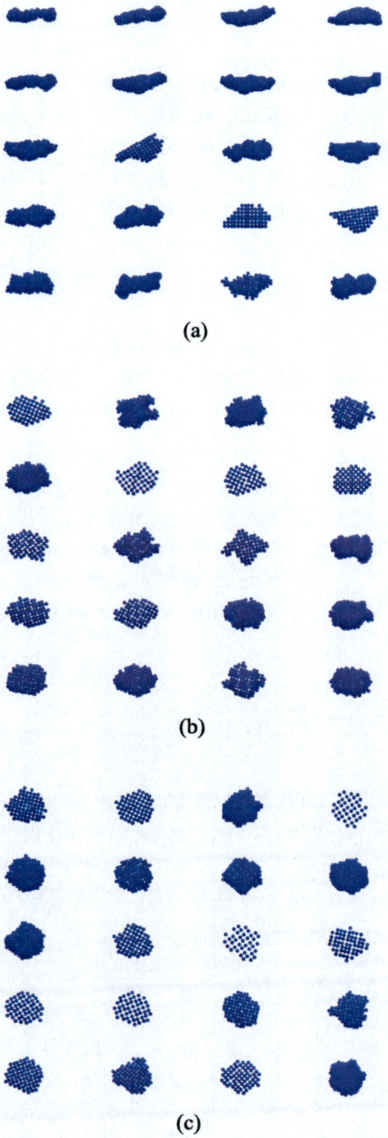

Fig. 10.14 Alignment outcomes for three shape categories. (**a**) Shape category 1: Rod. (**b**) Shape category 2: Ellipse. (**c**) Shape category 3: NearSphere. (Reprinted with permission from Sikaroudi et al. (2018))

where $(\tilde{\theta}_X^{(n)}, \tilde{\theta}_Y^{(n)})$ are the orientation angles of the two primary particles for the nth observation. Recall that $(\tilde{\theta}_X^{(n)}, \tilde{\theta}_Y^{(n)})$ are normalized to $[0, \pi/2]$.

We first looked at the angular correlation coefficient of $\tilde{\theta}_X^{(n)}$ and $\tilde{\theta}_Y^{(n)}$ for each aggregation group. Let $N_{k1,k2}$ denote the number of observations corresponding to the aggregation events involving shape categories $k1$ and $k2$. Following Fisher and Lee (1983), the angular correlation coefficient, $\rho_{k1,k2}$, is defined as,

$$\rho_{k1,k2} = \frac{\sum_{i,j\in N_{k1,k2}} \sin(\tilde{\theta}_X^{(i)} - \tilde{\theta}_X^{(j)}) \sin(\tilde{\theta}_Y^{(i)} - \tilde{\theta}_Y^{(j)})}{\sqrt{\sum_{i,j\in N_{k1,k2}} \sin^2(\tilde{\theta}_X^{(i)} - \tilde{\theta}_X^{(j)})}\sqrt{\sum_{i,j\in N_{k1,k2}} \sin^2(\tilde{\theta}_Y^{(i)} - \tilde{\theta}_Y^{(j)})}}.$$

The corresponding coefficient of correlation, $\rho^2_{k1,k2}$, is 0.1859 for Rod-Rod, 0.0273 for Rod-Ellipse, 0.0937 for Rod-NearSphere, 0.1195 for Ellipse-Ellipse, 0.0008 for Ellipse-NearSphere, 0.2252 for NearSphere-NearSphere. When $k1 = k2$, the coefficients were computed between $\min\{\tilde{\theta}_X^{(n)}, \tilde{\theta}_Y^{(n)}\}$ and $\max\{\tilde{\theta}_X^{(n)}, \tilde{\theta}_Y^{(n)}\}$. The small magnitudes of the coefficients imply that the two angles are weakly correlated.

Given the weak correlation of $\tilde{\theta}_X^{(n)}$ and $\tilde{\theta}_Y^{(n)}$ and a limited number of observations per group, we approximate the joint distribution of two angles with a product of the respective marginal distributions. Let $p_{k1|k2}(\tilde{\theta})$ denote the marginal density function of $\tilde{\theta}$ of shape category $k1$ when it aggregates with shape category $k2$. The marginal pdf, $p_{k1|k2}(\tilde{\theta})$, is assumed to be

$$p_{k1|k2}(\tilde{\theta}) = g(\tilde{\theta}; \gamma_{k1,k2}, \kappa_{k1,k2}).$$

The maximum likelihood estimation procedure, described in Sect. 10.7.2.1, was applied for $k1 = 1, 2$ and $k2 = 1, 2, 3$. We did not analyze the $k1 = 3$ case (i.e., the Near-Sphere cases) because that case is subject to relatively large estimation errors, as shown in the simulation study of Sect. 3.3 .

Let $\hat{\gamma}_{k1,k2}$ and $\hat{\kappa}_{k1,k2}$ denote the estimated $\gamma_{k1,k2}$ and $\kappa_{k1,k2}$. Figure 10.15 presents $p_{k1|k2}(\tilde{\theta})$ with $\hat{\gamma}_{k1,k2}$ and $\hat{\kappa}_{k1,k2}$.

The method described in Sect. 10.7.2.2 was applied for the goodness-of-fit test of the estimated density functions. For all cases, the estimated CDFs were very consistent with the corresponding empirical CDFs. The goodness-of-fit test showed no significant difference between them at the 95% significance level. Figure 10.16 presents the estimated CDFs (G), together with the empirical CDFs (G_n).

We also tested the hypothesis of whether there is a preferential orientation of shape category $k1$ when it aggregates with shape category $k2$. We used the method discussed in Sect. 10.7.2.3 to test

$$\begin{aligned} H_0 &: \kappa_{k1,k2} \le 0.5 \\ H_1 &: \kappa_{k1,k2} > 0.5. \end{aligned}$$

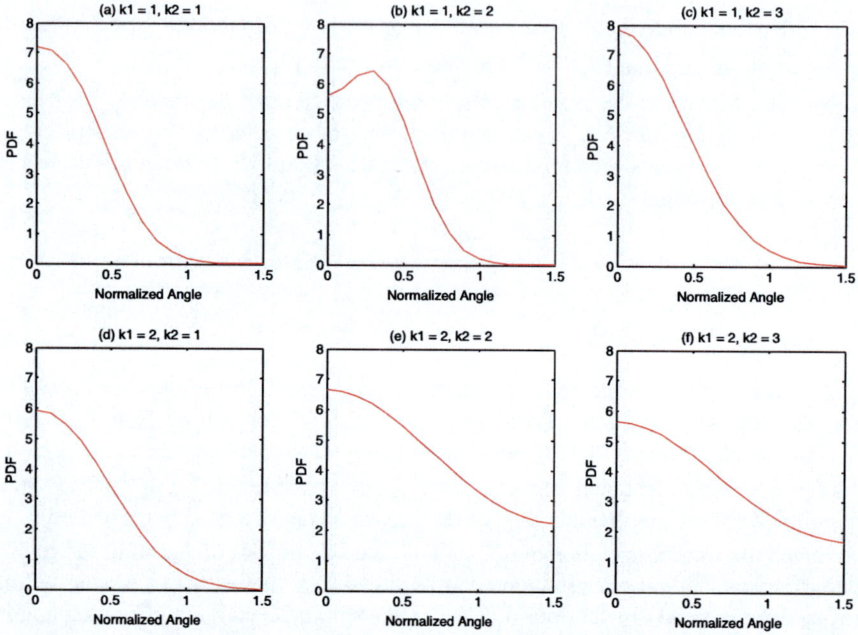

Fig. 10.15 Estimated probability density function of the orientation of shape category $k1$ when it aggregates with shape category $k2$. (Reprinted with permission from Sikaroudi et al. (2018))

At the 95% significance level, the null hypothesis was rejected for $(k1, k2) = (1, 1)$, $(1, 2)$, $(1, 3)$ $(2, 1)$, $(2, 2)$, and $(2, 3)$. The results suggest strong evidences that rod-like and ellipse-like nanoparticles have preferential orientations when they aggregate, respectively, with rod-like, ellipse-like, or near-sphere nanoparticles.

We performed a steered molecular dynamics (SMD) simulatio of a rod-to-rod particle aggregation (Welch et al. 2016), which allowed us to compute the energy barriers against aggregation for different orientations of rods. According to the simulation, when the major axes of two aggregating rods were not oriented toward the aggregation center, the compression of solvent monolayers at rod surfaces significantly increased when the rods became close to each other. The increase of the solvation force placed a large energy barrier against the aggregation of the two rods. The energy barrier was minimized when the major axes of both rods were oriented toward the aggregation center.

This implies the preferential orientation of a rod particle in its aggregate is zero (because the direction of the major axis is zero). To test whether our experimental observations are consistent with the simulation results, we formulated a hypothesis testing problem, which examines whether the mean orientation $\gamma_{1,1}$ for a Rod-Rod aggregation is zero,

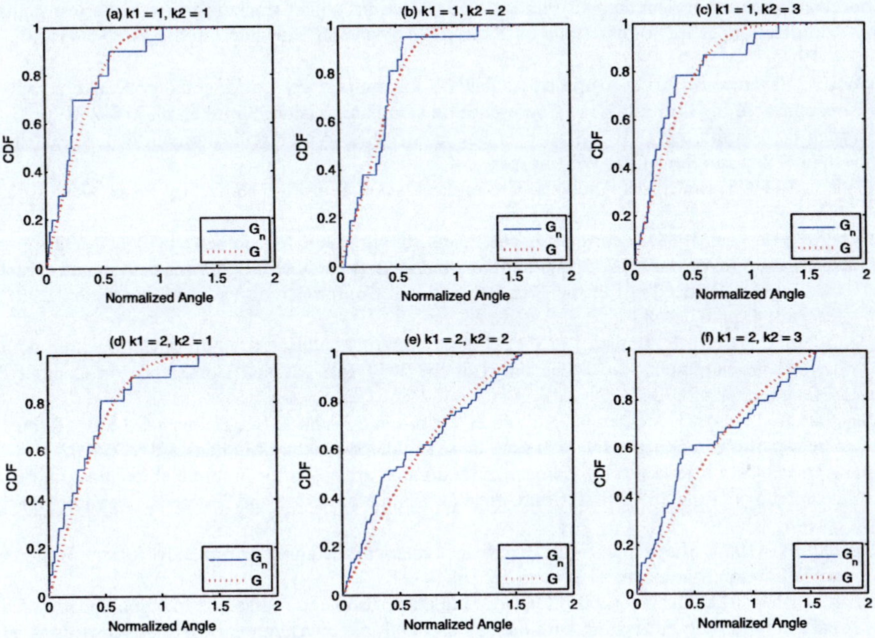

Fig. 10.16 Goodness-of-Fit Test: G is the estimated CDF and G_n is the empirical CDF. (Reprinted with permission from Sikaroudi et al. (2018))

$$\mathrm{H}_0 : \gamma_{1,1} = 0$$
$$\mathrm{H}_1 : \gamma_{1,1} \neq 0.$$

We employed the method in Section 10.7.2.4 to test the hypothesis. At the 95% significant level, the null hypothesis cannot be rejected. In other words, at a high significance (95%), the experimental observations are consistent with the outputs of the SMD simulation.

References

Ahujia R, Magnanti TL, Orlin JB (1993) Network flows: Theory, algorithms, and applications. New Jersey: Rentice-Hall

Anstreicher K (1999) Linear programming in $O(\frac{n^3}{\ln n}L)$ operations. SIAM Journal on Optimization 9(4):803–812

Arnold TB, Emerson JW (2011) Nonparametric goodness-of-fit tests for discrete null distributions. The R Journal 3(2):34–39

Arole V, Munde S (2014) Fabrication of nanomaterials by top-down and bottom-up approaches-an overview. Journal of Materials Science 1:89–93

Berclaz J, Fleuret F, Turetken E, Fua P (2011) Multiple object tracking using k-shortest paths optimization. IEEE Transactions on Pattern Analysis and Machine Intelligence 33(9):1806–1819

Choi W, Savarese S (2012) A unified framework for multi-target tracking and collective activity recognition. In: 12th European Conference on Computer Vision, Springer, pp 215–230

Dryden IL, Mardia KV (2016) Statistical Shape Analysis: With Applications in R, 2nd Edition. John Wiley and Sons Ltd., West Sussex, UK

Fisher NI (1995) Statistical Analysis of Circular Data. Cambridge University Press, New York, NY, USA

Fisher NI, Lee AJ (1983) A correlation coefficient for circular data. Biometrika 70:327–332

Genovesio A, Olivo-Marin JC (2004) Split and merge data association filter for dense multi-target tracking. In: Proceedings of the 17th International Conference on Pattern Recognition, 2004, Ieee, vol 4, pp 677–680

Henriques J, Caseiro R, Batista J (2011) Globally optimal solution to multi-object tracking with merged measurements. In: Proceedings of the 2011 International Conference on Computer Vision, Ieee, pp 2470–2477

Jaqaman K, Loerke D, Mettlen M, Kuwata H, Grinstein S, Schmid SL, Danuser G (2008) Robust single-particle tracking in live-cell time-lapse sequences. Nature Methods 5(8):695–702

Jiang H, Fels S, Little JJ (2007) A linear programming approach for multiple object tracking. In: Proceedings of the 2007 IEEE Conference on Computer Vision and Pattern Recognition, Ieee, pp 1–8

Kendall DG (1984) Shape manifolds, Procrustean metrics, and complex projective spaces. Bulletin of the London Mathematical Society 16(2):81–121

Khan Z, Balch T, Dellaert F (2006) MCMC data association and sparse factorization updating for real time multitarget tracking with merged and multiple measurements. IEEE Transactions on Pattern Analysis and Machine Intelligence 28(12):1960–1972

Kumar P, Ranganath S, Sengupta K, Weimin H (2006) Cooperative multitarget tracking with efficient split and merge handling. IEEE Transactions on Circuits and Systems for Video Technology 16(12):1477–1490

Li D, Nielsen MH, Lee JR, Frandsen C, Banfield JF, De Yoreo JJ (2012) Direction-specific interactions control crystal growth by oriented attachment. Science 336(6084):1014–1018

Mardia KV, Kent JT, Zhang Z, Taylor CC, Hamelryck T (2012) Mixtures of concentrated multivariate sine distributions with applications to bioinformatics. Journal of Applied Statistics 39(11):2475–2492

Martello S, Toth P (1987) Linear assignment problems. Surveys in Combinatorial Optimization 132:259–282

Memoli F (2007) On the use of Gromov-Hausdorff distances for shape comparison. In: Eurographics Symposium on Point-Based Graphics, The Eurographics Association, pp 81–90

Mémoli F, Sapiro G (2005) A theoretical and computational framework for isometry invariant recognition of point cloud data. Foundations of Computational Mathematics 5(3):313–347

Nemhauser G, Wolsey L (1988) Integer and Combinatorial Optimization. Jone Wiley & Sons, New York, NY, USA

Pardalos P, Vavasis S (1991) Quadratic programming with one negative eigenvalue is NP-hard. Journal of Global Optimization 1(1):15–22

Park C, Ding Y (2019) Automating material image analysis for material discovery. MRS Communications 9(2):545–555

Park C, Woehl TJ, Evans JE, Browning ND (2015) Minimum cost multi-way data association for optimizing multitarget tracking of interacting objects. IEEE Transactions on Pattern Analysis and Machine Intelligence 37(3):611–624

Perera AA, Srinivas C, Hoogs A, Brooksby G, Hu W (2006) Multi-object tracking through simultaneous long occlusions and split-merge conditions. In: Proceedings of the 2006 IEEE Conference on Computer Vision and Pattern Recognition, Ieee, vol 1, pp 666–673

Pirsiavash H, Ramanan D, Fowlkes CC (2011) Globally-optimal greedy algorithms for tracking a variable number of objects. In: Proceedings of the 2011 IEEE Conference on Computer Vision and Pattern Recognition, Ieee, pp 1201–1208

Polyak B (1987) Introduction to Optimization. Optimization Software Inc., New York, NY, USA

Rasmussen C, Hager GD (2001) Probabilistic data association methods for tracking complex visual objects. IEEE Transactions on Pattern Analysis and Machine Intelligence 23(6):560–576

Reid DB (1979) An algorithm for tracking multiple targets. IEEE Transactions on Automatic Control 24(6):843–854

Schrijver A (2003) Combinatorial optimization: polyhedra and efficiency, vol 24. Springer

Sergé A, Bertaux N, Rigneault H, Marguet D (2008) Dynamic multiple-target tracing to probe spatiotemporal cartography of cell membranes. Nature Methods 5(8):687–694

Sikaroudi AE, Welch DA, Woehl TJ, Faller R, Evans JE, Browning ND, Park C (2018) Directional statistics of preferential orientations of two shapes in their aggregate and its application to nanoparticle aggregation. Technometrics 60(3):332–344

Smeulders AW, Chu DM, Cucchiara R, Calderara S, Dehghan A, Shah M (2013) Visual tracking: An experimental survey. IEEE Transactions on Pattern Analysis and Machine Intelligence 36(7):1442–1468

Srivastava A, Klassen E, Joshi SH, Jermyn IH (2010) Shape analysis of elastic curves in Euclidean spaces. IEEE Transactions on Pattern Analysis and Machine Intelligence 33(7):1415–1428

Storlie CB, Lee TC, Hannig J, Nychka D (2009) Tracking of multiple merging and splitting targets: A statistical perspective. Statistica Sinica 19(1):1–52

Welch DA, Woehl T, Park C, Faller R, Evans JE, Browning ND (2016) Understanding the role of solvation forces on the preferential attachment of nanoparticles in liquid. ACS Nano 10(1):181–187

Woehl T, Evans J, Arslan I, Ristenpart W, Browning N (2012) Direct *in situ* determination of the mechanisms controlling nanoparticle nucleation and growth. ACS Nano 6(10):8599–8610

Younes L (1998) Computable elastic distances between shapes. SIAM Journal on Applied Mathematics 58(2):565–586

Yu Q, Medioni G (2009) Multiple-target tracking by spatiotemporal Monte Carlo Markov Chain data association. IEEE Transactions on Pattern Analysis and Machine Intelligence 31(12):2196–2210

Zhang L, Li Y, Nevatia R (2008) Global data association for multi-object tracking using network flows. In: Proceedings of the 2008 IEEE Conference on Computer Vision and Pattern Recognition, Ieee, pp 1–8

Zhang W, Crittenden J, Li K, Chen Y (2012) Attachment efficiency of nanoparticle aggregation in aqueous dispersions: Modeling and experimental validation. Environmental Science & Technology 46(13):7054–7062

Chapter 11
Super Resolution

11.1 Multi-Frame Super Resolution

One of the very first effort on the topic of super-resolution is credited to Tsai and Huang (1984), published more than 35 years ago. The phrase *multiframe* appeared in Tsai and Huang (1984), but the phrase *super-resolution* (SR) did not. The first time that super-resolution appeared in an academic work was most likely when Gross used it in the title of his Master's thesis at Tel Aviv University in 1986 (Gross 1986). A few years later, Irani and Peleg (1990) used super-resolution in their paper while publishing in an IEEE conference on pattern recognition (Irani and Peleg 1990). Since then, the phrase super-resolution has taken hold and the super-resolution image processing has become an active research field.

The phrase *resolution* in the expression of super-resolution, as discussed in this chapter, refers to the spatial resolution of an image, instead of the temporal resolution associated with a stream of images (i.e., video images). The temporal resolution is typically called the frame rate, the imaging rate, or simply, the rate, which is the number of frames captured per second. In terms of spatial resolution, a higher resolution is accomplished by having more pixels per unit area, capturing finer spatial details missed by a lower resolution image.

Super-resolution research starts off with using multiple frames of repeat-pass low-resolution images capturing the same scene. The idea is that each low-resolution image contains some information complementing the others and by combining the complementary information it is possible to enhance the quality of the low-resolution images and thus produce a higher-resolution one.

C. Park, Y. Ding, *Data Science for Nano Image Analysis*, International Series in Operations Research & Management Science 308,
https://doi.org/10.1007/978-3-030-72822-9_11

11.1.1 The Observation Model

There is a standard observation model linking a high-resolution image with its low-resolution counterpart in the super-resolution literature (Park et al. 2003; Tian and Ma 2011; Yue et al. 2016). Denote a high-resolution image by $\mathfrak{X}$ of pixel size of $M_h \times N_h$ and one of its corresponding low-resolution image by $\mathfrak{Y}$ of pixel size of $M_l \times N_l$. The ratios, M_h/M_l and N_h/N_l, are the respective down-sampling factors along each coordinate of the 2D image. For the same $\mathfrak{X}$, there may exist K low-resolution counterparts. To distinguish them, we put a subscript k on $\mathfrak{Y}$ as $\mathfrak{Y}_k$ for $k = 1, \ldots, K$.

Tian and Ma (2011) consider four operations when $\mathfrak{X}$ is converted into $\mathfrak{Y}_k$: geometric transformation (also known as warping), blurring, down-sampling, and addition of white Gaussian noises. If each of the first three operations can be approximated through a linear transformation, denoted by the matrix operator of $\mathbf{W}_k$ for warping, $\mathbf{B}_k$ for blurring, and $\mathbf{D}_k$ for down-sampling, then one can have the following relationship, which is essentially Eq. (1) in Tian and Ma (2011):

$$\mathfrak{Y}_k = \mathbf{D}_k\mathbf{B}_k\mathbf{W}_k\mathfrak{X} + \boldsymbol{\varepsilon}_k, \tag{11.1}$$

where $\boldsymbol{\varepsilon}_k$ represents the additive Gaussian noise. The variance of the elements in $\boldsymbol{\varepsilon}_k$ is assumed the same and denoted by $(\sigma_\epsilon)_k^2$ or simply σ_k^2. If σ_k^2 is the same for all k's, then it is denoted by σ_ϵ^2.

There are discussions about the order of the warping and blurring matrices in the above model. For instance, Chiang and Boult (2000) switch the order of warping and blurring in their observation model. Tian and Ma (2011, Page 331) claim that under certain assumptions on blurring, the operations of warping and blurring are actually commutable. Some researchers add additional operators into the above standard observation model. For example, the model in Yue et al. (2016, Eq. 1) has an additional operator for excluding the unobservable pixels from the low-resolution images.

The objective of multi-frame super-resolution is to exploit the high-to-low relationship in Eq. (11.1) for recovering the high resolution image $\mathfrak{X}$ from the set of low-resolution images of $\{\mathfrak{Y}_1, \ldots, \mathfrak{Y}_K\}$ capturing the same scene. The operation matrices are assumed known. In reality, it may be difficult to construct each of the operation matrices individually, and as such, researchers also choose to lump all of the matrices into a single transformation matrix, $\mathbf{H}_k$, such that,

$$\mathfrak{Y}_k = \mathbf{H}_k\mathfrak{X} + \boldsymbol{\varepsilon}_k. \tag{11.2}$$

The details of image operations, including the order of these operations, disappear in this aggregated expression. But the assumption of linear approximation is still essential for using the aggregated observation model in Eq. (11.2).

11.1.2 Super-Resolution in the Frequency Domain

When Tsai and Huang (1984) started the super-resolution research, they worked in the frequency domain by making use of certain properties of the continuous and discrete Fourier transforms associated with a series of shifted images. The basic assumption is that a low-resolution image is a shifted and down-sampled version of the high-resolution image. One can consider that the high-resolution image, $\mathfrak{X}(i, j)$, where $\mathfrak{X}(i, j)$ is the image pixel at the (i, j) location for $i = 1, \ldots, M_h$ and $j = 1, \ldots, N_h$, is first shifted by x and y, respectively, along each of its coordinates. A subscript is used to indicate the k-th shift in the multiple shifts, such as $\mathfrak{X}(i+x_k, j+y_k)$. With a downsampling, it yields the k-th low-resolution image as

$$\mathfrak{Y}_k(n_1, n_2) = \mathfrak{X}(n_1 \cdot T_1 + x_k, n_2 \cdot T_2 + y_k), \quad n_1 = 0, \ldots, M_l, \; n_2 = 0, \ldots, N_l, \tag{11.3}$$

where T_1 and T_2 are the sampling periods along each coordinate, respectively. The above expression suggests that the difference between the high-resolution and low-resolution images are entirely due to pixel shifting and downsampling and nothing else. This is of course a rather strong assumption.

Taking the Fourier transforms of both sides in Eq. (11.3) eventually yields a linear relationship like

$$\mathcal{D} = \boldsymbol{\Phi} C, \tag{11.4}$$

where $\mathcal{D}$ contains the coefficients resulting from performing discrete Fourier transforms on K low-resolution images and C is the sampled coefficients of the continuous Fourier transform of the high-resolution image but it is unknown. Tsai and Huang (1984) derive the expression of $\boldsymbol{\Phi}$ under the stated assumptions. Then, Eq. (11.4) is used to solve the inverse problem of estimating C from $\mathcal{D}$, thus accomplishing the objective of super-resolution. Here we omit the presentation of the detailed expressions of C, $\mathcal{D}$, and $\boldsymbol{\Phi}$, because this early method is not longer in active use. For those details, interested readers can refer to Park et al. (2003, pages 26–27).

It is not difficult to notice that this early work of super resolution assumed a rather restrictive relationship between the high-resolution image and its low-resolution counterparts. Later attempts were made to relax the assumptions, and thus make the problem setting more practical, by considering, for instance, noise and spatial blurring effects. But generally speaking, the frequency domain-based approach seems to be limited by the requirement of a global translation between images and the linear space-invariant blur during the image acquisition process, and was quickly superseded by other approaches.

11.1.3 Interpolation-Based Super Resolution

The basic idea of interpolation-based super-resolution works as follows. Suppose that one can register multiple low-resolution images on a high-resolution image grid. The high-resolution image grid is supposedly denser and finer than the low-resolution ones. After multiple low-resolution images are projected onto the high-resolution grid, many of the grid points do not have the corresponding pixels. Naturally, one can interpolate for the high-resolution grid points where there is no observation and reconcile for those grid points where there are more than one observations but of different values.

Based on the above description, it is not difficult to see that three steps are commonly employed in the interpolation-based super-resolution approaches. The first step is to align the low-resolution images and project them onto the high-resolution image, a step known as image registration. The second step is the interpolation step, as described above. The last step is a deblurring through some classical deconvolutional algorithms that remove image noise. Figure 11.1 illustrates the basic idea.

The focus of super-resolution is usually on the second step, i.e., interpolation, although image registration itself is often a tricky problem. But image registration alone is studied less frequently in the super-resolution literature because a separate subfield of image processing dedicates its effort to the problem of registration.

The simplest interpolation approach is to use the nearest neighbor idea, i.e., for each high-resolution grid point, its intensity value takes simply that of the nearest

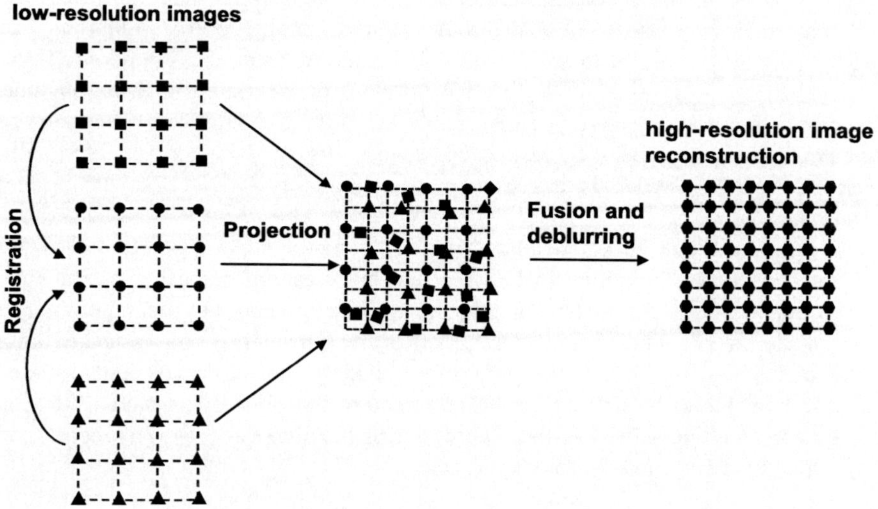

Fig. 11.1 Illustration of interpretation-based super-resolution (reprinted with permission from Tian and Ma (2011))

pixel. But other interpolation methods can be used, for instance, a weighted average of pixels in a neighborhood (kernel regression) or a simple average of the pixels of the k-nearest neighbors (k-NN).

Elad and Hel-Or (2001) present a mathematical framework, under certain assumptions, for Step 2 and 3 of the interpolation-based approaches. They assume the following:

- All the decimation operations are the same, i.e., $\mathbf{D}_k = \mathbf{D}$.
- All the blurring operations are linearly spatial invariant blurs and also the same. So $\mathbf{B}_k = \mathbf{B}$.
- All the warp operations correspond to pure translations.
- The additive noise is white noise and has the same variance for all frames, i.e., $\sigma_k^2 = \sigma_\epsilon^2$.

Under the first three assumptions, the operations of $\mathbf{B}$ and $\mathbf{W}_k$ are commutative. Therefore one can write

$$\mathfrak{Y}_k = \mathbf{DBW}_k\mathfrak{X} + \boldsymbol{\varepsilon}_k = \mathbf{DW}_k\mathbf{B}\mathfrak{X} + \boldsymbol{\varepsilon}_k = \mathbf{DW}_k\mathfrak{Z} + \boldsymbol{\varepsilon}_k, \tag{11.5}$$

where $\mathfrak{Z} = \mathbf{B}\mathfrak{X}$ is the blurred version of the high-resolution image. With Eq. (11.5), it is apparent that the interpolation is to estimate $\mathfrak{Z}$ and the deblurring is to recover $\mathfrak{X}$ from $\mathfrak{Z}$. Step 1, image registration, should have been performed before Eq. (11.5) takes effect.

Elad and Hel-Or (2001) provide an iterative estimation of $\mathfrak{Z}$ through

$$\mathfrak{Z}_{t+1} = \mathfrak{Z}_t + \vartheta\mathbf{BB}^T[\mathfrak{P} - \mathbf{R}\mathfrak{Z}_t], \tag{11.6}$$

where t is the iteration index, ϑ is the step-size variable, and $\mathfrak{P}$ and $\mathbf{R}$ are defined as follows:

$$\begin{aligned} \mathfrak{P} &= \sum_{k=1}^{K} \mathbf{W}_k^T\mathbf{D}^T\mathfrak{Y}_k, \\ \mathbf{R} &= \sum_{k=1}^{K} \mathbf{W}_k^T\mathbf{D}^T\mathbf{DW}_k. \end{aligned} \tag{11.7}$$

In this iterative procedure, the blur matrix $\mathbf{B}$ is still involved while estimating $\mathfrak{Z}$. But Elad and Hel-Or (2001) further claim that the steady-state solution to the above iteration, which they denote by $\mathfrak{Z}_\infty$, can be solved through a closed-form expression that does not depend on the blur matrix $\mathbf{B}$, i.e.,

$$\mathfrak{Z}_\infty = \mathbf{R}^{-1}\mathfrak{P}.$$

Using $\mathfrak{Z}_\infty$, the actions of interpolation and deblurring are de-coupled, so that they can be conducted sequentially.

11.1.4 Regularization-Based Super Resolution

Employing regularization to solve Eq. (11.2) should not come as a surprise because the super-resolution formulation is essentially an ill-posed inverse problem. The first attempt of regularization is the result of using a Bayesian framework and the associated *maximum a posterior* (MAP) estimation.

Denote by $f(\mathfrak{X}|\mathfrak{Y}_1, \ldots, \mathfrak{Y}_K)$ the posterior distribution of the high-resolution image, given the data from the set of low-resolution images. The MAP estimate of the high-resolution image is

$$\begin{aligned}\hat{\mathfrak{X}}^{\mathrm{MAP}} &= \arg\max_{\mathfrak{X}} f(\mathfrak{X}|\mathfrak{Y}_1, \ldots, \mathfrak{Y}_K)\\ &= \arg\max_{\mathfrak{X}} f(\mathfrak{Y}_1, \ldots, \mathfrak{Y}_K|\mathfrak{X}) f(\mathfrak{X}).\end{aligned} \tag{11.8}$$

The second line above is obtained by applying the Bayes rule, where a normalized constant is omitted—inclusion and exclusion of the constant do not affect the outcome of the maximization. The first term of the second line inside the optimization is the likelihood function and the second term is the prior distribution of the high-resolution image.

Using the observation model in Eq. (11.2), one can derive the likelihood function as

$$\begin{aligned}f(\mathfrak{Y}_1, \ldots, \mathfrak{Y}_K|\mathfrak{X}) &= \prod_{k=1}^{K} f(\mathfrak{Y}_k|\mathfrak{X})\\ &\propto \prod_{k=1}^{K} \exp\left(-\frac{1}{2\sigma_k^2}\|\mathfrak{Y}_k - \mathbf{H}_k\mathfrak{X}\|_2^2\right)\\ &= \exp\left(-\sum_{k=1}^{K}\frac{1}{2\sigma_k^2}\|\mathfrak{Y}_k - \mathbf{H}_k\mathfrak{X}\|_2^2\right).\end{aligned} \tag{11.9}$$

If one maximizes this likelihood function, i.e.,

$$\max_{\mathfrak{X}} f(\mathfrak{Y}_1, \ldots, \mathfrak{Y}_K|\mathfrak{X}),$$

the corresponding solution is the maximum likelihood estimation of the high-resolution image, namely $\hat{\mathfrak{X}}^{\mathrm{ML}}$. The MLE is equivalent to minimizing the power term without the minus sign, i.e.,

$$\min \sum_{k=1}^{K}\frac{1}{2\sigma_k^2}\|\mathfrak{Y}_k - \mathbf{H}_k\mathfrak{X}\|_2^2,$$

which is in fact the least-squares estimation without regularization.

Using the Bayesian framework and aiming at an MAP estimation, a regularization term arrives through the prior of the high-resolution image. The prior is typically defined using a Boltzmann distribution (Tian and Ma 2011, Eq. 11), such as

$$f(\mathfrak{X}) \propto \exp\left(-\alpha U(\mathfrak{X})\right), \tag{11.10}$$

where α is a hyperparameter, corresponding to the inverse temperature in physics, i.e., the reciprocal of Boltzmann's constant multiplying thermodynamic temperature in the original Boltzmann distribution, and $U(\mathfrak{X})$ is the so-called energy function. With this prior, Eq. (11.8) can be equivalently solved through a regularized least-squares estimation, i.e.,

$$\hat{\mathfrak{X}}^{\mathrm{MAP}} = \arg\min_{\mathfrak{X}} \left\{ \sum_{k=1}^{K} \|\mathfrak{Y}_k - \mathbf{H}_k\mathfrak{X}\|_2^2 + \lambda U(\mathfrak{X}) \right\}, \tag{11.11}$$

where the second term in the minimization is the regularization term and λ is the regularization coefficient, accounting for the effect of the noise variance, σ_k^2, and the hyperparameter, α, in the prior.

When the energy function, $U(\mathfrak{X})$, is modeled through a Gaussian Markov random field, it takes the following form (Yang and Huang 2010):

$$U(\mathfrak{X}) = \mathfrak{X}^T\mathbf{Q}\mathfrak{X}, \tag{11.12}$$

where $\mathbf{Q}$ is a precision matrix (the inverse of a covariance matrix), symmetric and positive-definite, characterizing spatial relations between adjacent pixels. For a Gaussian Markov random field, off-diagonal elements in the precision matrix, $\mathbf{Q}$, are non-zero only for the adjacent pixels in a defined neighborhood. The precision matrix can be furthermore factorized into $\boldsymbol{\Gamma}$ matrix of proper size, i.e.,

$$\mathbf{Q} = \boldsymbol{\Gamma}^T\boldsymbol{\Gamma}.$$

According to Yang and Huang (2010), $\boldsymbol{\Gamma}$ acts as some first or second derivative operator on the image of $\mathfrak{X}$. As such, the regularization term in Eq. (11.11) can be written as

$$\lambda U(\mathfrak{X}) = \lambda\|\boldsymbol{\Gamma}\mathfrak{X}\|_2^2,$$

which is known as the Tikhonov regularization in the field of image processing. The resulting regularization-based formulation,

$$\hat{\mathfrak{X}}^{\mathrm{MAP}} = \arg\min_{\mathfrak{X}} \left\{ \sum_{k=1}^{K} \|\mathfrak{Y}_k - \mathbf{H}_k \mathfrak{X}\|_2^2 + \lambda \|\boldsymbol{\Gamma}\mathfrak{X}\|_2^2 \right\},$$

is actually equivalent to the ridge regression in statistics (Hastie et al. 2009, Chapter 3).

11.2 Single-Image Super Resolution

Starting from around the turn of the twenty-first century, a new branch of super-resolution research emerged (Freeman et al. 2000) and eventually becomes the main stream of the super-resolution research until now. This new branch was initially referred to as single-frame super-resolution (Freeman et al. 2002) but the term quickly converged to single-image super-resolution (SISR) (Sun et al. 2003).

The basic idea of single-image super-resolution is to take a single low-resolution image and enhance its quality to match a higher resolution level by providing finer details that are not available in the original low-resolution image. Without any other additional information, this problem is ill-posed and technically unsolvable. Researchers, starting with Freeman et al. (2002), argue that one may get help from images unrelated to the low-resolution image in question because certain image features, like edges, and their impact on image quality, could have been learned from some unrelated image pairs. Using the learned relationship can boost the quality of the original low-resolution image.

To materialize the idea, one first needs to build a training dataset of many generic low-resolution and high-resolution image pairs and learn a relationship between the pairs. The training dataset is referred to as an *external* dataset, because they are not necessarily related to the low-resolution image to be processed. Building a connection between two whole images, and doing so more than what the observation model in Eq. (11.2) offers, is unrealistic. For this reason, the image relationship at the two resolution levels is learned between a pair of much smaller sized images, called *patches*, rather than between the original whole images. Image patches are usually of, say, 5×5 pixels, whereas the original images could be of 1024×1024 pixels or even larger. When a new low-resolution image arrives, it will first be chopped into small image patches. Each of the patches is to be enhanced, and then, the enhanced patches are to be stitched together to produce a whole high-resolution image. We want to stress that in the SISR research, the low-resolution images in the training set are typically generated from a physical high-resolution image though downsampling and blurring operations.

Let us use the following notations to describe the problem of SISR. Suppose that one has a training dataset having K pairs of images of different resolutions, and denote the images in the training set by a superscript, ‘tr’. The training set is hence denoted by

$$\mathcal{T} := \left\{(\mathfrak{Y}_1^{\text{tr}}, \mathfrak{X}_1^{\text{tr}}), \ldots, (\mathfrak{Y}_k^{\text{tr}}, \mathfrak{X}_k^{\text{tr}}), \ldots, (\mathfrak{Y}_K^{\text{tr}}, \mathfrak{X}_K^{\text{tr}})\right\}.$$

The low-resolution image to be enhanced is denoted by $\mathfrak{Y}^{\text{ts}}$, where the superscript, 'ts', means a test image. The goal is to produce a high-resolution counterpart, $\hat{\mathfrak{X}}^{\text{ts}}$, which does not yet exist. The superscript 'ts' on $\hat{\mathfrak{X}}^{\text{ts}}$ is usually dropped for notational simplicity, because there is no ambiguity where $\hat{\mathfrak{X}}$ (with the hat notation) belongs.

As discussed above, the images are further divided into small patches. These patches are not completely disjoint, but rather, have certain overlapping regions if two of them are adjacent to each other in the original image. Let us denote the low-resolution and high-resolution patches by, respectively, $\mathfrak{y}$ and $\mathfrak{x}$. For a given image pair, the low-resolution and high-resolution images in the training set, $\mathcal{T}$, have the same number of patches. If one puts the patches from all K training image pairs together, the low-resolution and high-resolution patches are still of the same number. Furthermore, they have a well-defined one-to-one correspondence, considering that the low-resolution images are generated from a corresponding high-resolution one through image operations. Let us index the patches from all images without necessarily being concerned with which image it belongs to. The patch pairs of all images in the training set can be denoted by

$$\wp := \left\{(\mathfrak{y}_1^{\text{tr}}, \mathfrak{x}_1^{\text{tr}}), \ldots, (\mathfrak{y}_j^{\text{tr}}, \mathfrak{x}_j^{\text{tr}}), \ldots, (\mathfrak{y}_{n_{\text{tr}}}^{\text{tr}}, \mathfrak{x}_{n_{\text{tr}}}^{\text{tr}})\right\}, \tag{11.13}$$

where n_{tr} is the total number of patches in the training set.

For the test image, $\mathfrak{Y}^{\text{ts}}$, it is divided into n_{ts} patches, which are denoted by $\{\mathfrak{y}_j^{\text{ts}}, j = 1, \ldots, n_{\text{ts}}\}$. It is understandable that the image patches in the test image are usually of the same pixel size as those in the training set. With these notations and background, one can state that an SISR method aims at producing the high-resolution image patches, $\{\hat{\mathfrak{x}}_j, j = 1, \ldots, n_{\text{ts}}\}$, which would lead to the construction of $\hat{\mathfrak{X}}$.

11.2.1 *Example-Based Approach*

Freeman and his colleagues are credited for having started the line of research on SISR (Freeman et al. 2000, 2002). The initial idea is as following. Given an observed low-resolution patch, $\mathfrak{y}^{\text{ts}}$, identify a single high-resolution patch from the training set of $\wp$ to replace it. Once this is done for every observed low-resolution patch, the set of $\{\mathfrak{x}_j^{\text{ts}}, j = 1, \ldots, n_{\text{ts}}\}$ is then attained, ready for constructing $\hat{\mathfrak{X}}$. Here $\mathfrak{x}_j^{\text{ts}}$ does not use the hat notation because it is the high-resolution patch itself rather than an estimate. The identified high-resolution patch is considered a local yet direct high-resolution *example* of the observed low-resolution counterpart, coming from the existing high-resolution images. Because of this, the resulting method was referred to as the example-based approach.

A key issue is how best to identify this high-resolution patch. A naive approach is to find the nearest neighbor of $\mathfrak{y}^{\text{ts}}$ in the set of $\wp$. This can be done by calculating the 2-norm distance between the observed patch, $\mathfrak{y}^{\text{ts}}$, and the candidate low-resolution patches, $\{\mathfrak{y}_i^{\text{tr}}, i = 1, \ldots, n_{\text{tr}}\}$, one at a time. The one with the smallest distance is the nearest neighbor. Provided the one-to-one connection between $\mathfrak{y}^{\text{tr}}$ and $\mathfrak{x}^{\text{tr}}$, once the best $\mathfrak{y}_j^{\text{tr}}$ is identified, the best $\mathfrak{x}_j^{\text{tr}}$ is found right away, which is subsequently used as the corresponding $\mathfrak{x}_j^{\text{ts}}$.

It turns out that this simple approach does not work well due to existence of noise as well as the ill-posed nature of super-resolution (Freeman et al. 2002, Fig. 4). Freeman et al. (2000, 2002) believe that a better approach is to use a broader neighborhood to account for spatial neighborhood effect.

Let us first consider randomly choosing n_{ts} high-resolution image patches, out of the n_{tr} patches in the training set, and then, matching them, one-by-one but arbitrarily, with the n_{ts} low-resolution patches partitioned from the observed low-resolution image. The neighborhood structure of the low-resolution patches in the observed image implies the neighborhood relationship among the matched high-resolution patches; see Fig. 11.2 for an illustration. Specifically, Freeman et al. (2002) choose their low-resolution and high-resolution patches to be of 7×7 and 5×5 pixels, respectively. The adjacent high-resolution patches have one pixel overlap. And the neighborhood of an image patch is defined as those having overlaps with it.

Recall that the j-th low-resolution image patch in the test image is denoted by $\mathfrak{y}_j^{\text{ts}}$. Its corresponding low-resolution candidate in the training set is labeled as $\mathfrak{y}_j^{\text{tr}}$,

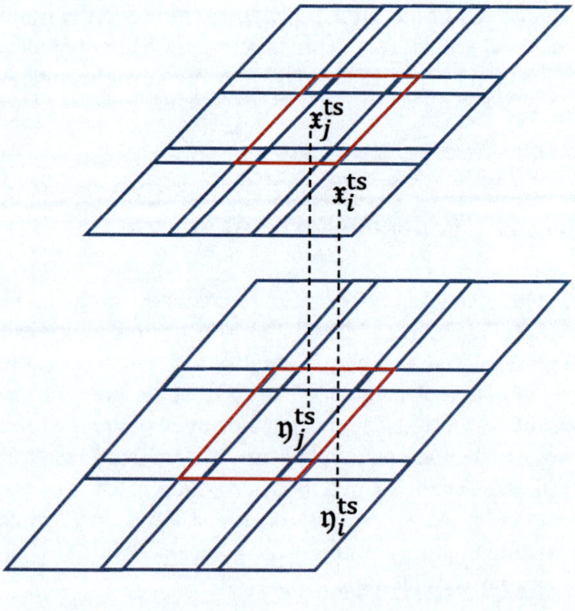

Fig. 11.2 Example-based super-resolution and the definition of the image patch neighborhood

and the corresponding high-resolution image patch is labeled as $\mathfrak{x}_j^{\text{tr}}$. The same $\mathfrak{x}_j^{\text{tr}}$ is used as a high-resolution image patch for super-resolution image reconstruction, so it is also labeled as $\mathfrak{x}_j^{\text{ts}}$. This high-resolution image patch is neighbored by a set of high-resolution image patches, denoted by $\mathfrak{x}_i^{\text{ts}} \in \mathfrak{N}_j$, where $\mathfrak{N}_j$ denotes the neighborhood set of image patch j.

Two compatibility functions are introduced to help quantify the spatial neighborhood effect. The function, $\psi(\mathfrak{x}_j^{\text{ts}}, \mathfrak{x}_i^{\text{ts}})$, characterizes the compatibility between two high-resolution patches, and $\phi(\mathfrak{y}_j^{\text{ts}}, \mathfrak{y}_j^{\text{tr}})$, characterizes the compatibility between an observed low-resolution patch and a candidate low-resolution patch in $\wp$. Freeman et al. (2000) choose $\psi(\mathfrak{x}_j^{\text{ts}}, \mathfrak{x}_i^{\text{ts}})$ as

$$\psi\left(\mathfrak{x}_j^{\text{ts}}, \mathfrak{x}_i^{\text{ts}}\right) = \exp\left(-\frac{\|\mathfrak{x}_j^{\text{ts}} - \mathfrak{x}_i^{\text{ts}}\|^2}{2\sigma^2}\right),$$

where $\|\cdot\|$ is an entry-wise matrix 2-norm, i.e., the Frobenius norm, if the image patches are represented via a matrix, and σ^2 is a tuning parameter. Note Freeman et al. (2002) specify that the entry-wise norm in the above equation is only computed in the overlapping regions between the two image patches. The function of $\phi(\mathfrak{y}^{\text{ts}}, \mathfrak{y}_j^{\text{tr}})$ takes a similar form but uses a different value for the tuning parameter σ^2.

To select the best high-resolution patch, Freeman et al. (2002) propose to optimize the posterior distribution of the high-resolution patch given all the observed low-resolution patches. The joint distribution is

$$f(\mathfrak{X}^{\text{ts}}|\mathfrak{y}_1^{\text{ts}}, \cdots, \mathfrak{y}_{n_{\text{ts}}}^{\text{ts}}) \propto \prod_{j=1}^{n_{\text{ts}}} \prod_{\mathfrak{x}_i^{\text{ts}} \in \mathfrak{N}_j} \psi(\mathfrak{x}_j^{\text{ts}}, \mathfrak{x}_i^{\text{ts}}) \cdot \prod_{j=1}^{n_{\text{ts}}} \phi(\mathfrak{y}_j^{\text{ts}}, \mathfrak{y}_j^{\text{tr}}). \tag{11.14}$$

Then, an optimization such as

$$\hat{\mathfrak{X}}^{\text{ts}} = \arg\max_{\mathfrak{X}^{\text{ts}}} \prod_{j=1}^{n_{\text{ts}}} \prod_{\mathfrak{x}_i^{\text{ts}} \in \mathfrak{N}_j} \psi(\mathfrak{x}_j^{\text{ts}}, \mathfrak{x}_i^{\text{ts}}) \cdot \prod_{j=1}^{n_{\text{ts}}} \phi(\mathfrak{y}_j^{\text{ts}}, \mathfrak{y}_j^{\text{tr}}),$$

is carried out, yielding the high-resolution patches with the best compatibility with adjacent patches as well as matching to the given low-resolution patch.

The brutal force to solve the above optimization is costly. Imagine the computational cost that one would choose n_{ts} high-resolution patches, out of the n_{tr} total, test all permutations that match the high-resolution patches with the observed low-resolution patches, and then exhaust all choose-n_{ts}-out-of-n_{tr} combinations. To expedite the optimization process, Freeman et al. (2002) suggest a one-pass algorithm. The procedure is a type of greedy algorithm, sequentially deciding the high-resolution patches one at a time. It proceeds in raster-scan order from left to right and top to bottom, keeps the previously selected high-resolution image patches, left to or above the current patch, unchanged, and selects the best high-resolution patch for the current position by computing patch compatibilities only for those already selected neighboring high-resolution patches.

11.2.2 Locally Linear Embedding Method

Researchers soon realize that the strategy of selecting a single best high-resolution image patch is too restrictive (Yang and Huang 2010, Page 18). Instead, one can possibly connect an observed low-resolution image patch with a set of candidate low-resolution patches via a linear combination, much like in linear regression but it is referred to as *linear embedding* in the image processing language. Chang et al. (2004) materialize this idea. Rather than learning a linear combination of all possible image patches, Chang et al. (2004) limit the set of candidate low-resolution patches in a local neighborhood and thus label their method as *locally* linear embedding.

Specifically, Chang et al. (2004) propose to find the k-nearest neighborhood among $\mathfrak{y}^{\text{tr}}$'s in $\wp$ for each observed low-resolution patch, $\mathfrak{y}_j^{\text{ts}}$, and designate those training images as constituting $\mathfrak{y}_j^{\text{ts}}$'s neighborhood, $\mathfrak{N}_j$. Then, they solve the following minimization problem,

$$\begin{aligned} \hat{\mathbf{w}} &= \arg\min_{\mathbf{w}} \left\| \mathfrak{y}_j^{\text{ts}} - \sum_{\mathfrak{y}_i^{\text{tr}} \in \mathfrak{N}_j} w_i \mathfrak{y}_i^{\text{tr}} \right\|^2, \\ &\text{such that} \sum_{\mathfrak{y}_i^{\text{tr}} \in \mathfrak{N}_j} w_i = 1, \end{aligned} \tag{11.15}$$

where $\mathbf{w}$ is the vector of reconstruction weights to be learned from the observed low-resolution image patch and the candidate patches in $\wp$. The reconstruction weights are then employed to produce the corresponding high-resolution image patch through

$$\hat{\mathfrak{x}}_j = \sum_{\mathfrak{y}_i^{\text{tr}} \in \mathfrak{N}_j} \hat{w}_i \mathfrak{x}_i^{\text{tr}}. \tag{11.16}$$

One may have noticed that the expression in Eq. (11.15) looks similar to a kernel regression, which is a localized, nonparametric regression method (Hastie et al. 2009, Chapter 6). The difference is that the weight in a kernel regression is defined through a kernel function and its controlling parameter, the bandwidth, which defines the size of the neighborhood, is learned from the training data. Here in super-resolution, the weights are not regularized by any function but learned directly from the data. The neighborhood size is defined in a preceding, but separate, step in a k-NN fashion. In this sense, the locally linear embedding idea in super-resolution combines the features of both kernel regression and k-NN methods.

11.2.3 Sparse Coding Approach

Similar to Chang et al. (2004), Yang et al. (2010) develop another machine learning-based approach to learn the relationship of an observed low-resolution image patch as a linear combination of the candidate patches in the training set $\wp$. Unlike Chang et al. (2004), Yang et al. (2010) do not limit the candidate patches to a prescribed neighborhood. Instead their method is derived from the compressive sensing theory, meaning that they mean to learn a sparse representation of the observed low-resolution image patch using a low-resolution image patch dictionary, denoted by $\mathfrak{D}_l$, and reconstruct the corresponding high-resolution image patch using the corresponding high-resolution image patch dictionary, denoted by $\mathfrak{D}_h$.

The basic idea can be described as follows. The low- and high-resolution image patch dictionaries are assumed given. A straightforward way for constructing those dictionaries is to use the raw image patches and their corresponding downsampled counterparts as the elements (called atoms in the compressed sensing literature) in the respective dictionaries, i.e.,

$$\begin{aligned}\mathfrak{D}_l &= \left[\mathfrak{y}_1^{\text{tr}}, \quad \mathfrak{y}_2^{\text{tr}}, \quad \cdots, \quad \mathfrak{y}_{n_{\text{tr}}}^{\text{tr}}\right], \\ \mathfrak{D}_h &= \left[\mathfrak{x}_1^{\text{tr}}, \quad \mathfrak{x}_2^{\text{tr}}, \quad \cdots, \quad \mathfrak{x}_{n_{\text{tr}}}^{\text{tr}}\right].\end{aligned} \tag{11.17}$$

Given an observed low-resolution patch, $\mathfrak{y}^{\text{ts}}$, Yang et al. (2010) seek to find its sparse representation by solving the ℓ_1-minimization problem:

$$\begin{aligned}&\min \|\boldsymbol{\alpha}\|_1 \\ &\text{such that} \quad \|\mathfrak{y}^{\text{ts}} - \mathfrak{D}_l \boldsymbol{\alpha}\|_2^2 \leq \epsilon,\end{aligned} \tag{11.18}$$

where ϵ is the allowance for fitting error and $\boldsymbol{\alpha}$ is the sparse representation to be learned. The constrained minimization problem above is usually solved in the following equivalent unconstrained form, which is in fact the formulation for Lasso regression:

$$\hat{\boldsymbol{\alpha}} = \arg\min_{\boldsymbol{\alpha}} \|\mathfrak{y}^{\text{ts}} - \mathfrak{D}_l \boldsymbol{\alpha}\|_2^2 + \lambda \|\boldsymbol{\alpha}\|_1, \tag{11.19}$$

where λ is the regularization coefficient, playing the same role as in Eq. (11.11). Then, the corresponding high-resolution patch is recovered by

$$\hat{\mathfrak{x}} = \mathfrak{D}_h \hat{\boldsymbol{\alpha}}. \tag{11.20}$$

The actual algorithm of Yang et al. (2010) considers a number of practical concerns within the aforementioned framework in Eq. (11.19).

- **Feature extraction operator**. Yang et al. (2010) state that fitting directly to the image pixels is not the best practice in image processing. Rather people recognize the importance of fitting the most relevant part of the low-resolution image. This goal is carried out by adding an operator, called the feature extraction operator and denoted by **F**, in front of the dictionary. In other words, the fitness term in Eq. (11.19) becomes

$$\|\mathbf{F}\mathfrak{y}^{\mathrm{ts}} - \mathbf{F}\mathfrak{D}_l\boldsymbol{\alpha}\|_2^2. \tag{11.21}$$

The feature extraction operator, **F**, is typically some kind of high-pass filter, because viewers of images are more sensitive to the high-frequency content of the image. Therefore, by focusing on the high-frequency content, the machine learning estimate of $\boldsymbol{\alpha}$ is more perceptually meaningful. Yang et al. (2010) suggest using the first-order and second-order derivatives of the image patches as the feature extraction operators.
- **Enforce the compatibility between adjacent patches**. Yang et al. (2010) realize that solving Eq. (11.19) individually for each patch does not ensure the compatibility between adjacent patches. To address this issue, Yang et al. (2010) borrow the idea of one-pass algorithm (Freeman et al. 2002), mentioned at the end of Sect. 11.2.1, in which the patches are processed in raster-scan order, from left to right and top to bottom. To enforce the compatibility between adjacent patches, Yang et al. (2010) add a second constraint, in addition to that already in Eq. (11.18), which is to force the super-resolution reconstruction, $\mathfrak{D}_h\boldsymbol{\alpha}$, to agree closely with the previously computed high-resolution patches in its neighborhood. Thus, Eq. (11.18) is augmented as

$$\begin{aligned} \min \|\boldsymbol{\alpha}\|_1, \quad \text{such that} \quad & \|\mathbf{F}\mathfrak{y}^{\mathrm{ts}} - \mathbf{F}\mathfrak{D}_l\boldsymbol{\alpha}\|_2^2 \le \epsilon_1 \\ & \|\mathbf{F}\hat{\mathfrak{x}}_* - \mathbf{P}\mathfrak{D}_h\boldsymbol{\alpha}\|_2^2 \le \epsilon_2 \end{aligned} \tag{11.22}$$

where $\hat{\mathfrak{x}}_*$ represents the previously reconstructed high-resolution images overlapping with the current patch under processing and **P** extracts the region of overlaps between the current patch and previously reconstructed high-resolution image. By defining a few new notations, such as

$$\tilde{\mathfrak{D}} = \begin{bmatrix} \mathbf{F}\mathfrak{D}_l \\ \\ \mathbf{P}\mathfrak{D}_h \end{bmatrix}, \quad \tilde{\mathfrak{y}} = \begin{bmatrix} \mathbf{F}\mathfrak{y}^{\mathrm{ts}} \\ \\ \mathbf{F}\hat{\mathfrak{x}}_* \end{bmatrix},$$

Eq. (11.19) can be reformulated as

$$\hat{\boldsymbol{\alpha}} = \arg\min_{\boldsymbol{\alpha}} \|\tilde{\mathfrak{y}} - \tilde{\mathfrak{D}}\boldsymbol{\alpha}\|_2^2 + \lambda\|\boldsymbol{\alpha}\|_1. \tag{11.23}$$

Once $\hat{\boldsymbol{\alpha}}$ is obtained, the high-resolution reconstruction is still carried out using Eq. (11.20).

- **Learning the dictionary pair**. When we explain the basic framework of the sparse representation work, the dictionaries are the collection of raw image patches. Yang et al. (2010) believe that such a strategy is not most effective and will result in large dictionaries and expensive computation. They instead propose to learn compact dictionaries also through an ℓ_1-regularization. Suppose that $\mathfrak{x}_c$ and $\mathfrak{y}_c$ are a collection of sampled raw image patches of high-resolution and low-resolution, respectively. Then the dictionary learning is through

$$\min_{\{\mathfrak{D}_h, \mathfrak{D}_l, \mathbf{z}\}} \left\{ \frac{1}{M_0} \|\mathfrak{x}_c - \mathfrak{D}_h \mathbf{z}\|_2^2 + \frac{1}{N_0} \|\mathfrak{y}_c - \mathfrak{D}_l \mathbf{z}\|_2^2 + \lambda \left(\frac{1}{N_0} + \frac{1}{M_0} \right) \|\mathbf{z}\|_1 \right\}, \tag{11.24}$$

where M_0 and N_0 are the dimensions of the high-resolution and low-resolution image patches in vector form and $\mathbf{z}$ is the coefficients of the sparse representation, like the $\boldsymbol{\alpha}$ used above.

11.2.4 Library-Based Non-local Mean Method

Sreehari et al. (2017) present a library-based (LB) non-local mean (NLM) method for handling a single low-resolution image. Their work is also one of the first applying the super-resolution technique to electron microscopic images. The objective function used by Sreehari et al. (2017) is

$$\hat{\mathfrak{X}}^{\text{MAP}} = \arg\min_{\mathfrak{X}} \{-\log f(\mathfrak{Y}|\mathfrak{X}) - \lambda \log f(\mathfrak{X})\}, \tag{11.25}$$

which is in principle the same as Eq. (11.8). One notes that there is a single low-resolution image here, corresponding to $K = 1$ in Eq. (11.8).

Sreehari et al. (2017) propose to solve Eq. (11.25) using the ADMM method (Boyd et al. 2011). In doing so, Sreehari et al. (2017) introduce an $\mathfrak{X}_*$ to be used as input in the second term of Eq. (11.25), with an enforcer that $\mathfrak{X}_* = \mathfrak{X}$, in order to split the decision variable. Then, Sreehari et al. (2017) express the constrained optimization problem in the form of an augmented Lagrangian, such as

$$\min_{\mathfrak{X}, \mathfrak{X}_*, \mathfrak{U}} \left\{ -\log f(\mathfrak{Y}|\mathfrak{X}) - \lambda \log f(\mathfrak{X}_*) + \frac{1}{2\sigma^2} \|\mathfrak{X} - \mathfrak{X}_* + \mathfrak{U}\|_2^2 - \frac{1}{2\sigma^2} \|\mathfrak{U}\|_2^2 \right\}, \tag{11.26}$$

where σ is the augmented Lagrangian parameter and $\mathfrak{U}$ is the dual variable. To solve Eq. (11.26) via ADMM, the following specific steps are to be taken:

- Introduce a couple of new notations: $\tilde{\mathfrak{X}} = \mathfrak{X}_* - \mathfrak{U}$ and $\tilde{\mathfrak{X}}_* = \mathfrak{X} + \mathfrak{U}$.

- Initialize $\hat{\mathfrak{X}}_*$ and $\mathfrak{U}$. Let $\hat{\mathfrak{X}}_*$ be a baseline image reconstruction, for instance, via a bicubic interpolation (Keys 1981) and let $\mathfrak{U} = \mathbf{0}$.
- Iterate through the following steps.
  1. $\tilde{\mathfrak{X}} \longleftarrow \hat{\mathfrak{X}}_* - \mathfrak{U}$.
  2. Solve the following optimization to update $\hat{\mathfrak{X}}$.

$$\begin{aligned} \hat{\mathfrak{X}} &= \arg\min_{\mathfrak{X}} \left\{ -\log f(\mathfrak{Y}|\mathfrak{X}) + \frac{1}{2\sigma^2} \|\mathfrak{X} - \tilde{\mathfrak{X}}\|_2^2 \right\}, \\ \Rightarrow \quad \hat{\mathfrak{X}} &= \arg\min_{\mathfrak{X}} \left\{ \frac{1}{2\sigma_\epsilon^2} \|\mathfrak{Y} - \mathbf{H}\mathfrak{X}\|_2^2 + \frac{1}{2\sigma^2} \|\mathfrak{X} - \tilde{\mathfrak{X}}\|_2^2 \right\}, \end{aligned} \tag{11.27}$$

  where the observation model in Eq. (11.2) is used to get the second minimization in the above equation.
  3. $\tilde{\mathfrak{X}}_* \longleftarrow \hat{\mathfrak{X}} + \mathfrak{U}$.
  4. Update $\hat{\mathfrak{X}}_*$ through the library-based non-local mean method. Here "library" refers to an external set of high-resolution image patches, much like $\wp$ mentioned above. But in Sreehari et al. (2017)'s library, there is no low-resolution image patch. Instead, Sreehari et al. (2017) use $\tilde{\mathfrak{X}}_*$ to create the patches to be matched with that in the library.

  Denote by $\tilde{\mathfrak{x}}_*(s)$ a patch in $\tilde{\mathfrak{X}}_*$, which is of $N_p \times N_p$ pixels and centered at position s. The l-th patch in the high-resolution library is denoted by $\mathfrak{x}_l^L$, where the library is the same as the training set mentioned earlier and the superscript 'L' bears a similar meaning as 'tr' before. Then, the LB-NLM weight is computed by

$$w_{s,l} \leftarrow \exp\left\{ \frac{-\|\tilde{\mathfrak{x}}_*(s) - \mathfrak{x}_l^L\|_2^2}{2N_p^2\sigma_n^2} \right\} \quad \text{and} \quad w_{s,l} \leftarrow \frac{w_{s,l}}{\sum_{l=1}^{n_L} w_{s,l}}, \tag{11.28}$$

  where $\sigma_n = \sqrt{\lambda}\sigma$ is interpreted by Sreehari et al. (2017) for being the noise standard deviation in the denoising operation and n_L is the number of patches in the library. Then, the pixel at position s can be updated by

$$\left(\hat{\mathfrak{X}}_*\right)_s = \sum_{l=1}^{n_L} w_{s,l} z_l,$$

  where z_l is the center pixel of patch $\mathfrak{x}_l^L$.
  5. $\mathfrak{U} \longleftarrow \mathfrak{U} + (\hat{\mathfrak{X}} - \hat{\mathfrak{X}}_*)$.

11.2.5 Deep Learning

In recent years, the deep learning methods (Wang et al. 2021) have been adopted for SISR. Deep learning methods train an end-to-end deep neural network, which, once trained, can take in an low-resolution image and produce a reconstructed image, supposedly of higher resolution. In the 2017 single-image super-resolution challenge, `NTIRE 2017` (Timofte et al. 2017), all the competitive algorithms adopt the deep learning approach.

One of the first deep convolutional neural networks is SRCNN (Dong et al. 2016); the acronym stands for the super-resolution convolutional neural network. The default setting of SRCNN has three (hidden) layers, as illustrated in Fig. 11.3. The three layers fulfill, respectively, the tasks of patch extraction and representation, nonlinear mapping, and reconstruction.

At each layer, there are two important parameters to specify the network setting: the filter size and the number of filters. Dong et al. (2016) choose their basic network setting as the following—filter size: $a_1 = 9$, $a_2 = 1$, and $a_3 = 5$, and the number of filters: $n_1 = 64$, $n_2 = 32$, and $n_3 =$ the number of channels in the image. This network is denoted using the convention of 9-1-5, which represents the spatial size of the filters on the three layers. The first two layers perform a standard convolution with the rectified linear unit (ReLU), of the form

$$\mathbf{F}_1(\mathfrak{Y}) = \max(\mathbf{0}, \mathbf{W}_1 \otimes \mathfrak{Y} + \mathbf{B}_1) \quad \text{and} \quad \mathbf{F}_2(\mathfrak{Y}) = \max(\mathbf{0}, \mathbf{W}_2 \otimes \mathbf{F}_1(\mathfrak{Y}) + \mathbf{B}_2),$$

where $\mathbf{F}_1(\cdot)$ and $\mathbf{F}_2(\cdot)$ denote the operation on the respective layer, $\mathbf{W}_1$ and $\mathbf{W}_2$ are the filters, and $\mathbf{B}_1$ and $\mathbf{B}_2$ are the biases, and $\otimes$ denotes a convolution operation. The third layer does not apply ReLU but simply the filter and bias, i.e.,

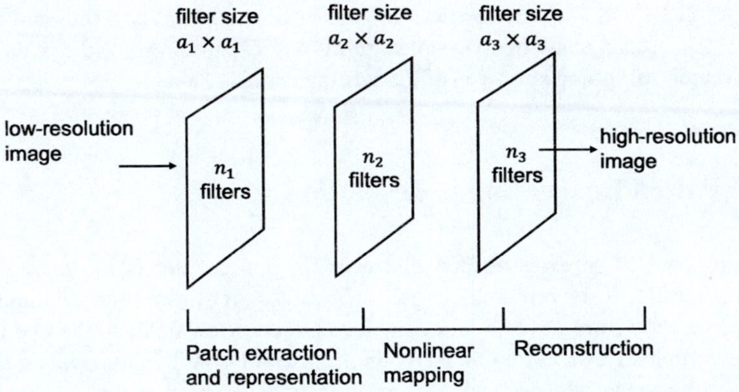

Fig. 11.3 Schematic of the three-layer convolutional neural network developed by Dong et al. (2016) for super-resolution

$$\mathbf{F}(\mathfrak{Y}) = \mathbf{W}_3 \otimes \mathbf{F}_2(\mathfrak{Y}) + \mathbf{B}_3.$$

The network parameters are collectively denoted by $\boldsymbol{\Theta} = \{\mathbf{W}_1, \mathbf{W}_2, \mathbf{W}_3, \mathbf{B}_1, \mathbf{B}_2, \mathbf{B}_3\}$. The loss function used in SRCNN is a mean-squared error loss, expressed as

$$L(\boldsymbol{\Theta}) = \frac{1}{K}\sum_{k=1}^{K} \|\mathbf{F}(\mathfrak{Y}_k^{\mathrm{tr}}; \Theta) - \mathfrak{X}_k^{\mathrm{tr}}\|^2, \tag{11.29}$$

where $\mathbf{F}(\mathfrak{Y}_k^{\mathrm{tr}}; \Theta)$ means the k-th reconstructed image. One thing worth noting is that when using SRCNN, the low-resolution image input is first upscaled to the desired size using bicubic interpolation. In other words, $\mathfrak{Y}^{\mathrm{tr}}$ in Eq. (11.29) is the up-scaled low-resolution image of the same pixel amount as its corresponding $\mathfrak{X}^{\mathrm{tr}}$.

With three layers, SRCNN is not really a deep network. Dong et al. (2016) conduct experiments, exploring how the number of layers may help with the super-resolution objective. Their conclusion at that time was that "*All these experiments indicate that it is not 'the deeper the better'*. . . ." Their final model is a three-layer, 9-5-5 network, which performs only slightly better than their basic 9-1-5 network.

Kim et al. (2016) develop a very deep convolutional network for super-resolution (VDSR). They argue that "*increasing depth significantly boosts performance*." VDSR uses 20 layers for super resolution and their filter size is 3×3 on all layers. All the intermediate layers use 64 filters for each color channel while the last layer uses a single filter for each color channel. Their loss function for training is still the mean squared error loss.

There are other newer deep learning networks such as the enhanced deep-residual networks super-resolution (Lim et al. 2017, EDSR), developed by the same research team who developed VDSR, and the residual channel attention networks (Zhang et al. 2018, RCAN). EDSR and VDSR are two of the most competitive methods in the `NTIRE 2017` challenge. RCAN was published later and thus not part of the `NTIRE 2017` challenge. Deep learning-based super-resolution is an active research area still going strong as of the writing of this chapter.

11.3 Paired Images Super Resolution

The third kind of super-resolution discussed in this chapter is the paired image super-resolution. This refers to a pair of images of different resolutions both obtained directly from physical imaging devices. The view fields of the two images overlap completely, or more precisely, the high-resolution image covers a smaller subset of the view field of the low-resolution image. Suppose that the resolution ratio of the two images is 2:1 and the two images have the same amount of pixels, i.e., $M_h = M_l$ and $N_h = N_l$. Then, the high-resolution image covers only 25% of the area that is covered by the low-resolution image.

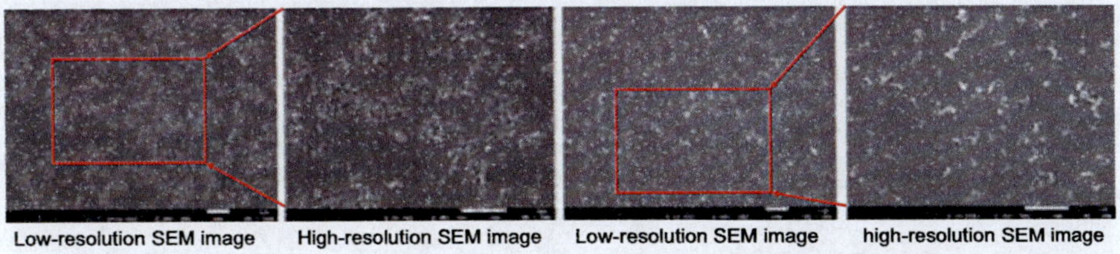

Fig. 11.4 Two pairs of low-resolution and high-resolution SEM images. The red rectangles in the low-resolution images are the areas corresponding to the high-resolution images (*source*: Qian et al. (2020))

Two such examples presented by Qian et al. (2020) are reproduced in Fig. 11.4. The pair of images in each example are obtained by the same SEM in one experimental setting but through two actions. First, the SEM is set at a low magnification level and takes the low-resolution image. Then, with the same sample still in the platform, the SEM is adjusted to a higher magnification level, i.e., it is zoomed in, taking the high-resolution image. The objective of super-resolution is to boost the quality of the low-resolution image to be of $2M_l \times 2N_l$ pixels over the whole area of the view field.

The paired image super-resolution is clearly different from the multi-frame super-resolution research, as one only deals with a single low-resolution image. In the sense of using a single low-resolution image, this line of research is closer to the branch of single-image super-resolution. But in SISR, the image pairs in the training set are external to the low-resolution image to be super resolved. Almost invariably in the current practice, the low-resolution images in the training set are the blurred and downsampled version of the high-resolution images, rather than physical images directly acquired by an imaging device.

Qian et al. (2020) argue that the properties of physical low-resolution images do not necessarily satisfy the observation model used in the synthetic image-based super-resolution, e.g., Eq. (11.2). Their demonstration of the differences is presented in Fig. 11.5, in which a physical low-resolution image is compared with a synthetic image, downsampled from their commonly paired high-resolution image. The right most figure of Fig. 11.5 shows the difference between the two images, which is rather pronounced. The reason of discrepancy are the noise in the high-resolution image, different contrast levels between the paired images, and/or different natures and degrees of local distortion from individual image-capturing processes. Consequently, the SISR methods thus far developed may not necessarily be effective to handle the paired image problem.

The use of external image datasets for training, as done in the current SISR, may not be the best practice in handling the paired image problem. Being "external," it means that the image pairs in the training set are unrelated to the image to be super resolved. That setting is understandable when one only has a low-resolution image without its high-resolution counterpart. For the paired image cases, given the complete overlap, albeit a subset of the view field, between a low-resolution image

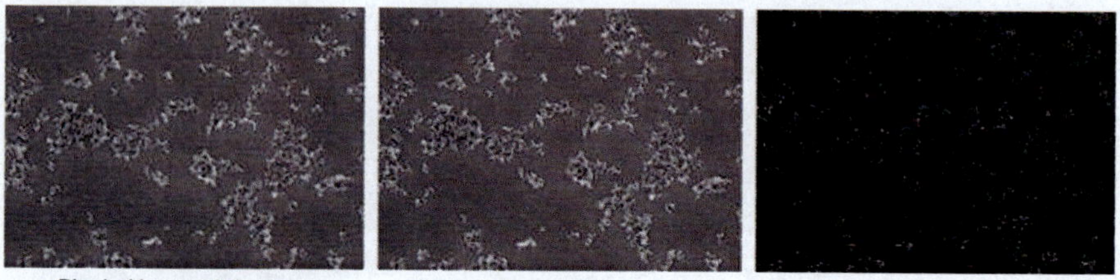

Fig. 11.5 Comparison of a physical low-resolution SEM image and the synthetic downsampled image from their commonly paired high-resolution image (*source*: Qian et al. (2020))

and its high-resolution counterpart, one would think that a relationship learned directly from this specific pair is the best for boosting the resolution for the rest of the area uncovered by the high-resolution image.

Paired images are not very common in public image databases because taking them needs special care and specific setup. This may explain that this particular type of problem has not received enough attention as much as the other two branches (Zhang et al. 2019; Qian et al. 2020). Paired images, on the other hand, are rather common in scientific experiments, especially in the material and manufacturing research, as well as in medical imaging (Trinh et al. 2014).

The use of electron microscopes exacerbates the need for handling paired image for super-resolution. The particles generating images in electron microscopes are electrons, much heavier than photons (a photon's static mass is zero). Unlike optical photographing, the imaging process using an electron microscope is not non-destructive. Imaging using high-energy electron beams can damage sample specimen and must be carefully administrated. Consequently, researchers tend to use low-energy beams or subject the samples to a short duration of exposure. The results are of course low-resolution images. But clear images are much desired for, and instrumental to, discovery and material designs. When a greater area of material samples is imaged using low-energy beams, an effective super-resolution method that can subsequently boost these low-resolution images to a higher resolution definitely has a significant impact on scientific fields relying on electron imaging.

Qian et al. (2020) look into the issue of paired image super-resolution. They proceed with two schools of approaches for handling the paired electron images. The first school is to apply the current popular SR methods, specifically the sparse-coding approaches explained (Yang et al. 2010; Trinh et al. 2014) and deep-learning methods (Kim et al. 2016; Lim et al. 2017; Zhang et al. 2018), but using the physical electron image pairs as input, rather than using the downsampled synthetic images. To handle the uniqueness of paired EM images, Qian et al. (2020) explore different training strategies. The second school is to devise a simpler super-resolution method that uses an LB-NLM filter with a paired library, a revision to the method presented in Sect. 11.2.2. The common preprocessing step in both schools is to register the high-resolution and low-resolution physical images. For that, Qian et al. (2020) devise a global and local registration procedure. The global registration is applied to the whole image, so that this step is common to all super-resolution methods. The local registration is applied to the image patches and thus common only to the sparse coding methods and the LB-NLM methods. The deep learning methods take the whole images as input of their networks, to which only the global registration is applied, and conduct an end-to-end super-resolution.

11.3.1 Global and Local Registration

Qian et al. (2020) start off by upsampling the low-resolution image, $\mathfrak{Y}$, through bicubic interpolation and by a factor of two; this produces $\mathfrak{Y}^{\text{up}}$. Then, Qian et al. (2020) estimate a shift transform (x, y) and a rotation transform (θ) by comparing the overlapping area of $\mathfrak{X}$ and $\mathfrak{Y}^{\text{up}}$. The mean squared error (MSE) between $\mathfrak{X}$ and $\mathfrak{Y}^{\text{up}}$ are calculated in their overlapping area (see Eq. (11.33) for the definition of MSE between two images) and is to be minimized to produce the estimate of the global registration parameters of (x, y, θ). To accelerate the matching process, Qian et al. (2020) first downsample both images and estimate (x, y, θ) with a grid search, and then refine the estimation by searching the neighborhood of the initial estimates using the original image of $\mathfrak{Y}^{\text{up}}$ and $\mathfrak{X}$. The registered upsampled image is then denoted by $\mathfrak{Y}^{\text{reg}}$. This completes the procedure of global registration.

For the purpose of local registration, Qian et al. (2020) segment the matched $\mathfrak{X}$ and $\mathfrak{Y}^{\text{reg}}$ into overlapping patches of $N_p \times N_p$ pixels. Let $\mathfrak{x}(i, j)$ and $\mathfrak{y}^{\text{reg}}(i, j)$ denote the patches centered at (i, j) in $\mathfrak{X}$ and $\mathfrak{Y}^{\text{reg}}$, respectively. For notional simplicity, the superscript, 'reg', is often omitted if it is clear which image or image patch is referred to in the context.

Qian et al. (2020) search the neighborhood of (i, j) to find the best shift that registers the two images, through solving the following optimization problem:

$$\min_{i^*, j^*} \frac{\langle \mathfrak{x}(i^*, j^*), \mathfrak{y}(i, j) \rangle}{\|\mathfrak{x}(i^*, j^*)\|_2 \cdot \|\mathfrak{y}(i, j)\|_2}, \tag{11.30}$$

where $\langle \cdot \rangle$ denotes a matrix inner product, and $|| \cdot ||_2$ is the Frobenius norm. Qian et al. (2020) state that they prefer the use of an inner product over the use of Euclidean distance for matching two patches as the former is insensitive to the contrast difference between the two image patches.

Qian et al. (2020) only apply the local registration to the patches with rich texture, which can be selected by deeming the variance of pixel values of $\mathfrak{y}(i, j)$ larger than a certain threshold. For the electron microscopic images at their disposal, Qian et al. (2020) set the threshold as 100. Figure 11.6 presents one example of local registration, where the red arrows illustrate the displacements, $(i^* - i, j^* - j)$, between the matched patches from $\mathfrak{X}$ and $\mathfrak{Y}^{\text{reg}}$. The magnitudes and directions of the displacements vary significantly across the image, showing a complex and irregular pattern of local distortions, which would not have been adjusted by a global registration alone.

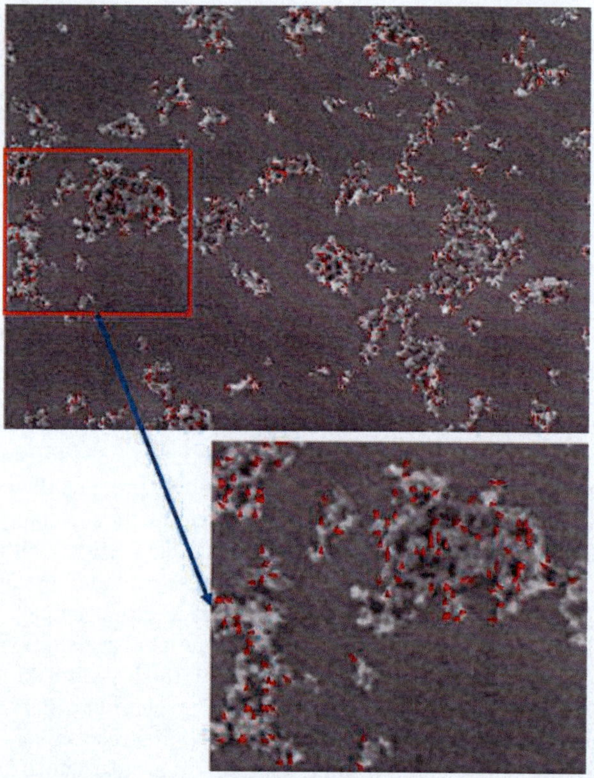

Fig. 11.6 The results of the local registration. The bottom image is magnified from the red rectangle in the top image, in which the red arrows indicate the displacements $(i^* - i, j^* - j)$ between the matched patches (*source*: Qian et al. (2020))

11.3.2 Existing Super-Resolution Methods Applied to Paired Images

Qian et al. (2020) test two main approaches: the sparse-coding methods and the deep learning methods. When using the sparse-coding methods, Qian et al. (2020) removed the back-projection step after the super-resolution reconstruction. The back-projection step was included in the original sparse-coding method under the assumption that by downsampling the super-resolution result, one can get the same image as the low-resolution input. This assumption is not valid for the paired images; see the example presented in Fig. 11.5.

When using the deep-learning methods, Qian et al. (2020) are mindful of the small sample size of the paired images. The small number of electron images is a result of the expensiveness to prepare material samples and operate electron microscopes. In reality, one can expect a handful, to a few dozens at best, of such paired electron images. To prevent overfitting, Qian et al. (2020) adopt two techniques: data-augmentation and early-stopping. A larger dataset is created by flipping each image pair row-wise and column-wise, rotating them by 90, 180 and 210 degrees, and downsizing them by the factors 0.7 and 0.5. By calculating the accuracy of validation data, Qian et al. (2020) state that they also discover that the training achieves the best performance before its 30th epochs and should be stopped accordingly.

There is the question of how to train a super-resolution model. Suppose that one has a total of m_{pr} pairs of electron images, each of which has a low-resolution image and its corresponding high-resolution image. In the particular study reported in the later section of Case Study, $m_{\mathrm{pr}} = 22$ and they are all SEM images. The size of both types of images is 1280×944 pixels. Through image registration, Qian et al. (2020) identify the overlapping areas of each pair and carve out the corresponding low-resolution image, which is of 640×472 pixels. The 1280×944-pixel high-resolution image and the 640×472-pixel low-resolution image are what Qian et al. (2020) use to train the model and do the testing. The non-overlapping area of the low-resolution image is not used in the experimental analysis because there is no ground truth for that area to be tested.

To mimic the practical applications where the super-resolution method is to be applied to the area where there is no corresponding high-resolution images, Qian et al. (2020) partition the low-resolution and high-resolution images in each pair into 3×4 subimages. Qian et al. (2020) treat $m_{\mathrm{pp}}^{\mathrm{tr}}$ subimages as the training images and keep the remaining $m_{\mathrm{pp}}^{\mathrm{ts}}$ subimages unused in the training stage and treat them as the out-of-sample test images. In the later Case Study, the number of training images per pair is $m_{\mathrm{pp}}^{\mathrm{tr}} = 9$ and the number of test images per pair is $m_{\mathrm{pp}}^{\mathrm{ts}} = 3$. The size of a high-resolution subimage is 320×314, where the size of a low-resolution subimage is 160×157, still maintaining the 2:1 resolution ratio. The training and test subimages of two SEM image pairs are shown in Fig. 11.7.

There are naturally two training strategies. To reconstruct the test subimages from Pair $i = 1, \ldots, m_{\mathrm{pr}}$, one can use the training subimages coming from the same pair to train the model. As such, there will be m_{pr} individual models trained. In the phase of testing, each model is used individually for the specific image pair from which

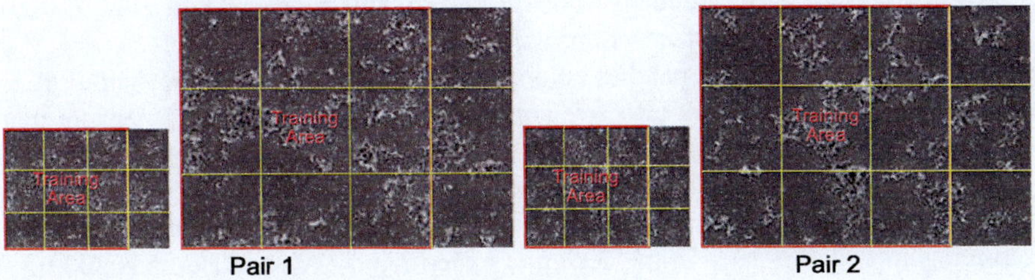

Fig. 11.7 The overlapping areas of two pairs of SEM images. The left 75% is the training area and the right 25% is the test area. The yellow lines partition each image into 3×4 subimages

the model is trained. Each model is trained by $m_{\mathrm{pp}}^{\mathrm{tr}}$ pairs of subimages and evaluated on $m_{\mathrm{pp}}^{\mathrm{ts}}$ pairs of subimages. Qian et al. (2020) refer to this strategy as self-training.

Alternatively, one can pool all the training sample pairs together and train a single model. In the phase of testing, this single model is used for reconstructing the test images for all image pairs. Qian et al. (2020) refer to this strategy as pooled-training. Under this setting, there are a total of $m_{\mathrm{pr}} \times m_{\mathrm{pp}}^{\mathrm{tr}}$ pairs of training images and $m_{\mathrm{pr}} \times m_{\mathrm{pp}}^{\mathrm{ts}}$ pairs of test images. In the Case Study, the training sample size in the pooled-training is 198 pairs of subimages and the test sample size is 66 pairs of subimages, much greater than the sample sizes used in self training.

The conventional wisdom, especially when deep learning approaches are used, is that the $m_{\mathrm{pp}}^{\mathrm{tr}}$ training images, which are nine in the Case Study, are too few to be effective. The popular strategy is to use the pooled training. For the paired electron images, however, Qian et al. (2020) find that using self-training in fact produces the best super-resolution results, despite the relatively small sample size used. This appears something unique for the paired electron image problem—the pairing in the images makes using training samples internal to a specific image pair a better option than using more numerous external images.

11.3.3 Paired LB-NLM Method for Paired Image Super-Resolution

Qian et al. (2020) devise a simple but effective super-resolution method for the paired electron images, based on the LB-NLM method originally used by Sreehari et al. (2017), but made a few changes to the original LB-NLM method.

After the local registration, Qian et al. (2020) save the matched patches from $\mathfrak{x}$'s and $\mathfrak{y}$'s into a paired library. Including $\mathfrak{y}$'s in the library, i.e., the upsampled physical low-resolution image patches, is a key difference between Qian et al. (2020)'s method and Sreehari et al. (2017)'s method, as the library in Sreehari et al. (2017) uses the high-resolution patches only. Sreehari et al. (2017) also propose to create the library with dense sampling. But the training area of each pair of electron microscopic images has about one million overlapping patches and many of them are of low texture and redundant information. For this reason, Qian et al. (2020) want to reduce the library size to improve the learning efficiency.

As a large portion of the patches belongs to the background, random sampling is understandably not the most effective approach for patch selection. To ensure that different categories of image patches are adequately included, such as foreground, background, and boundaries, Qian et al. (2020) devise a k-means clustering method to build the paired library, which is, in spirit, similar to the stratified random sampling approach as used in the design of experiments (Wu and Hamada 2009).

Assume that one would like to build a library with n_L pairs of image patches. Qian et al. (2020) first randomly sample a total of $G \times n_L$ high-resolution patches from $\mathfrak{x}$'s. Then they apply the k-means method to classify the high-resolution

Fig. 11.8 Demonstration of a paired library with 800 patches of 9×9, classified into 10 categories. Left: the selected high-resolution patches, where each row makes up one category; middle: the corresponding upsampled low-resolution patches, right: the histogram of the original patches (*source*: Qian et al. (2020))

patches into g categories according to the pixel's intensity. After that, Qian et al. (2020) randomly sample n_L/g high-resolution patches from each category, and save them and their matched upsampled patches $\mathfrak{y}$'s in the library.

Qian et al. (2020) denote each pair of the patches by $\mathfrak{x}_l$ and $\mathfrak{y}_l$, for $l = 1, \cdots, n_L$. They state that when a large enough G, say $G = 10$, is used, there are usually more than n_L/g patches in each category. But just in case that the number of patches in one category is fewer then n_L/g, Qian et al. (2020) use all the patches in that category. The library size will then become smaller than n_L, but that is fine. In their implementation, Qian et al. (2020) set $G = 10$ and $g = 10$.

Figure 11.8 demonstrates a library with 800 paired patches of 9×9. Figure 11.8, the right most panel, presents the histogram of patches in the ten categories of the original image data. One can see that the first, fourth and fifth categories account for a large portion of the randomly sampled patches and these categories correspond to the patches in the background area. After the selection, there will be 80 patches in each category equally. The background patches make up only 30% of the selected ones in the library, whereas the other 70% are the patches with rich texture. Comparing the two figures on the left, one observes that the noise and contrast levels are represented with a good balance in both high-resolution and low-resolution image patches.

With the paired library in place, Qian et al. (2020) reconstruct a high-resolution image for the low-resolution image's coverage area. The real impact is on the area of the low-resolution image where there is no high-resolution image correspondence. In doing so, Qian et al. (2020) again start with the upsampled low-resolution image, $\mathfrak{Y}^{\text{up}}$. Then a revised LB-NLM filter, based on the paired library established thus far, is applied to $\mathfrak{Y}^{\text{up}}$ for obtaining the reconstructed image $\hat{\mathfrak{X}}$.

Specifically, for each pixel (i, j) in $\mathfrak{Y}^{\text{up}}$, Qian et al. (2020) extract an $N_p \times N_p$ patch $\mathfrak{y}^{\text{up}}$ centered at (i, j). Then a weighting vector, $\mathbf{w}$, is calculated by comparing $\mathfrak{y}^{\text{up}}(i, j)$ and the upsampled patches $\mathfrak{y}_l$'s in the paired library as

$$w_l(i, j) = \exp\left\{-\frac{\|\mathfrak{y}^{\text{up}}(i, j) - \mathfrak{y}_l\|_2^2}{2N_p^2\sigma_n^2}\right\}, \qquad (11.31)$$

where σ_n follows the same interpretation as in Eq. (11.28) (Sreehari et al. 2017). This equation is similar to Eq. (11.28) but a key difference is that the low-resolution patches are used here, whereas in Eq. (11.28), high-resolution patches are used.

After $\mathbf{w}$ is normalized (see Eq. (11.28)), the intensities of the filtered image's pixels are then calculated as the weighted average of the intensity of the center pixel of the high-resolution patch, $\mathfrak{x}_l$, in the paired library, such that

$$\hat{\mathfrak{X}}(i, j) = \sum_{l=1}^{n_L} w_l(i, j) z_l, \tag{11.32}$$

where z_l is the center pixel of the high-resolution patch $\mathfrak{x}_l$.

Since the patches in the library have been classified into g categories, Qian et al. (2020) accelerate the LB-NLM filter by calculating only the weights of the category closest to the current patch $\mathfrak{y}^{\text{up}}(i, j)$. As the weights are calculated by an exponential function, their values are close to zero when a category is dissimilar enough to the current patch. Qian et al. (2020) compare the average value of the $\mathfrak{y}_l$ patches in each category with $\mathfrak{y}^{\text{up}}(i, j)$ for finding the closest category. The selection is based on the shortest Euclidean distance between the two patches. Then the pixel intensity of the reconstructed image, $\hat{\mathfrak{X}}(i, j)$, is calculated by the weighted average of the patches only in the respective category. This approach can reduce the computational cost by g times.

The scale parameter, σ_n, can affect the LB-NLM filtering outcomes. For a small σ_n, $\hat{\mathfrak{X}}(i, j)$ is determined by a few closest patches, yielding a similar result as the locally linear embedding method that is described in Sect. 11.2.2. This choice of σ_n helps reconstruct image details in the foreground area. When σ_n is large, LB-NLM averages a large number of patches in the library, decreasing the noise carried over from the training high-resolution images. As electron microscopic images usually have a high noise level, especially in the background area, a default setting $\sigma_n = 1.0$ can provide a good trade-off between enhancing signals and de-noising.

A key difference between the original LB-NLM in Sreehari et al. (2017) and the revised LB-NLM method in Qian et al. (2020) is the following. In Sreehari et al. (2017), both the weights and intensity are calculated from the same high-resolution patches $\mathfrak{x}_l$, whereas in Qian et al. (2020), the intensity of the center pixel, z_l, still comes from the high-resolution patch, $\mathfrak{x}_l$ but the weight, w_l, is calculated from $\mathfrak{y}_l$, which is the matched and upsampled low-resolution patch.

The paired LB-NLM is summarized in Algorithm 11.1.

11.4 Performance Criteria

To measure the performance of a super-resolution method, the standard approach is to consider the high-resolution image as the ground truth, and compare it with the reconstructed image. Two most popular quantitative metrics are the peak signal-to-

Algorithm 11.1 Paired library-based NLM method. Inputs: high-resolution image, $\mathfrak{X}$, low-resolution image, $\mathfrak{Y}$, $N_p = 9$, $g = 10$, $G = 10$, $\sigma_n = 1.0$, and n_L is from a few thousand to tens of thousand. Outputs: super-resolution image reconstruction, $\hat{\mathfrak{X}}$

Registration:

1. Enlarge $\mathfrak{Y}$ by a factor of two via a bicubic interpretation for attaining an upsampled image. Denote the resulting image by $\mathfrak{Y}^{\text{up}}$.
2. Search a shift transform (x, y) and a rotation transformation θ. After the transformation, the overlapping parts on $\mathfrak{Y}^{\text{up}}$ and $\mathfrak{X}$ are supposed to have the smallest mean squared error. The registered image of $\mathfrak{Y}^{\text{up}}$ is denoted by $\mathfrak{Y}^{\text{reg}}$.
3. For each $N_p \times N_p$ image patch, $\mathfrak{x}(i, j)$, search a local (x_{ij}, y_{ij}) to minimize the MSE between $\mathfrak{x}(i, j)$ and $\mathfrak{y}(i + x_{ij}, j + y_{ij})$. The matched patch is denoted by $\mathfrak{y}(i^*, j^*)$.

Paired library building:

4. Sample $G \times n_L$ patches from $\mathfrak{x}$'s, then use the k-means method to classify them into g categories.
5. Sample n_L/g patches from each category to obtain a library with n_L high-resolution patches.
6. Add the matched upsampled patches of $\mathfrak{y}$'s into the library. Each pair is denoted by $\{\mathfrak{x}_l, \mathfrak{y}_l\}_{l=1}^{n_L}$.

LB-NLM filtering:

7. Upsample $\mathfrak{Y}$ by a factor of two using bicubic interpolation. This produces $\mathfrak{Y}^{\text{up}}$.
8. Obtain an $N_p \times N_p$ patch $\mathfrak{y}^{\text{up}}(i, j)$ from $\mathfrak{Y}^{\text{up}}$.
9. Denote the weighting vector by $\mathbf{w}(i, j) = \{w_1(i, j), w_2(i, j), \cdots, w_{n_L}(i, j)\}$. For $l = 1, 2, \cdots, n_L$, the weight $w_l(i, j)$ is calculated by:

$$w_l(i, j) = \exp\left\{-\frac{\|\mathfrak{y}^{\text{up}}(i, j) - \mathfrak{y}_l\|_2^2}{2N_p^2\sigma_n^2}\right\}.$$

10. Normalize $\mathbf{w}$ by

$$w_l(i, j) = \frac{w_l(i, j)}{\sum_{l=1}^{n_L} w_l(i, j)}.$$

11. $\hat{\mathfrak{X}}(i, j)$ is the weighted average of $\mathfrak{X}$ as:

$$\hat{\mathfrak{X}}(i, j) = \sum_{l=1}^{n_L} w_l(i, j) z_l,$$

where z_l is the center pixel of the high-resolution patch $\mathfrak{x}_l$ and $\hat{\mathfrak{X}}(i, j)$ is the pixel intensity of the reconstructed image at position (i, j).
12. Reconstruct the super-resolved image $\hat{\mathfrak{X}}$ by combining $\hat{\mathfrak{X}}(i, j)$ for all (i, j) positions.

noise ratio (PSNR) and the structural similarity (SSIM) index. The closer the two images are, the higher PSNR and SSIM will be. The two metrics are defined as follows.

To define PNSR, let us define the MSE for a pair of images first. For two images, $\mathfrak{X}$ and $\mathfrak{Y}$ of the same size of $M \times N$, the MSE between the two images is defined as

$$\text{MSE} = \frac{1}{MN}\sum_{i=1}^{M}\sum_{j=1}^{N}(\mathfrak{X}(i,j) - \mathfrak{Y}(i,j))^2 . \tag{11.33}$$

Then, PSNR is defined as

$$\text{PSNR} = 10 \cdot \log_{10}\left(\frac{\text{MAX}^2}{\text{MSE}}\right) = 20 \cdot \log_{10}\text{MAX} - 10 \cdot \log_{10}\text{MSE}, \tag{11.34}$$

where MAX is the maximum possible pixel value of the baseline image, $\mathfrak{X}$. When the pixels are represented using eight bits per sample, then MAX = 255. Generally, when samples are represented with B bits per sample, $\text{MAX} = 2^B - 1$.

SSIM is defined as follows,

$$\text{SSIM} = \frac{(2\mu_{\mathfrak{X}}\mu_{\mathfrak{Y}} + c_1)(2\sigma_{\mathfrak{X}\mathfrak{Y}} + c_2)}{(\mu_{\mathfrak{X}}^2 + \mu_{\mathfrak{Y}}^2 + c_1)(\sigma_{\mathfrak{X}}^2 + \sigma_{\mathfrak{Y}}^2 + c_2)}, \tag{11.35}$$

where

- $\mu_{\mathfrak{X}}$ and $\mu_{\mathfrak{Y}}$ are, respectively, the average of the pixel values in $\mathfrak{X}$ and $\mathfrak{Y}$.
- $\sigma_{\mathfrak{X}}^2$ and $\sigma_{\mathfrak{Y}}^2$ are, respectively, the variance of the pixel values in $\mathfrak{X}$ and $\mathfrak{Y}$.
- $\sigma_{\mathfrak{X}\mathfrak{Y}}$ is the covariance between the pixel values of $\mathfrak{X}$ and $\mathfrak{Y}$.
- $c_1 = (b_1\text{MAX})^2$ and $c_2 = (b_2\text{MAX})^2$ are two variables to stabilize the division with weak denominator. By default, $b_1 = 0.01$ and $b_2 = 0.03$.

In the comparison studies of Sect. 11.5, what is reported is the delta in PSNR or the delta in SSIM, which is the change made in respective metrics by a method over the bicubic interpolation baseline. Furthermore, Qian et al. (2020) propose to segment the nanomaterial clusters (foreground) and the host material (background) through image binarization and evaluate the improvements in PSNR and SSIM separately for foreground and background. The foreground improvement reveals how well a super-resolution method enhances the details of the image texture, whereas the background improvement points to a better de-noising capability.

Qian et al. (2020) also propose to add a third metric to measure more directly the impact made by a super-resolution method on electron microscopic images. They argue that the main objective of material science imaging is to increase the ability of material characterization; for instance, increase the accuracy of morphology analysis. But PSNR and SSIM do not necessarily reflect a change in this capability. Qian et al. (2020) propose to check whether the reconstructed images are able to facilitate a better detection of a nanomaterial object's boundary. They use Canny's edge detector (Canny 1986) to identify the boundaries and textures of the nanomaterial clusters and label the detected edges in a binary map. Let $\mathcal{B}_{\text{HR}}$ denote the binary edge map detected from the original high-resolution image (ground truth) and $\mathcal{B}_{\text{SR}}$ denote the binary map detected from the reconstructed image resulting from a super-resolution method. The similarity between them is defined as:

$$\text{sim} = 1 - \frac{\|\mathcal{B}_{\text{HR}} \neq \mathcal{B}_{\text{SR}}\|_1}{\|\mathcal{B}_{\text{HR}}\|_1 + \|\mathcal{B}_{\text{SR}}\|_1}, \tag{11.36}$$

where $\mathcal{B}_{\text{HR}} \neq \mathcal{B}_{\text{SR}}$ produces an indicator matrix whose element is 1 where $\mathcal{B}_{\text{HR}}$ and $\mathcal{B}_{\text{SR}}$ have different values and 0 otherwise, and $\|\cdot\|_1$ is the entry-wise matrix 1-norm. A high sim indicates a better performance.

11.5 Case Study

This section applies various super-resolution methods to paired SEM images and compare their performances.

11.5.1 VSDR Trained on Downsampled Low-Resolution Images

Qian et al. (2020) first demonstrate that the default setting of the current SISR methods, which use the downsampled low-resolution images in training, is ineffective for handling paired image problems. They use VDSR as a representative of the existing SISR methods.

First, Qian et al. (2020) choose 539 high-resolution optical images from the IAPR TC-12 Benchmark (Grubinger et al. 2006) and sample 256 patches from each image. They train a VDSR with 20 layers on these patches with a scale factor of two. Other parameters, such as the training rate and the maximum number of iterations, are set as the default values (Kim et al. 2016). Next, Qian et al. (2020) collect 539 high-resolution nanoimages captured by electron microscopes and again sample 256 patches per image. With the same neural network structure and parameter setting, they train another VDSR using the electron microscopic images. Once given the high-resolution images (in both circumstances), the corresponding low-resolution images are automatically created by downsampling and blurring the high-resolution ones. Qian et al. (2020) denote the VDSR trained on optical images by *Net_Optical*, and the VDSR trained on electron images by *Net_EM*.

For testing, Qian et al. (2020) prepare two datasets: synthetic electron images downsampled from the high-resolution images and the physical low-resolution electron images corresponding to the same high-resolution images. Either image option is used as the input for enhancement via the resulting VDSR. Then, Qian et al. (2020) compare the reconstructed images, presumably enhanced, with the actual high-resolution images and calculate the delta in PSNR, relative to bicubic interpolation. The results are presented in Fig. 11.9.

In Fig. 11.9, left panel, when applying to the downsampled images, both versions of VDSR improve upon bicubic interpolation, but the improvement made by

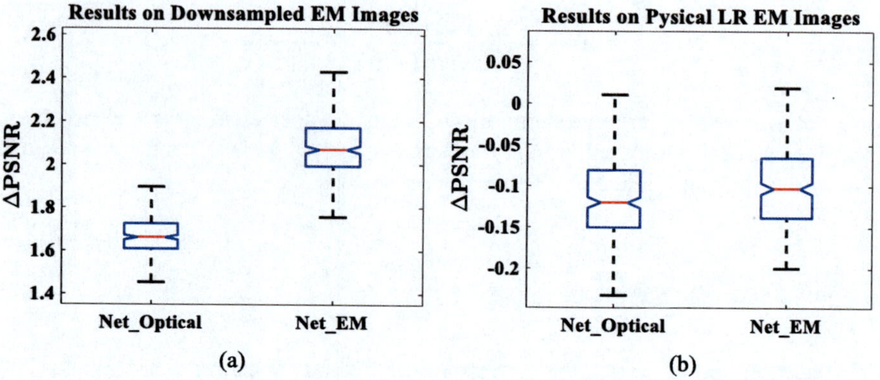

Fig. 11.9 The performance of VDSR when its two versions, *Net_Optical* and *Net_EM*, are applied to the downsampled and physical low-resolution electron images (*source*: Qian et al. (2020))

Net_EM is appreciably greater than that made by *Net_Optical*. This provides one example supporting the conjecture that VDSR trained by optical images become less effective when the resulting VDSR is applied to electron images. In Fig. 11.9, right panel, when applying to the physical low-resolution electron images, both versions of VDSR perform worse relative to bicubic interpolation. The message is that material scientists cannot simply grab an existing pre-trained super-resolution model for processing their unique paired electron images.

11.5.2 Performance Comparison

Qian et al. (2020) collect 22 pairs of SEM images with low and high resolutions. The resolution of high-resolution images double that of the low-resolution ones, meaning that the scale ratio is 2:1. The size of both types of images and the training/test sample preparation have been described in Sect. 11.3.2. Two image pair examples are presented in Fig. 11.7.

With the two training options, self-training versus pooled-training, Qian et al. (2020) compare a total of seven methods in three categories:

- Two sparse-representation methods: the method proposed by Yang et al. (2010) is nicknamed ScSR (spare-coding based super resolution) and the method proposed by Trinh et al. (2014) is nicknamed SRSW (super-resolution by sparse weight).
- Three deep learning-based super-resolution methods: VDSR (Kim et al. 2016), EDSR (Lim et al. 2017), and RCAN (Zhang et al. 2018).
- Two versions of LB-NLM methods: the original LB-NLM (Sreehari et al. 2017) and the paired LB-NLM (Qian et al. 2020).

The 22 image pairs are partitioned into 198 in-sample subimages and 66 out-of-sample subimages. For ScSR, $L = 80,000$ paired patches of size 9×9 are randomly sampled to train a paired dictionary of size $1,024$. The same paired patches also make up the library for SRSW. VDSR, EDSR and RCAN are trained with their default settings with the data-augmentation and early-stopping options. For the original and paired LB-NLM methods, a paired library with the same size as in SRSW is constructed using the corresponding portion of code in Algorithm 11.1.

Table 11.1 presents the average improvement of PSNR and SSIM by these super-resolution methods as compared with the bicubic interpolation baseline.

The first observation is that for the paired image problem, self-training is a better strategy, despite the relatively small image sample size used. For all methods, self-training outperform pooled-training in terms of the out-of-sample ΔPSNR. For most methods, the self-training also produces a better out-of-sample ΔSSIM while for some methods the pooled-training's ΔSSIM is better. But either way, the difference in ΔSSIM is marginal. As argued earlier, using the learned relationship specific to a particular image pair pays off when that relationship is used for reconstruction. This pair-specific type of information does not exist in the general SISR when an external training set is used.

Among the methods in comparison, ScSR is not competitive when it is applied to the paired electron images. The lack of competitiveness of ScSR can be explained by certain options used in its training process. ScSR extracts the high-frequency features from low-resolution images. As the physical low-resolution images contain heavy noisy, those high-frequency features do not adequately represent the image information. Also, ScSR assumes the reconstructed high-resolution patches sharing

Table 11.1 The improvements of PSNR and SSIM of the reconstructed SEM images after applying different super-resolution methods, relative to the performance of bicubic interpolation (*source*: Qian et al. (2020))

Methods		Self-training		Pooled-training	
		In-sample	Out-of-sample	In-sample	Out-of-sample
ScSR	ΔPSNR	0.26 dB	0.23 dB	0.18 dB	0.19 dB
	ΔSSIM	0.019	0.015	0.012	0.014
SRSW	ΔPSNR	1.41 dB	1.17 dB	0.28 dB	0.31 dB
	ΔSSIM	0.026	0.026	0.019	0.022
VDSR	ΔPSNR	2.22 dB	2.07 dB	1.24 dB	1.25 dB
	ΔSSIM	0.052	0.051	0.044	0.047
EDSR	ΔPSNR	2.16 dB	2.06 dB	1.56 dB	1.35 dB
	ΔSSIM	0.052	0.052	0.050	0.051
RCAN	ΔPSNR	2.24 dB	2.07 dB	1.84 dB	1.59 dB
	ΔSSIM	0.053	0.050	0.051	0.051
Original LB-NLM	ΔPSNR	0.46 dB	0.45 dB	0.23 dB	0.28 dB
	ΔSSIM	0.016	0.016	0.017	0.018
Paired LB-NLM	ΔPSNR	3.75 dB	1.67 dB	0.87 dB	0.78 dB
	ΔSSIM	0.132	0.037	0.034	0.031

the same mean and variance as the input low-resolution patches, which is again not true for the physically captured image pairs. SRSW, on the other hand, obtains much better results by directly using the original patches. However, the randomly sampled library used in SRSW retains too many background patches with very little useful information. Such construction of the image library hampers SRSW's effectiveness.

Trained from the physically captured low-resolution images, the performance of VDSR improves significantly as compared to the results in Sect. 11.5.1. In terms of both ΔPSNR and ΔSSIM, the three deep-learning methods yield very similar results with self-training. Using pooled-training, the most advanced RCAN achieves the best performance but still is outperformed by its self-training counterpart. A possible reason is that while EDSR and RCAN can benefit from their complex architectures in pooled-training, this advantage disappears in self-training. Considering the training time cost, VDSR under self-training appears to be the most competitive super-resolution approach yet for the paired electron images.

The simple, paired LB-NLM method achieves rather competitive performances and outperform the original LB-NLM, ScSR and SRSW. There are certain similarities between paired LB-NLM and SRSW. The paired LB-NLM method accounts for more factors behind the difference between a pair of physical images acquired at different resolutions, whereas SRSW primarily deals with the noise aspect. Both paired LB-NLM and SRSW show an obvious better performance when applied to the in-sample images under self-training, while for ScSR, the deep-learning methods, and the original LB-NLM, their in-sample and out-of-sample performance difference is much less pronounced.

The out-of-sample performance of the paired LB-NLM method under self-training reaches 80% accuracy of the deep-learning methods under the same setting. Considering the simplicity of the paired LB-NLM method, it is difficult to imagine that a simple method like that is able to achieve such a performance, relative to deep learning methods, on the general SISR problems; these results highlight the uniqueness of the super-resolution problem for paired physical images.

Let us take a closer look at the reconstruction results. Figures 11.10 and 11.11 the original low-resolution images, bicubic interpolated images, the reconstructed images by the VDSR (both self-training and pooled-training), the reconstructed images by the paired LB-NLM method (self-training only), and the high-resolution images (ground truth). Here VDSR is again used as a representative of the three deep-learning methods, since their respective best performances are similar. In each figure, four images are shown. The four images in Fig. 11.10 are in-sample subimages, whereas those in Fig. 11.11 are out-of-sample subimages. VDSR and the paired LB-NLM method both give a clear foreground and a less noisy background. The visual results of the LB-NLM method are comparable to those of VDSR under self-training. The visual comparison between the images under VDSR (self-training) and those under VDSR (pooled-training) highlights the benefit of using the self-training strategy—the benefit of using self-training is particularly noticeable on the last two images, namely the last two rows of Figs. 11.10 and 11.11.

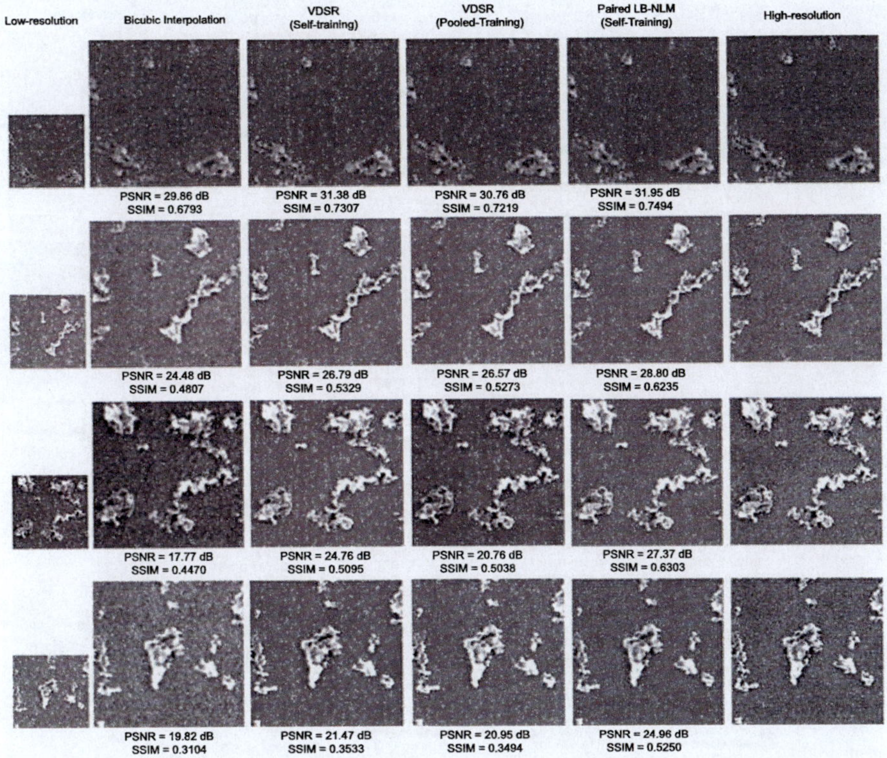

Fig. 11.10 The low-resolution images, the bicubic interpolated results, the image reconstruction results using VDSR (self-training and pooled-training), the paired LB-NLM method (self-training), and the ground truth (high-resolution images) for four in-sample subimages (*source*: Qian et al. (2020))

11.5.3 Computation Time

Qian et al. (2020) present the computation time of training and inference for five methods: three deep-learning methods, SRSW and the paired LB-NLM. SRSW is considered here as it is the better one of the two sparse-representation methods. The three deep-learning methods, implemented by `PyTorch`, are trained at Texas A&M University on one of its High Performance Research Computing (HPRC) cluster with GPUs. The other two methods are trained on personal computers with an `MATLAB` implementation. Because of the differences, in terms of both hardware and software, the computation times are not directly comparable. The exercise here is to let readers have a feeling of the computational demand of each method. Table 11.2 presents the average training and inference time when analyzing the 22 paired SEM images.

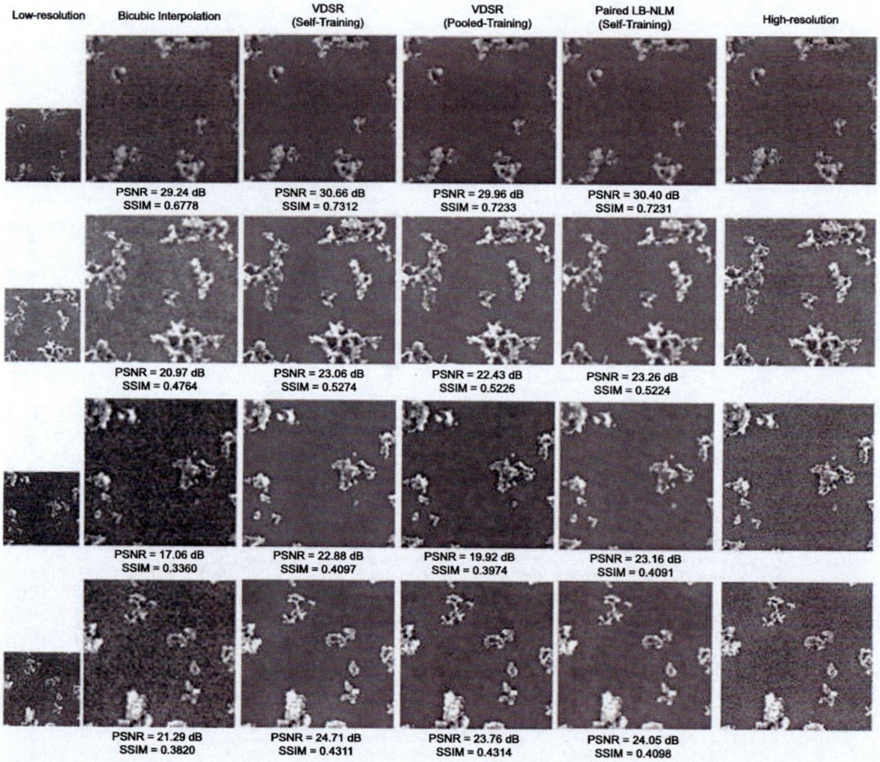

Fig. 11.11 The low-resolution images, the bicubic interpolated results, the image reconstruction results using VDSR (self-training and pooled-training), the paired LB-NLM method (self-training), and the ground truth (high-resolution images) for four out-of-sample subimages (*source*: Qian et al. (2020))

With the aid of high computing power on HPRC, the deep learning methods still need a relatively long training time, especially when the pooled-training strategy is used. Training those models on a regular laptop computer without GPUs is not practical. Both SRSW and the paired LB-NLM methods can be trained and used on regular personal computers but SRSW suffers from a much longer inference time because it solves an optimization problem for each input low-resolution patch. The paired LB-NLM method performs is simpler, much faster, and presents itself as a competitive alternative that can be easily implemented on ordinary laptop computers.

Table 11.2 Computation time of training and inference of some super-resolution methods

	PyTorch on HPRC cluster			MATLAB on regular computer	
	VDSR	EDSR	RCAN	SRSW	Paired LB-NLM
Training	∼30 min (self)	∼30 min (self)	∼10 h (self)	∼5 min (both)	∼5 min (both)
	∼2 h (pooled)	∼5 h (pooled)	∼40 h (pooled)		
Inference	0.21 s	0.17 s	2.66 s	∼20 min	∼30 s

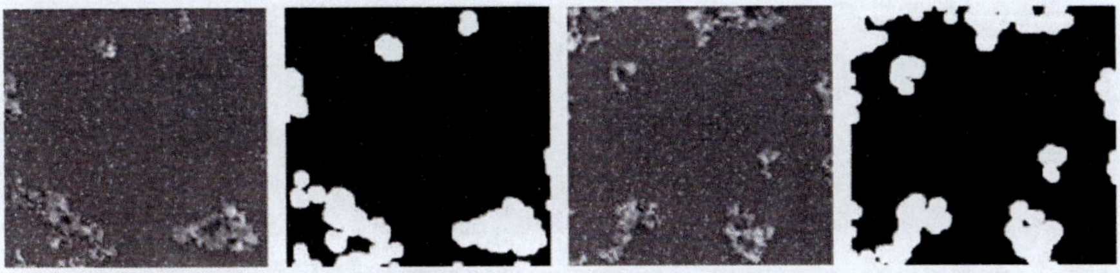

Fig. 11.12 The foreground and background masks of an in-sample and an out-of-sample SEM subimage. In the 2nd and 4th panels, the white areas indicate the nanomaterial (foreground), whereas the black areas indicate the host material (background) (*source*: Qian et al. (2020))

Table 11.3 Changes in PSNR calculated separately for foreground and background for different super-resolution results (*source*: Qian et al. (2020))

		VDSR		SRSW	Paired LB-NLM
		Self-training	Pooled-training	(Self-training)	(Self-training)
In-sample	Foreground	1.21 dB	0.53 dB	0.03 dB	4.27 dB
	Background	2.86 dB	1.68 dB	2.27 dB	3.52 dB
Out-of-sample	Foreground	0.97 dB	0.48 dB	−0.25 dB	0.23 dB
	Background	2.83 dB	1.75 dB	2.15 dB	2.65 dB

11.5.4 Further Analysis

Qian et al. (2020) provide further quantitative analysis using the new criteria mentioned in Sect. 11.4—the separate foreground and background analysis and the edge detection analysis.

Qian et al. (2020) segment the SEM images by using Otsu's method (Otsu 1979) to highlight the separation of foreground from background and remove the isolated noise points in the foreground. Figure 11.12 shows in two image examples of which the binary masks indicate the foreground.

Table 11.3 presents the changes in PSNR, calculated separately for the foreground and the background, for three methods: VDSR (both self-training and pooled-training), SRSW (self-training), and the paired LB-NLM method (self-training). It is apparent that all these methods denoise the background much more than they enhance the foreground for the out-of-sample images. The main advantage of VDSR is its ability to improve the foreground better than the paired LB-NLM.

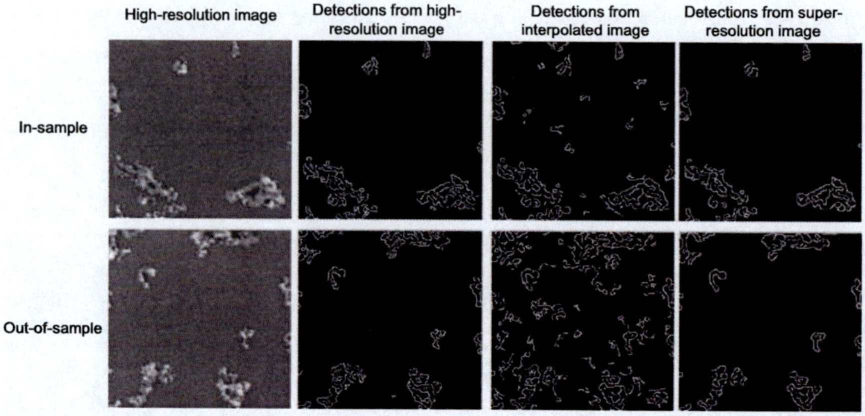

Fig. 11.13 The results of applying Canny's edge detection to the high-resolution images and the reconstructed images by using bicubic interpolation and the paired LB-NLM method, respectively (*source*: Qian et al. (2020))

This is not entirely surprising because the non-local-mean methods were originally designed as an image de-noising tool. It is also observed that the self-training VDSR is better than the pooled-training VDSR more so in terms of a stronger de-nosing capability over the background. SRSW does a similar job in terms of denoising the background. But there is a slight decrease in terms of PSNR for the foreground, which suggests that the particular mechanism used in SRSW, especially the mechanism to create its library, is not effective for enhancing the foreground signals in the physical electron images.

Qian et al. (2020) then apply the Canny's edge detector to the high-resolution images, the bicubic interpolated images, and the reconstructed images. The parameter of Canny's edge detector is set at 0.2. Figure 11.13 demonstrates the detection results for both an in-sample and out-of-sample subimage. The visual inspection shows clearly that the paired LB-NLM's reconstruction results facilitate more accurate edge detections than the bicubic interpolated images when both are compared with the boundaries detected in the high-resolution images (the ground truth).

To quantify the improvement in edge detection accuracy, Table 11.4 presents the results of the similarity index, sim, as defined in Eq. (11.36). Except for SRSW, all methods can improve, as compared with the bicubic interpolation baseline, the edge detection accuracy by around 50% on the out-of-sample images. The self-training VDSR achieves the largest improvement, although its sim is just slightly higher than that of the pooled-training VDSR and the paired LB-NLM. This result is consistent with the foreground PSNR improvements made by the four method in Table 11.3.

Table 11.4 Results of sim for different super-resolution methods and bicubic interpolation (*source*: Qian et al. (2020))

	Bicubic Interpolation	VDSR		SRSW	Paired LB-NLM
		Self-training	Pooled-training	(Self-training)	(Self-training)
In-sample	0.25	0.39	0.37	0.33	0.56
Out-of-sample	0.24	0.37	0.35	0.24	0.33

References

Boyd S, Parikh N, Chu E, Peleato B, Eckstein J (2011) Distributed optimization and statistical learning via the alternating direction method of multipliers. Foundations and Trends in Machine Learning 3(1):1–122

Canny J (1986) A computational approach to edge detection. IEEE Transactions on Pattern Analysis and Machine Intelligence 8(6):679–698

Chang H, Yeung DY, Xiong Y (2004) Super-resolution through neighorhood embedding. In: Proceedings of the 2004 IEEE Computer Society Conference on Computer Vision and Pattern Recognition (CVPR), Washington, DC

Chiang MC, Boult TE (2000) Efficient super-resolution via image warping. Image and Vision Computing 18:761–771

Dong C, Loy CC, He K, Tang X (2016) Image super-resolution using deep convolutional networks. IEEE Transactions on Pattern Analysis and Machine Intelligence 38(2):295–307

Elad M, Hel-Or Y (2001) A fast super-resolution reconstruction algorithm for pure translational motion and common space-invariant blur. IEEE Transactions on Image Processing 10(8):1187–1193

Freeman WT, Pasztor EC, Carmichael OT (2000) Learning low-level vision. International Journal of Computer Vision 40(1):25–47

Freeman WT, Jones TR, Pasztor EC (2002) Example-based super-resolution. IEEE Computer Graphics and Applications 22(2):56–65

Gross D (1986) Super Resolution from Sub-pixel Shifted Pictures. Master's Thesis, Tel Aviv University

Grubinger M, Clough P, Müller H, Deselaers T (2006) The IAPR TC-12 benchmark: A new evaluation resource for visual information systems. In: International Conference on Language Resources and Evaluation, Genoa, Italy. [online] https://www.imageclef.org/photodata

Hastie T, Tibshirani R, Friedman J (2009) The Elements of Statistical Learning: Data Mining, Inference, and Prediction, 2nd edn. Springer, New York

Irani M, Peleg S (1990) Super resolution from image sequences. In: Proceedings of the 10th International Conference on Pattern Recognition, Atlantic City, NJ, pp 115–120

Keys R (1981) Cubic convolution interpolation for digital image processing. IEEE Transactions on Acoustics, Speech, and Signal Processing 29(6):1153–1160

Kim J, Lee JK, Lee KM (2016) Accurate image super-resolution using very deep convolutional networks. In: Proceedings of the 2016 IEEE Conference on Computer Vision and Pattern Recognition (CVPR), Las Vegas, NV, pp 1646–1654

Lim B, Son S, Kim H, Nah S, Mu Lee K (2017) Enhanced deep residual networks for single image super-resolution. In: Proceedings of the IEEE Conference on Computer Vision and Pattern Recognition Workshops (CVPRW), Honolulu, HI, USA, pp 136–144

Otsu N (1979) A threshold selection method from gray-level histograms. IEEE Transactions on Systems, Man, and Cybernetics 9(1):62–66

Park SC, Park MK, Kang MG (2003) Super-resolution image reconstruction: A technical overview. IEEE Signal Processing Magazine 20(3):21–36

Qian Y, Xu J, Drummy LF, Ding Y (2020) Effective super-resolution method for paired electron microscopic images. IEEE Transactions on Image Processing 29:7317–7330
Sreehari S, Venkatakrishnan S, Bouman KL, Simmons JP, Drummy LF, Bouman CA (2017) Multi-resolution data fusion for super-resolution electron microscopy. In: Proceedings of the IEEE Conference on Computer Vision and Pattern Recognition Workshops (CVPRW), Honolulu, HI, pp 1084–1092
Sun J, Zheng NN, Tao H, Shum HY (2003) Image hallucination with primal sketch priors. In: Proceedings of the 2003 IEEE Computer Society Conference on Computer Vision and Pattern Recognition (CVPR), Madison, WI
Tian J, Ma KK (2011) A survey on super-resolution imaging. Signal, Image and Video Processing 5(3):329–342
Timofte R, Agustsson E, Van Gool L, Yang MH, Zhang L (2017) NTIRE 2017 challenge on single image super-resolution: Methods and results. In: Proceedings of the 2017 IEEE Conference on Computer Vision and Pattern Recognition Workshops (CVPRW), Honolulu, HI, pp 1110–1121
Trinh DH, Luong M, Dibos F, Rocchisani JM, Pham CD, Nguyen TQ (2014) Novel example-based method for super-resolution and denoising of medical images. IEEE Transactions on Image Processing 23(4):1882–1895
Tsai RY, Huang TS (1984) Multiframe image restoration and registration. In: Huang TS (ed) Advances in Computer Vision and Image Processing, vol 1, JAI Press Inc., Greenwich, CT, pp 317–339
Wang Z, Chen J, Hoi SCH (2021) Deep learning for image super-resolution: A survey. IEEE Transactions on Pattern Analysis and Machine Intelligence, in press
Wu CFJ, Hamada M (2009) Experiments: Planning, Analysis, and Parameter Design Optimization, 2nd edn. Wiley Series in Probability and Statistics, John Wiley & Sons, New York
Yang J, Huang TS (2010) Image super-resolution: Historical overview and future challenges. In: Milanfar P (ed) Super-Resolution Imaging, Chapman & Hall/CRC Press, Boca Raton, FL, pp 3–35
Yang J, Wright J, Huang TS, Ma Y (2010) Image super-resolution via sparse representation. IEEE Transactions on Image Processing 19(11):2861–2873
Yue L, Shen H, Li J, Yuan Q, Zhang H, Zhang L (2016) Image super-resolution: The techniques, applications, and future. Signal Processing 128:389–408
Zhang X, Chen Q, Ng R, Koltun V (2019) Zoom to learn, learn to zoom. In: Proceedings of the IEEE Conference on Computer Vision and Pattern Recognition (CVPR), IEEE, Long Beach, CA, pp 3762–3770
Zhang Y, Li K, Li K, Wang L, Zhong B, Fu Y (2018) Image super-resolution using very deep residual channel attention networks. In: Proceedings of the European Conference on Computer Vision (ECCV), Munich, Germany, pp 286–301

Index

C. Park, Y. Ding, *Data Science for Nano Image Analysis*, International Series in Operations Research & Management Science 308,
https://doi.org/10.1007/978-3-030-72822-9

T

U

V

W

X